Participation Based Intelligent Manufacturing

Participation Based Intelligent Manufacturing: Customisation, Costs, and Engagement

EDITED BY

BIRUTĖ MOCKEVIČIENĖ

Mykolas Romeris University, Lithuania

United Kingdom – North America – Japan – India – Malaysia – China

Emerald Publishing Limited
Emerald Publishing, Floor 5, Northspring, 21-23 Wellington Street, Leeds LS1 4DL.

First edition 2025

British Library Cataloguing in Publication Data
A catalogue record for this book is available from the British Library

ISBN: 978-1-83797-363-7 (Print)
ISBN: 978-1-83797-362-0 (Online)
ISBN: 978-1-83797-364-4 (Epub)

Contents

List of Figures and Tables *xi*

Abbreviations *xv*

About the Editor *xvii*

About the Contributors *xix*

Chapter 1 Introduction *1*
Birutė Mockevičienė
A Problem for Customised Manufacturing That Deserves a Response *1*
A Solution in Under 30 Minutes? *2*
Rationality of the Research *3*
The Purpose of the Book *3*
Relevance and Challenges *4*

Chapter 2 Intelligent Manufacturing From A Theoretical Perspective: The Technological Revolution and Social Participation *7*
Birutė Mockevičienė
A Historical Perspective for Industry 4 *7*
Smart Manufacturing: Measuring, Managing, and Advancing Processes *9*
Industry 4.0: The Concept *10*
The Role of Smart Manufacturing in Industry 4.0 *16*
Digitalisation in Smart Manufacturing *20*
Big Data *20*
Data Mining and Big Data Analytics in Manufacturing *21*
Applications of Artificial Intelligence *26*
Machine Learning in Manufacturing *32*
Digital Platforms for Manufacturing *34*
Small and Medium Enterprises' Readiness to Embrace Industry 4.0 Possibilities *37*
Factors That Are Linked with the Success of Industry 4.0 *38*

SME-specific Maturity and Readiness Models *39*
Barriers to the Implementation of Industry 4.0 *40*

Chapter 3 Furniture Industries: Challenges of Regionalisation, Customisation and New Paradigm of Pricing *47*
Birutė Mockevičienė and Tomas Vedlūga
Sustainability Trends in the Furniture Industry *48*
Regionalisation and Frugal Innovation in Manufacturing *52*
Regionalisation *52*
The Concept of FI *53*
Impact of FI on Customised Furniture Manufacturing *55*
Customised Manufacturing: Challenges and Opportunities *56*
Basic Principles of Customised Manufacturing *56*
Customer Integration *61*
Knowledge-based Engineering for Customised Product Lifecycle Management *64*
ERP and Customisation *66*
Empowering People in Manufacturing Industries: Participation-Based Management *76*
Stages of Participation *77*
Instrumental Approach Towards Participation *78*
Higher Scale of Participation and Engagement *78*
Employee Participation in the Era of Industry 4.0 *79*
Cost Estimation Approaches *81*
Traditional Costing Methods as Life Cycle Costing *81*
Managerial Approach to Costing *85*
New Paradigm of Cost Evaluation *88*
A Machine Learning Approach to Estimate Early Costs of New Product *90*
Expert Judgement in Price Estimation *91*

Chapter 4 Methodological Implications Seeking To Solve Cost Estimation Issues For Customise Production Process *101*
Birutė Mockevičienė
Introduction *101*
Outlining of the Methodology *104*
Phase I. Analysis of Customised Production Processes *104*
Phase II. Historical Manufacturing Data Collection and Modelling of Manufacturing Processes Based on Machine Learning Algorithms *106*
Phase III. Inclusive Governance and Modelling of Expert Decisions *108*

Companies Selected as Case Studies for the Empirical Data 110
Case 1 (Company A) 111
Case 2 (Company B) 114

Chapter 5 How The Small Country's Furniture Sector Builds Its International Competitiveness (Survey of Lithuania, Latvia and Estonia) 117
Julija Moskvina and Birutė Mockevičienė
Furniture Manufacturing Sector: Similarities and Differences Between the Baltic Countries 117
Estonia 119
Latvia 120
Lithuania 120
Prevailing Organisational Structure in Furniture Manufacturing Companies 121
Competences of the Manufacturing Team 125
Tools for Data Managing and Exchange 127
Production and Experts 128
Cost Estimation Practice 130
Price Evaluation Strategy 130
Attitudes of Manufacturers Towards IT and Other Decision Support Instruments 131
IT Adoption Factors and Organisational Features 131
Technology Usage and Sector Specificity 134
Assessment of Baltic Manufacturers' Attitudes Towards IT 137
In Conclusion 145

Chapter 6 Frugal Innovation as Intersection Between Complexity of Early Cost Estimation, Machine Learning and Expert-Based Decision System 151
Julija Moskvina, Anca Hanea, Tomas Vedlūga and Birutė Mockevičienė
Complexity Management in Customised Furniture Manufacturing 152
Complexity in Manufacturing 152
Specificity of Complexity Management in Customised Furniture Manufacturing 155
Early Cost Estimation: Managerial Perspective 158
Managerial Challenges Related to Uncertainty 158
Role of Data in Early Price Estimation 160
Methodology of Qualitative Analysis 162
Findings from the Selected Enterprises 167
Pricing Processes in Made-to-Order Manufacturing Businesses 176

Strategies for Using Furniture Manufacturing Data to Promote Industry 4.0 — 180
The State of the Art of Data Analytics in Furniture Manufacturing — 182
Methodological Considerations in the Search for a Data Usability Strategy — 186
Factual Data Representing Furniture Manufacturing Companies — 188
Possible Strategies for Using Data Based on the Level of Complexity of the Tasks Customised Manufacturing — 193
In Conclusion — 198
Early Cost Estimation by Means of Machine Learning with Data Visualisation — 200
Historical Production Data in Furniture Companies — 200
Early Cost Estimation by Means of Machine Learning with Data Visualisation — 203
Empowering and Engaging Industrial Workers Using Structured Expert Judgment — 217
Cost Estimation and Expert Based Knowledge — 219
Methodology for SEJ Application — 223
From Experiments to SEJ Benefits of Employees' Engagement for Early Cost Estimation — 224
In Conclusion — 231

Chapter 7 Conceptional Knowledge Management Tool for Early Furniture Cost Estimation (Integrated Early Price Assessment System) — 239
Birutė Mockevičienė
Communication Flow Diagram for the Operation of the Prototype — 239
Types of Knowledge — 240
Knowledge Management Cycle — 240
Prototype Functionality and Structure — 243
Functionality — 243
Structure — 244
Back End of the Prototype — 245
Front End — 247
Prototype Users — 247
Structural Elements — 248
User Interface — 248
Design of User Interface — 249
Algorithm for Verification of Performance Under Laboratory Conditions (Combining ML with SEJ) — 249

Prototype Operating Principle for Estimating the Cost of a Piece of Furniture 249
A Case Study on the Application of the Price Evaluation Methodology 253
Prototype Testing Conditions and Challenges 258
Preconditions 258
Selecting the Company 259
Data for Testing 260
Refining the Prototype 261

Chapter 8 Epilogue 263
Birutė Mockevičienė
Customised Furniture Manufacturing in the Era of Industry 4.0 263
Regionalisation Issues 263
Attitudes Towards Digitalisation and Data Usability of Furniture's Companies 264
Attitudes Towards Pricing Strategies of Furniture's Companies 264
Preconditions for Making the Pricing Paradigm Shift Happen 265
Uniqueness of Solution of Proposed Integrated Early Price Assessment System 266
Transferability of Findings 267
Future Research 268

Annex 1 269

Annex 2 279

Annex 3 281

Annex 4 285

Index 287

List of Figures and Tables

Figures

Fig. 2.1. The Emergence of Industry 4.0 as a New Phenomenon. 8
Fig. 2.2. The Content and Components of Industry 4.0. 15
Fig. 2.3. Theoretical Framework of Industry 4.0 Technologies According to Various Technologies (Frank et al., 2019). 17
Fig. 2.4. Classification of Big Data Analytics and Used Methods for Manufacturing (Belhadi et al., 2019). 24
Fig. 2.5. Generic Machine Learning Pipeline with Quality Attributes. 28
Fig. 2.6. Machine Learning Models (Techniques) (Based on Sarker, 2021). 29
Fig. 2.7. In Classification, Two Classes are Distinguished by A Dashed Line Representing the Separation, While in Regression, the Model is Used to Represent the Axis of the Relationship Between Variables. 31
Fig. 2.8. Architecture of a Typical Data Mining System (Han & Kamber, 2006). 35
Fig. 3.1. Furniture Industry Sustainability. 50
Fig. 3.2. The Benefits of Customisation Intersect with the Challenges That Have Arisen, and Industry 4.0 Offers Practical Solutions to Address Them. 61
Fig. 3.3. Lifecycle of Knowledge Pattern. 65
Fig. 3.4. Key Features and Aspects of ERP Systems. 67
Fig. 3.5. The Concept of Employee Participation. 80
Fig. 3.6. The Concept of the Activity-based Costing Approach. 85
Fig. 3.7. Case-based Reasoning Flowchart. 92
Fig. 4.1. A Flowchart of the Methodology of an Interdisciplinary Study Combining the Disciplines of Custom Furniture Manufacturing, Data Science and Inclusive Management. 103
Fig. 4.2. An Example of the Products that were Ordered and Produced by the Company, Which Represents Case 1 (Photo of the Company). 112
Fig. 4.3. An Example of Design Produced by the Company, Which Represents Case 1 (Photos of the Company). 112
Fig. 4.4. A Two-dimensional Classification of Products (Case 1) (Provided by the Company). 113

Fig. 5.1. Levels of Organisational Structure at Furniture Manufactures. 123
Fig. 5.2. Structure of Surveyed Companies with Divisions/ Departments by Country. 125
Fig. 5.3. Order Price Calculation Methods by Country. 130
Fig. 5.4. Employees of Companies Involved in the Calculation of the Order Price Per cent, Multiple Choice). 132
Fig. 5.5. The Approach Towards Survey's Questions Superposition of IT Adoption Factors with Pricing Strategy and Technology Usage. 138
Fig. 5.6. Purchase of Services (Per cent, Multiple Choice). 139
Fig. 5.7. Changes in the Performance Indicators of Some Companies Over the Past Year, Compared to The Previous Year, Per cent. 141
Fig. 5.8. Manufacturing Activities, Where Specialised IT Process Management Tools Are Applied (Per cent, Multiple Choice). 142
Fig. 5.9. The Model of Factors of IT Adoption in Furniture Manufacturing, Linking Data Management Levels. 144
Fig. 6.1. The Diagram of the Qualitative Data Categorisation. 163
Fig. 6.2. Word Frequency Cloud. 163
Fig. 6.3. Pricing Processes in Made-to-Order Manufacturing Businesses. 177
Fig. 6.4. Hierarchical Category Tree and Category Group Clusters. 178
Fig. 6.5. Data Sample (Modified). 189
Fig. 6.6. Customised Manufacturing Data Specificity. 190
Fig. 6.7. Final Prices of all Furniture Products Manufactured by Company A During the Analysis Period. 191
Fig. 6.8. Three Data Utilisation Goals for Customised Manufacturing. 195
Fig. 6.9. Preliminary Cost Estimation: (a) Manual-intensive Process, (b) Proposed Machine Learning-based Approach (Kurasova et al., 2021). 204
Fig. 6.10. Relationship Between the True Price (Y True) and the Predicted Price (Y Predicted) Obtained by Linear Regression (Kurasova et al., 2021). 207
Fig. 6.11. Data Visualisation Process (Kurasova et al., 2021). 210
Fig. 6.12. Price Distribution (Kurasova et al., 2021). 212
Fig. 6.13. Distribution of Different Parts (Kurasova et al., 2021). 213
Fig. 6.14. Data Visualised by Two Principal Components (Kurasova et al., 2021). 213
Fig. 6.15. Data Visualised Using the Multidimensional Scaling Method (Kurasova et al., 2021). 214
Fig. 6.16. Data Visualised by the t-SNE Method (Kurasova et al., 2021). 214
Fig. 6.17. Data Clustered by Louvain's Algorithm and Visualised by t-SNE (Kurasova et al., 2021). 215
Fig. 6.18. Mapping of Cluster C6 (Kurasova et al., 2021). 216

Fig. 6.19. Data Visualised by an Autoencoder Neural Network (Kurasova et al., 2021). 216
Fig. 6.20. Data Subset Clustered by Louvain's Algorithm and Visualised by t-SNE (Kurasova et al., 2021). 217
Fig. 6.21. Results of Scoring Experts on (a) 20 Seed Questions, and on (b) 11 Seed Questions. 226
Fig. 6.22. Results of Cost Estimation for Four Products (Solid Line – the Real Cost of the Product, DOTTED Line – Machine Learned Cost Estimation). 228
Fig. 6.23. Results of Scoring Experts on 14 Seed Questions and the Equally and Differentially Weighted Combinations of Experts. 228
Fig. 6.24. Results of Scoring Experts on Different Subsets of Seed Questions & and the Corresponding Equally and Differentially Weighted Combinations of Experts. 229
Fig. 6.25. Example of 3 Seed Questions and the Experts Estimates. PW is the Performance-Based Aggregation Based on the 23 Seeds, Calculated with Power Reduced to Half. The '['Represents the Fifth Percentiles, ']' Represents the 95th Percentile and the '*' Represents the Median. The '#' Represents the True Value. 230
Fig. 7.1. The Mapping of the Knowledge Management Process to the Operational Phases of the Cost Estimation/ Prognosis Prototype. 241
Fig. 7.2. Communication Flow Diagram for Prototyping. 242
Fig. 7.3. Integrated Early Price Assessment System Prototype Operating Diagram (Prognostic Approach and Interface Diagram). 246

Tables

Table 2.1. The Most Advanced Technologies, Associated with Industry 4.0 (VTT & Consult, 2016). 16
Table 5.1. Rotated Component Matrix. 123
Table 5.2. Turnover and Nature of Production (Crosstab). 126
Table 5.3. Summarising the Factors of the Level of IT Adoption Within Industrial Company. 136
Table 6.1. Interview Questions Made in Compliance with the Research Instrument Scheme. 164
Table 6.2. Classification of the Interviewed Representatives of Furniture Manufacturing Businesses. Organisational Levels Are Identified as Described by Navickienė and Mikulskienė (2019). 166
Table 6.3. Hypothetical Grouping of Furniture Products by Price, Features, and Customer Order Size. 192
Table 6.4. Data Usability Strategies for Furniture Manufacturing. 199

Table 6.5. Forecasting Results. 206
Table 6.6. Attributes of Production Data. 211
Table 7.1. The List of Notions. 251
Table 7.2. Answers to Questions for Expert Calibration (Expert was Asked to Give His/Her Best Guess (The Median) and the 5% and 95% Confidence Bounds and, Respectively, for Each of the Variables). 255
Table 7.3. Breaking Down the Experts' Predictions into Ranges. 255
Table 7.4. Averages of the Experts' Predictions Falling Within the Specified Ranges. 256
Table 7.5. Calculation of the Calibration Score. 256
Table 7.6. Calculation of Lower and Upper Limits. 256
Table 7.7. Calculation of the Information Estimate (Per Expert, Per Question). 256
Table 7.8. Calculation of the Average Information Values. 257
Table 7.9. Calculation of the Weights of the Experts. 257
Table 7.10. Estimation of the Final Expert Price. 257
Table 7.11. Estimation of the Blended Price. 258

Abbreviations

ACSI	American Customer Satisfaction Index
AI	Artificial intelligence
ANN	Artificial neural networks
AR	Augmented reality
ATO	Assembly to order
B2B	Business-to-business
BA	Business analyst
BD	Big data
BDA	Big data analytics
CBR	Case-based reasoning
CEO	Chief executive officer
CM	Cloud computing
CNC machinery	Computer numerical control machinery
CPS	Cyber-physical system
CRM	Customer relationship management
DARPA	Defense Advanced Research Projects Agency
DM	Direct materials
ERP	Enterprise resource planning
ETO	Engineer to order
FBA	Forecasting by analogy
FI	Frugal innovation
HMI	Human–machine interaction
ICT	Information and communication technology
IDS	Intrusion detection systems
IL	Indirect labour
IM	Indirect materials
IoT	Internet of Things
IPNs	Industry peer networks
IT	Information technology
KBE	Knowledge based engineering
KPIs	Key performance indicators
MDS	Multidimensional scaling
MES	Manufacturing execution systems
MFA	Multifactor authentication
ML	Machine learning
MOH	Manufacturing overhead

MTO	Manufacturing to order
MTS	Manufacturing from stock
PCA	Principal component analysis
PLM	Product lifecycle management
PSS	Product and service systems
RBAC	Implement role-based access control
RBF	Rule-based forecasting
RF	The Random Forest
SAPA	Semantic annotation and processing agent
SEJ	Structured expert judgement
SME	Small and medium enterprise
VBP	Value based pricing

About the Editor

Birutė Mockevičienė is the author of the idea to use machine learning to address the problem of customised furniture production orders, the initiator and the leader of all the research work carried out in order to prove this innovative approach. She developed both the concept of the innovative solution and the research methodology combining multidisciplinary approach. She also initiated and carried out the empirical data collection (interview process, survey process, process of collecting, compiling and updating historical data, process of testing the prototype under real conditions). The prototype design was developed and a list of functionalities of the prototype for the price evaluation was prepared.

She is a Professor of Management since 2017. She started her career in the field of Social Sciences in 2006, and currently works at Mykolas Romeris University, Institute of Management and Policy Studies, as a Director. Her basic education (Ph.D. in Physics) has allowed her to work at the frontier of scientific fields, bridging the approaches of mathematics, physics, management and economics.

About the Contributors

Anca Hanea is a renowned structured expert judgement (SEJ) expert. In this work, she has ensured the validity of the method of involving employees in price forecasting, the technicalities of integrating the SEJ methodology into the prototype and the processing and interpretation of the experimental data. Her extensive experience enabled the precise implementation of the SEJ method.

She is a probabilistic modeller interested in uncertainty quantification using data (when available) or SEJ (when data are sparse or missing). She was instrumental in building a COST European network for SEJ elicitation and aggregation, and related standards for the European Food Safety Authority (EFSA). Together with colleagues from the University of Melbourne, she has developed, tested and co-authored guidelines for the IDEA protocol for SEJ. She has taken part and facilitated numerous expert elicitations in various projects undertaken by World Health Organization, EFSA, the European Framework Seven Programme, Defense Advanced Research Projects Agency (DARPA), Cawthron Institute, the Australian Department of Agriculture, Fisheries and Forestry and many universities around the world.

Julija Moskvina was responsible for analysing the furniture sector in Baltic countries, and the sociological interpretation of the interview and survey data working in a team with Birutė Mockevičienė. The qualitative data analysis performed by her covered transcribing interviews, data cleaning, developing categories, data coding, interpretation and report drafting.

She is a Senior Researcher at the Labour Market Research Division of the Lithuanian Centre for Social Sciences. She received her Ph.D. in Sociology from Vilnius University. While participating in national and international research projects, she has developed thorough experience in qualitative and quantitative analysis of labour market processes and employment policies with a particular focus on vulnerable groups in employment.

Tomas Vedlūga is a Doctor of social sciences, specialising in performance management, performance evaluation systems in modelling, selection of indicators and their measurement. While participating in national and international scientific conferences, he gained experience in process evaluation, data analysis and prototyping. In his research, he pays a lot of attention to the modelling and evaluation of prototypes in terms of sustainability.

–He assisted in the justification of sustainability or further use of the collected data for the functioning of the prototype and the transfer of ERP logic in the design of the prototype.

Chapter 1

Introduction

Birutė Mockevičienė

Mykolas Romeris University, Lithuania

Abstract

This chapter aims to introduce the book's research field and the primary challenges faced by the modern manufacturing sector in terms of operational planning and process cost reduction. It will focus on the impact of technology and social technologies, such as digital employee involvement in process assessment, which facilitates organisational innovation through ICT and high-value collaborative technologies. In other words, this chapter presents solutions to the issues that arise from using current production data and employee collaboration to develop future expert predictive solutions tailored to the furniture manufacturing process.

Keywords: Customised manufacturing; structured expert judgement; machine learning; furniture industry; frugal innovations

A Problem for Customised Manufacturing That Deserves a Response

Worldwide, engineering contributes positively to national GDP growth. However, the manufacturing sector is facing significant challenges around the world: increasing global competition and consumers' desire for customised solutions that prioritise design. This phenomenon has reoriented the existing supply-driven market towards *consumer-driven production*. The success of customised production now relies on efficient planning. This planning should consider the required amount of materials, the time and effort needed to complete the project, and the evaluation of all the client's requirements. It is also important to choose the

Participation Based Intelligent Manufacturing:
Customisation, Costs, and Engagement, 1–5

Published under exclusive licence by Emerald Publishing Limited
doi:10.1108/978-1-83797-362-020241001

best strategies, creative design, and project implementation. Production process for construction. Customised production requires completely new technological solutions, long and careful prototyping and testing. This increases *the cost of the product*, lengthens the *production time* and leads to persistent errors in *product quality*. Therefore, the customised manufacturing sectors are intensively looking for new solutions with quite specific requirements: to perform the predictive evaluation of a customised order quickly (*in less than 30 minutes*) and reliably (to reduce the likelihood of prediction errors). Following the global trends in the furniture industry and considering the production organisation solutions used in this field (Amarilli & Spreafico, 2010), in order to reconcile a number of different objectives at the same time, the need for rapid information transfer and efficient decision-making is becoming more and more critical (Kucharska et al., 2015). In customised furniture production, where design is an added value to the product, the preparation of design drawings and estimates to determine the cost and lead time of an order can take from four working days to two weeks and account for as much as 30% of the company's total working time. Unfortunately, even with this amount of time, errors due to the complexity of the structural elements and the innovativeness of the design elements can reach up to 70%. This situation creates *tensions between designers, engineers, and managers* at various levels, and the company starts to avoid complex and innovative orders. Therefore, to preserve the sustainability of the company, support instruments are needed (Bagchi et al., 2003), which allow to react in an increasingly shorter period of time to the content of the user's customised order, and to properly assess and plan the production processes, costs and deadlines (Gawroński, 2012; Møller, 2006). As different employees in a company have different experiences in product evaluation (designers anticipate design strategy, engineers are better at predicting lead times, managers are better at 'feeling' the pulse of the market – acquisitions and consumer expectations), an instrument based on the *involvement of different levels of employees* is needed (Wilkinson et al., 2010), with a reactive (learning from past mistakes and the growing experience of the employees) production process evaluation system.

A Solution in Under 30 Minutes?

This issue calls for innovative solutions that can revolutionise how we manufacture products to meet the unique needs of our customers. For the sake of accuracy, the time needed for the specialists to prepare the initial start-up of the system is not included in the formulation of the performance indicator for the cost evaluation module, that is, the time needed to collect the historical data, the time required to get used to working with the system are not taken into account. Also, the time spent working with the client to understand the client's needs is not part of the working time of the price evaluation module. The 30 minutes is the suggested timeframe for the manager to gather the required information for price evaluation: log in, access the order details, execute the task, choose the experts, and request them to conduct the evaluation.

Why would it seems important for a manufacturing company to have an order price forecasting system that is capable of solving the task at hand (analysing the order price) in some specific time frame, for example, the time frame we have proposed, which would be shorter than 30 minutes? We have to look for the answer in innovation theory. Those 30 minutes can be used as a baseline to measure the effectiveness of the innovation. This time period must be so short that it can demonstrate a significant change in the practice of price evaluation and initiate the development of a new paradigm. In 2018, when this idea was submitted as a project application for peer review, the idea of fixed 30-minute processes sounded unlikely given the IT technologies used at the time.

Today, from a historical perspective, it turns out that this was not a completely random guess. The application of the latest Industry 4.0 technologies and approaches has succeeded in demonstrating this, confirming a paradigm shift towards cost in a traditional industry such as furniture. Analogous situations related to the paradigm shift can be found in other sectors. Quite illustrative is the experience of Rolls Royce (Rodríguez et al., 2020). The Rolls Royce case is interesting because it coincided with a desire to set the price in an unconventional way (without linking it to the sector's usual indicators, as in the furniture industry with the usual resource volumes). We are talking about Rolls Royce's 'Power by the hour' programme (Smith, 2013), in which the pricing mechanism of the contract is based on the number of hours the aircraft engine is running, as opposed to the traditional method where the customer buys the equipment and is billed according to the results of the support services that keep the engine running.

Rationality of the Research

Custom furniture companies, large and small, in both developed and developing regions, suffer from a detailed, cost-modelling, early pricing process for new products at the product order stage. This process is demanding in terms of specialist effort and man-hours, but it does not avoid both costing, planning and quality errors in production processes. There is therefore a need for a symbiosis between science and practice which, in addition to theoretical assumptions, examines scientific progress in the context of Industry 4.0 and addresses the practical problems of furniture production in the early pricing of newly developed products. Industry 4.0 offers a broad perspective on the complex challenges of manufacturing business. Industry 4.0 tools (e.g. robotics, sensors, and big data analytics) fully address manufacturing issues only when applied in an integrated manner. That is, solutions often require an interdisciplinary approach combining social context with technological capabilities. There is still, no ready-made solution.

Therefore, the book will present an evidence-based innovative digital solution to address pricing issues in early new product development, presenting methodological and production embedding aspects. This innovation will be referred to hereafter as the *Integrated early price assessment system*.

The Purpose of the Book

The purpose of the book is to review existing knowledge in the field of furniture manufacturing, combining capabilities of artificial intelligence and discuss the availability of human involvement into rigid prognostic process of price estimation. At the same time, we want to present the innovation approach we created to predict the early price of customised product development, which, being cheap, can be accessible to a wide range of users, including small businesses.

The dimensions of issues solved here, can be distinguished as following:

- *Manufacturing* itself constantly raises many managerial questions, many of which can be solved by innovative technological solutions of the industry (robotisation and digitalisation). Unfortunately, solutions for all manufacturing sectors are not easily available due to the large initial investment in innovation and the availability and accessibility of big data. The furniture sector, which is largely composed of small and medium-sized enterprises, faces greater challenges due to low levels of innovation and financial capacity.
- *Customisation.* Production customisation comes with its issues when it is necessary to control the customer's needs, the desire to get involved and create a unified product together. Management challenges arise in controlling time and quality while maintaining a competitive price.
- *Regionalisation.* The challenges posed by regionalisation are related to customisation, when an individual product must be developed locally, close to the customer.

The motivation to write a book is related to the completion of a 4-year project supported by the Lithuanian state called Participation Based Intelligent System to Estimate Customise Production Process (this project has received funding from the European Regional Development Fund [project No 01.2.2-LMT-K-718-01-0076], duration 2018–2021 under a grant agreement with the Research Council of Lithuania). The support scheme is aimed at increasing applied scientific research and commercialisation capabilities. This book presents the theoretical material that underpinned the methodology and research, which was not presented in the report due to its applied nature. We believe it will contribute to a better understanding of the field of both scholars and practitioners who would like to deepen their knowledge in the field.

In this book, *we fill the gap* for frugal innovation that helps manufacturing changes adapt to the market, describing the possibilities of interdisciplinary integration to create real-world working and science-based smart instruments for direct business use. This book can be seen as a synthetic work that presents a unique innovation. The described innovation was tested as a prototype. Its verification in practice is valuable in a scientific sense. The content of the book is extremely new and innovative, because when solving an old problem (determining the forecast price of a new product), innovatively combining approaches from several sciences. It is a completely new way to manage production processes and use the available historical data together with the knowledge of the employees.

Relevance and Challenges

The relevance of this study is underlined by the fact that it has been launched at the same time as preparatory research initiatives to assess the potential of Industry 4.0 are being launched across Europe. One of the noteworthy projects is titled 'Industry 4.0 for SMEs – Smart Manufacturing and Logistics for SMEs in an X-to-order and Mass Customisation Environment'. This project is funded by the European Commission H2020 through the MSCA Research and Innovation Staff Exchange (RISE) program. The fact that initiatives have been launched in a totally uncoordinated and spontaneous way in different parts of the world, raising very similar issues, shows the timeliness of the problem.

In fact, 2017–2018 was seen as still so early in the perspective of the introduction of Industry 4.0 tools. Many initiatives have been limited to the identification of needs and problems.

Our idea was to look deeper and propose operational ideas for manufacturing, specifically for the SME sector, which is not able to do it on its own.

In tackling this scientific question, we had to face paradigmatic difficulties. In today's dynamic market, relying solely on manufacturing costs to evaluate product pricing is no longer sufficient. To stay competitive, it's crucial to take a fresh approach to pricing that considers various factors such as market trends, consumer demand, and the overall value of the product. By doing so, businesses can ensure that their pricing strategy is not only cost-effective but also effective in meeting the needs of their customers. Therefore, the proposed solution of using machine learning algorithms to evaluate price destroys the existing perception and requires a change in business behaviour. Making such transformations can be mentally demanding and time-consuming.

References

Amarilli, F., & Spreafico, A. (2010). Tisettanta case study: The interoperation of furniture production companies. In *Evolving towards the internetworked enterprise*. Springer-Verlag, Inc.

Bagchi, S., Kanungo, S., & Dasgupta, S. (2003). Modeling use of enterprise resource planning systems: A path analytic study. *Journal of Information Systems*, *12*(2), 142–158.

Gawroński, T. (2012). Optimization of setup times in the furniture industry. *Annals of Operations Researchs*, *201*(1), 169–182.

Kucharska, E., Grobler-Dębska, K., Gracel, J., & Jagodziński, M. (2015). Idea of impact of ERP-APS-MES systems integration on the effectiveness of decision making process in manufacturing companies. In *International conference* S. Kozielski et al. (Eds.)*: Beyond databases, architectures and structures* (pp. 551–564).

Møller, C. (2006). On managing the enterprise information systems transformation: Lessons learned and research challenges. *Research and Practical Issues of Enterprise Information Systems*, *205*, 307–317.

Rodríguez, A. E., Pezzotta, G., Pinto, R., & Romero, D. (2020). A comprehensive description of the product-service systems' cost estimation process: An integrative review. *International Journal of Production Economics*, *221*, 1074–1081. https://doi.org/10.1016/j.ijpe.2019.09.002

Smith, D. J. (2013). Power-by-the-hour: The role of technology in reshaping business strategy at Rolls-Royce. *Technology Analysis & Strategic Management*, *25*(8), 987–1007. https://doi.org/10.1080/09537325.2013.823147

Wilkinson, A., Gollan, P. J., Marchington, M., & Lewin, D. (Eds.). (2010). Conceptualizing employee participation in organizations. In *The Oxford handbook of participation in organizations* (June 2017, pp. 1–24). https://doi.org/10.1093/oxfordhb/9780199207268.003.0001

Chapter 2

Intelligent Manufacturing From A Theoretical Perspective: The Technological Revolution and Social Participation

Birutė Mockevičienė

Mykolas Romeris University, Lithuania

Abstract

This chapter is designed to provide an overview of the challenges facing Industry 4.0, focussing on the manufacturing sector, and highlighting the specifics of small to medium-sized enterprises. Recent technologies for data science, analysts, robotics, and other smart manufacturing trends are discussed, and the opportunities, difficulties, and limitations for breakthrough development are highlighted.

Keywords: Industry 4.0; digitalisation; big data; customisation; smart manufacturing; engagement; SME

A Historical Perspective for Industry 4

The year 2011 marks the beginning of a major transformation in manufacturing. The industrial environment is starting to change radically, and this is linked to a new concept of manufacturing based on technology, in particular IT capabilities and the development of the internet. Since 2011, Industry 4.0 has been discussed, which refers to three areas of activity: manufacturing, IT, and the Internet.

The Industry 4.0 paradigm marks a substantial transformation in manufacturing, presenting a novel approach to products and their consumers. In this approach,

Participation Based Intelligent Manufacturing:
Customisation, Costs, and Engagement, 7–45

Published under exclusive licence by Emerald Publishing Limited
doi:10.1108/978-1-83797-362-020241002

the consumer assumes the dual role of both customer and designer within the product development process. The manufacturer is then tasked with identifying and satisfying the consumer's needs accurately and efficiently. This approach differs greatly from the one that led to Henry Ford's success in 1914, where he famously stated 'Any customer can have a car painted any colour that he wants, as long as it is black' became a prerequisite for Ford Motors to reduce costs, while ensuring the quality of the car manufacturing (Ford & Crowther, 1922). Today, however, this phrase is a matter of historical artifice, as more and more instruments are being developed to meet the needs of the consumer in ever greater depth.

The concept of Industry 4.0 was launched in Germany at the Hannover Fair in 2011 and has become a concept for a broad change in manufacturing. At the same time, similar movements started in the USA with Smart Manufacturing programmes. In 2010, the Smart Manufacturing Leadership Coalition (SMLC) brought together a group of more than 50 industry leaders and published a report on the goals of smart manufacturing (Smart Manufacturing Leadership Coalition, 2011). Later South Korea loaded a program for Smart Factory (Park, 2015) and Japan recognised the value of Smart Manufacturing in 2015 (Nishioka, 2015). It is not long since the European Union announced its strategic direction in this area as the 'Factories of the Future' (Davies, 2015) initiative. The emergence of Industry 4.0 is summarised in Fig. 2.1.

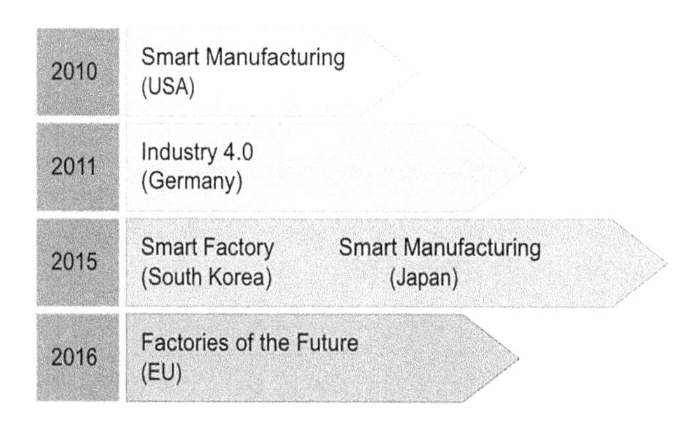

Fig. 2.1. The Emergence of Industry 4.0 as a New Phenomenon.

The importance of the Industry 4.0 concept has also been reflected in the strategies of every other national state, highlighting that the technological shift towards robotics, online live processes, and smart analytics is an inevitability that will be the subject of coherent efforts, both at the level of the manufacturing company and the state. All countries have moved along this path, recognising the need to accelerate digitalisation, as the realisation has become more than clear that innovation and change are right on the doorstep and that delay could cost regional competitiveness. And here we see a whole series of movements:

> Platform Industry 4.0 in Germany, Catapult in UK, Fabbrica Digitale in Italy, Made Different in Belgium, Industry du Futur in France, Produktion 2030 in Sweden, Made in Denmark, Smart Industry in Netherlands, Produtech in Portugal, Industria Conectada 4.0 in Spain, Production of the Future in Austria, Průmysl 4.0 in Czech Republic, Smart Industry SK in Slovakia and many others. (Matt et al., 2020)

These actions to strengthen Industry 4.0 have spread globally and continue to be reinforced in Europe, the USA, and elsewhere. The strategies being developed continue to face significant challenges, but there are also success stories (Yang & Gu, 2021). Let's take a look at Denmark. The Danish case is characterised by the significant role played by the state, not only in networking but also in funding. The state has provided 50 million dollars in funding from 2014 to 2019. The story has started in 2013, when the Danish Academy of Manufacturing (MADE) was launched as a bottom-up initiative. The MADE brings together Danish manufacturing companies, five universities, and three research and technology organisations to work together without the need for government intervention. This risk-taking approach has paid off, and the companies have increased their revenues.

Worth to list the success of France as well. Discussions that the French industry is suffering significant losses due to under-investment in digitisation have led the government to take measures to encourage investment in this area and to take steps to provide funding itself. In September 2013, the 'New Industrial France' and 'Investing for the Future' programmes were launched. New programmes covered basic research, innovation, and technology transfer (Dosso, 2020).

Smart Manufacturing: Measuring, Managing, and Advancing Processes

The manufacturing revolution (Industry 4.0) has become an inevitable stage of development. Smart manufacturing opportunities have led to new strategies in the whole production and supply chain of the component process. The approach to the product life cycle has started to change, allowing us to provide more functionalities and customisation options to customers. In addition, there is an increasing degree of flexibility in production processes and more transparency in supply chains.

Even mass customisation offers the possibility to transform into a customer-specific orientation (De Backer & De Stefano, 2021). The implementation of significant changes necessitates further restructuring in the governance layers. For instance, it is imperative to transition from traditional hierarchical and centralised organisational structures to more flexible and adaptable flat structures. Collaboration within and beyond the organisation is now a critical aspect that plays a vital role in ensuring the excellence and sustainability of supply chains.

We will then discuss the innovations of Industry 4.0 and how they are shaping manufacturing.

Industry 4.0: The Concept

The concept of Industry 4.0 covers the whole transformation along the entire chain of activities, from manufacturing, services, and the consumer market, which is based on maximum digitisation. Digitalisation will lead first to the emergence of advanced manufacturing and then inevitably embrace of supply channels, the necessary flows of all resources and values (Pech & Vrchota, 2022).

When starting a discussion on Industry 4.0, it is important to define it. Although it is a new field, it is evolving so rapidly that new and more advanced and complementary technologies are emerging every year, making it tempting to revise the definition of Industry 4.0 annually.

In 2015, the EU has defined Industry 4.0 in its strategic documents as a set of technologies and enabling mechanisms. According to the European Commission (EC, 2015), Industry 4.0 is defined as follows:

- *Information and communication technology* (ICT) to digitise information and integrate systems at all stages of product creation and use (including logistics and supply), both inside companies and across company boundaries.
- *Cyber-physical systems* use ICTs to monitor and control physical processes and systems. These may involve embedded sensors, intelligent robots that can configure themselves to suit the immediate product to be created, or additive manufacturing (3D printing) devices.
- *Network communications*, including wireless and internet technologies serve to link machines, work products, systems, and people, both within the manufacturing plant and with suppliers and distributors.
- *Simulation, modelling, and virtualisation* in the design of products and the establishment of manufacturing processes.
- *Big data analysis* and exploitation, either immediately on the factory floor or through cloud computing.
- *Digital assistance systems* for human workers, including robots, augmented reality, and intelligent aid.

The goals of Industry 4.0 are defined by leveraging the advantages and tools of digitisation. It aims to cover various *aspects*, such as (Yang & Gu, 2021):

- ensuring seamless mass production by adopting IT;
- managing the production chain automatically and flexibly;
- tracking the necessary components for production;
- creating an automated connection between parts, products, and machines, and
- applying human-machine interaction paradigms.

Additionally, it aims to achieve optimisation of production in smart factories through the Internet of Things and provide new types of services and interaction business models along the value chain.

These objectives can be achieved by focussing on five areas of *activity (values)*, namely:

1. interoperability;
2. virtualisation;
3. decentralisation;
4. real-time capabilities; and
5. service orientation and modularity.

Modern technologies are now available to digitise all stages of production. This includes the use of sensors in almost all production components, analysing sensor data, and automating and robotising systems.

Industry 4.0 as a concept was made possible by the identification of five highly interdisciplinary key areas of activity that unite for the benefit of the manufacturing future. These are:

1. digitisation of production, optimisation, and customisation;
2. automation and customisation;
3. human-machine interaction (HMI);
4. value-added services and business; and
5. automated data exchange and communication.

Here, we review the most common list of components of Industry 4.0:

Cyber-Physical System

Industry 4.0 can be interpreted and understood as the Cyber-Physical System (CPS). It is a secure and interoperable smart grid system that combines and integrates cyber and physical components such as sensors, control, processing and computing, communications and execution elements. The main task of such a system is to interact in real-time with the physical world and with human users (e.g. a Building Management System, or a Smart Grid covering an entire country) (Abiri & Kundur, 2020). In manufacturing, these systems connect physical objects (production lines) to virtual models with the help of sensors. It is important that the CPS performs in a stable and meaningful way when coupled with large artificial intelligence tools, as it becomes a source of big data generation itself (Tay et al., 2018).

The expanding functionality of the CPS gives importance to both the development of the Internet of Things (IoT), which can expand its scope and make it easier for companies to build global networks, and the diversity of the Internet of Services (IoS).

Internet of Things

The Internet of Things (IoT) is about physical objects or groups of objects with sensing and communication capabilities that connect and exchange data over networks. It is the interconnectedness of everyday objects and things, often with ubiquitous intelligence. The main aim is to integrate the object system into the web space so that such an object can be continuously connected, with unique identification on the network (Caputo et al., 2016).

When it comes to manufacturing, we bring the concept of the Industrial Internet of Things (IIoT) (Le Mouël & Carrillo, 2023), which combines the IoT and CPS in the context of Industry 4.0. The IIoT enables machines/devices to communicate and share information, creating a smooth industrial ecosystem in manufacturing that requires reliable interoperability to function. IIoT interoperability is commonly referred to machine-to-machine/device-to-device connections, where multiple devices can exchange data. It therefore becomes important to establish reliable and ubiquitous communication between heterogeneous embedded devices in the network.

Internet of Services

The Internet of Services (IoS) acts as a component of the IOF and is particularly significant for the manufacturing sector. It is the communication of information for knowledge management in the service domain. The IoS is initiated through the transfer of data by information technology, as 'service dealers' provide services over the Internet according to service types, such as digitisation services. These services are available and on-demand according to business models, partners or any set-up. This is how value-added services are designed, so that Service Providers deliver and bundle services.

Big Data and Analytics

Industry 4.0 generates big data from a large variety of channels, industrial sensors, financial transactions, log files, video/audio, network activity, and social media feeds. This data is becoming very interesting for decision-making in various areas, such as product quality management, meeting user needs, or strategic decisions. The need to collect and analyse big data is therefore driving the search for new methods of analysis, as the traditional ones are no longer able to process such volumes.

In defining what constitutes big data, we refer to the classification of big data according to its volume, variety, value, and velocity (Witkowski, 2017).

Augmented Reality

Augmented reality (AR) technologies are radically improving the perception of the real world by providing its users with additional information such as graphics, text, video, and 2D and 3D models. Increasingly, AR experiences can be mediated by simple devices such as smartphones and tablets or by more sophisticated tools such as smart glasses (Aquino et al., 2023). Augmented Reality as a technology, can provide tremendous support for maintenance work in manufacturing by reducing the time spent on maintenance work and potential errors in maintenance activities (Tay et al., 2018).

Autonomous Robots

Autonomous robots are automated machines designed to perform repetitive actions at high speed and with high precision. In addition, such robots can work where human work is limited (dangerous or difficult for humans to access) (Yang & Gu, 2021). In this case, industrial robots are often controlled using augmented reality. Depending on the different needs, small robots or even a whole assembly line, a vehicle-like robot, an android or a patrolling robot with legs are developed. They have penetrated into the areas of chemical processing, pharmaceutical manufacturing, food, and beverage production. In addition to automation robots, collaborative robots are being designed. The first are designed to automate operational processes, while the second aims to work alongside humans to help them perform tasks that increase human flexibility and productivity (Frank et al., 2019).

3D Printing

3D printing, officially known as additive manufacturing in Industry 4.0 terminology, allows the construction of a variety of complex geometries from metal or plastic. 3D is only made possible by IT technologies and is fundamentally changing the principles of manufacturing, encouraging the search for new materials to exploit the possibilities of the 3D principle. 3D printing can be applied to a wide variety of products. One of the more popular areas of development is customised manufacturing. Additive manufacturing promotes sustainable production because it requires only one process and generates less waste than traditional manufacturing (Frank et al., 2019). With the growing trend towards digitisation and the requirement to shorten the life cycle of products, more and more areas are discovering the benefits of 3D printing. In addition, it helps to reduce the weight of components and reduce waste, making it particularly useful for the automotive and aerospace industries. It is important to recognise that 3D printing allows flexibility in the choice of the production site, thus reducing transport costs and inventory. 3D printing has led to the development of related technologies (deposition melting or selective laser melting) and, as customisation is becoming a trend, it is expected that more and more customised 3D printing services will become available in the coming years (Yang & Gu, 2021).

Cloud Computing

Cloud computing (CM) is a system that provides the user with a huge amount of storage space and the ability to move and access computing services over the internet (access to networks, servers, storage, applications, and services) (Chen, 2016). Such big data storage is becoming cheap and affordable for manufacturing companies, which are starting to generate more and more data. Thedesignofnewdatacentresystemsisgraduallyshiftingtowardsthistypeofstorage. Cloud manufacturing is becoming much more flexible and can involve many actors from all stages of the production process. As a result, cloud computing makes it possible to incorporate customer requirements and engineering concerns into the development phase of a new product, which significantly reduces delivery times. In addition to design, there are aspects of cloud computing, such as software-as-a-service, customer relationship management, data analysis, collaboration and business planning platforms (Yang & Gu, 2021).

Simulations

Simulation is a technique for reproducing a real or virtual process or system in order to clarify or estimate the output. This simulation is more advanced with digital manufacturing, which offers the possibility to consider data from all the virtualised objects on the shop floor and then simulate the operational processes, taking into account several parameters that can affect production. Simulations are performed using real-time data to represent the real-world simulation model, and the models include workforce, product, and machinery interactions (Rosin et al., 2020). Operators can therefore optimise machine settings in a virtual simulation before implementing them in the physical world. This reduces machine setup time and improves product quality. A breakthrough in simulation has come from the application of the virtual factory concept.

In an attempt to summarise and group all emerging new technologies aimed at improving industrial performance, one of the most representative ways is to group them into the following five areas (see Fig. 2.2):

1. Internet of Service:
   - Delivery channels.
   - Application programming interface economy.
   - Integration across value chains.
2. Internet of Things:
   - Ability to connect and manage devices.
   - Real-time data collection.
   - Insight of what is happening.
   - New business model.
3. Flexible manufacturing. Flexible Manufacturing Systems (FMS) serves these capabilities by engaging programmable automation for machining and material handling. A flexible manufacturing system usually consists of computer numeric control (CNC) machine tools, interconnected by automated material handling and storage systems, and control by an integrated computer system.
   - Flexible machine.
   - 3D printing.
   - Mixed human robotics.
   - New protocols.
   - Vertical and horizontal integration.
4. Autonomous system
   - Smart and networked products.
   - Self diagnosis.
5. Analytics
   - Embedded in equipment products and services.
   - Predict what may happen.

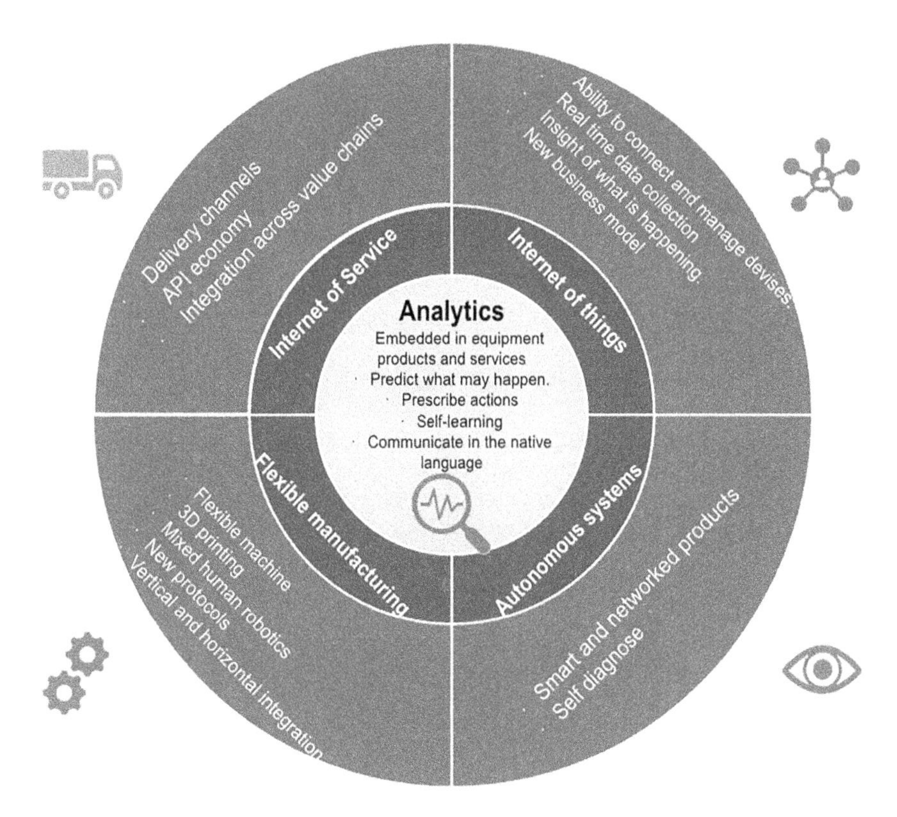

Fig. 2.2. The Content and Components of Industry 4.0.

- Prescribe actions.
- Self learning.
- Communicate in native language.

This classification is quite convenient for the manufacturing business as it identifies an area of performance that can be potentially strengthened.

However, this grouping has its own limitations due to the constant emergence of so-called *front-end technologies (smart manufacturing, smart product, smart working, and so on* (Frank et al., 2019)) to address the challenges of smart manufacturing.

The greatest benefits of Industry 4.0 will be generated mainly in these production fields:

- Speed of production – production processes have become faster as a result of IT, the Internet and data processing.
- Product quality – monitoring and sensors have made production quality easier to achieve and parameters more precisely adjusted.
- Flexibility of production – new IT technologies allow new developments and design solutions to be implemented right here and now, simulation capabilities allow ideas to be tested at an early stage with minimal cost.

- Customisation – flexibility of production allows customisation of production, which becomes a prerequisite for changes in market needs. The market for customised products is therefore growing rapidly.

Summarising, the possibilities of Industry 4,0 triggered diverse human activities, such as *business models, data manipulation, legal issues concerning intellectual property rights, quality standards, and the competencies and skills of professionals* (Matt et al., 2020).

The Role of Smart Manufacturing in Industry 4.0

Industry 4.0 has always been associated with smart manufacturing, which is still a crucial part of the Industry 4.0 framework. As new technologies emerge, the discussion about the significance and relevance of smart manufacturing continues. It is essential to understand each technology's intended purpose and role in the production process to make the most of them. Smart manufacturing involves advanced technologies such as machining, cutting, and forming. In Table 2.1 presented a frequent, though not definitive, list of available technologies.

However, to gain a deeper understanding of the role of the concept of Industry 4.0 to Smart Manufacturing, a more comprehensive and conceptual categorisation of technologies is required. A simple and clear conceptual model was proposed by Frank and co-authors by differentiating between two groups of technologies, base technologies and front-end technologies and demonstrating how Smart Manufacturing plays its part in combining technologies.

This conceptual model analyses Smart manufacturing as one of the key components of Industry 4.0, along with Smart Products, Smart Work, Smart Supply Chain, and enabling base technologies (see Fig. 2.3).

Table 2.1. The Most Advanced Technologies, Associated with Industry 4.0 (European Commission, Directorate-General for Internal Market, Industry, Entrepreneurship et al. 2016).

High Performance Manufacturing Technologies	**ICT-Enabled Technologies**
- Industrial robots/ handling systems - Automated Warehouse Management Systems - Technologies for safe human-machine cooperation improved usability - Manufacturing micromechanical components - Additive manufacturing - Photonics (other than additive) Processes specific to Advanced Materials - Nanomanufacturing - Processes for bio-manufacturing - High-performance machinery	- Virtual reality/simulation in production reconfiguration - Virtual reality/simulation in product design, Digital design technologies, Design platforms for modular, adaptable manufacturing - Supply chain management with suppliers/customers, Network-centric production, Optimisation of production networks - Product Lifecycle Management Systems, Product Data Management Systems

Table 2.1. (*Continued*)

High Performance Manufacturing Technologies	ICT-Enabled Technologies
- Modular and adaptable (interoperable) machines - Cutting and machining techniques for rapid prototyping equipment manufacture - Rapid time-to-market enabling technologies - Self-adaptive production lines—Printed electronics/roll-to-roll processes - Silicon-on-chip, heterogeneous circuits, embedded systems, and Integrated photonic circuits - Microelectromechanical systems (MEMS) and sensor devices - Nanoelectronics materials and patterning - Nanoimprint (process and equipment), Precision manufacturing and metrology	- Enterprise Resource Planning - Technologies that depend on the use and coordination of information, automation, computation, software, sensing, and networking - Mass customisation (three-dimensional printing, direct digital manufacturing) - Cyber-physical (production) systems, intelligent components - Cloud manufacturing other

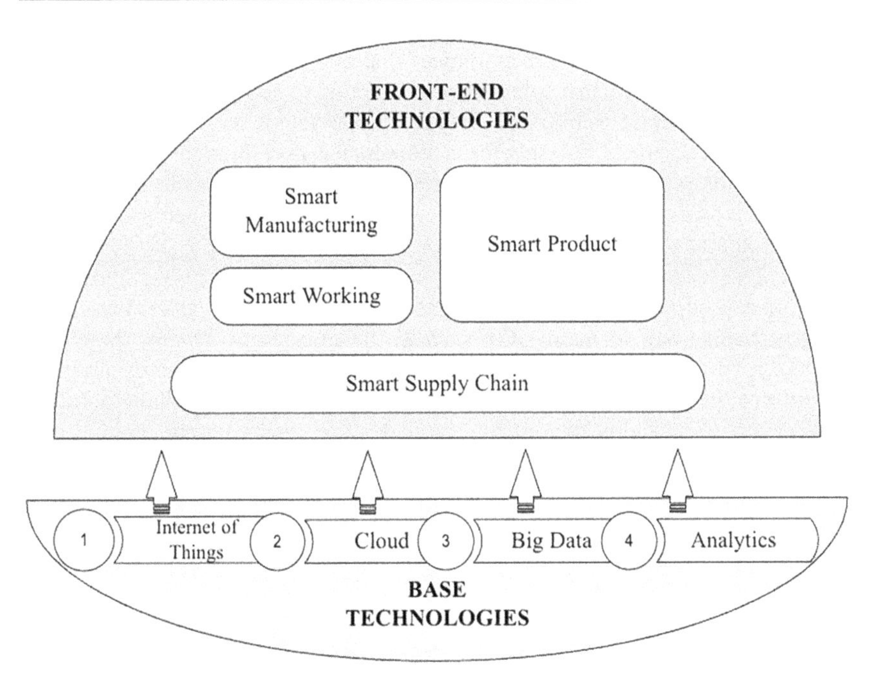

Fig. 2.3. Theoretical Framework of Industry 4.0 Technologies According to Various Technologies (Frank et al., 2019).

The concept distinguishes between the 'smartness' dimension, which focusses on operational and market needs, and the technologies that provide connectivity and intelligence. Front-end technology groups primarily serve the enterprise value chain and, due to their complexity, are the most challenging to realise. These combinations of technologies are what differentiate Industry 4.0 from other industry sectors.

Here is the list of front-end technologies:

- *Smart manufacturing*: Initially, the term Industry 4.0 was associated with the concept of Smart manufacturing. It described a system that could adapt and adjust production processes on flexible assembly lines to accommodate different product types and changing conditions. Smart manufacturing is then described as one that enables quality, productivity and flexibility, so that it can realise customised products on a large scale with efficiently managed resources. Smart manufacturing involves technologies for product development and production (production system), as distinct from smart product technology, which refers to new product features—technologies related to product innovation.
- *The smart supply chain*: Ensures the functioning of the smart manufacturing by synchronising production with suppliers through intelligent IT systems (reducing information distortion). The opportunity to specialise in a particular field is created by sharing resources in a collaborative way through the supply chain.
- *Smart working*: Smart working as a concept deserves to be categorised in its own category from smart manufacturing, as smart working contributes to the social dimension and ensures the efficiency of production operations. It is about those Industry 4.0 technologies that allow workers to perform tasks more productively and in a coherent way, adapting to new production methods.
- *Smart product*: Base technology changes and develops the product, creating more product features not only for customers but also for smart manufacturing itself, for example, the product provides feedback on the use of its features and this information can be used to develop new products. Such product features fundamentally change the business models themselves.

Another group of technologies could be called *base technologies*. Base technologies include new forms of ICT such as the Internet of Things, the cloud, big data, and analytics. Each group is characterised by its unique capabilities. The Internet of Things has become increasingly important in the manufacturing industry due to its ability to connect objects through new communication links. Additionally, the cloud provides an extensive range of possibilities, while big data and analytics provide access to intellectual applications, allowing for increased precision and unexpected analytical findings.

After identifying the key features of Smart Manufacturing, it is important to group the chosen technologies based on the most significant challenges. Six main objectives can be identified (Frank et al., 2019):

1. *Vertical integration*: Vertical integration reduces decision-making levels and eliminates human intervention. This is possible by digitising processes

(installing sensors and controllers). The sensor system requires its own data collection system (supervisory control and data acquisition system) to manage the sensors, to provide on-site control and diagnostics. These systems are connected to the Manufacturing Execution System (MES) and provide production status data to the Enterprise Resource Planning (ERP) system. Therefore, vertical integration provides greater transparency and control of the production process and helps to improve decision-making on the shop floor through machine-to-machine communication.

2. *Virtualisation*: In digital manufacturing, virtualisation enables direct communication between machines. This technology plays a vital role in streamlining the manufacturing process. Various technologies exist for simulating processes using sophisticated artificial intelligence techniques. These technologies are designed to replicate real-world scenarios and generate insights that can be used to optimise processes, reduce errors and improve performance.
3. *Automation*: Technology has the potential to enable maximum automation, as robots can perform tasks with greater accuracy and speed than humans. Any spelling, grammar, or punctuation errors have been corrected. Only processes that automate production without human intervention are considered robotic. Utilising artificial intelligence can facilitate automation by analysing activities such as monitoring and predicting equipment failures, scheduling preventive maintenance, and detecting product nonconformities during the development phase. This can significantly elevate quality control, decrease production costs, and offer predictive data to ERP systems.
4. *Traceability*: Traceability becomes an added benefit and value for both the producer and the user. We can trace deviations in raw materials and the product itself at any stage of production, from raw material in the warehouse to the point of consumption. We can also improve products at an early stage of development by integrating information gathered from the user or the production line, thus contributing to flexibility.
5. *Flexibility*: Modular machine systems allow flexible production lines that can be easily modified to fit customised production needs, increasing product variety without sacrificing quality or productivity.
6. *Energy management*: Smart manufacturing inherently involves sustainability, which comprises factory efficiency and energy management. Leveraging existing IT to ensure traceability through monitoring energy consumption data and virtualisation by applying different consumption patterns based on price is a mandatory requirement.

By categorising and emphasising the primary goals for operational modifications, the classification of technology types enables us to comprehend the principal motivators for implementing Industry 4.0. This approach also indicates that it is feasible to adapt and transform through specific levels of maturity. Research indicates (Frank et al., 2019) that once a company reaches a particular level of maturity in its front-end technologies, smart manufacturing technologies become complementary rather than substitutable. Once a mature technology is introduced, it is used and improved if necessary, and an emerging technology is adopted to tackle new challenges. Furthermore, certain technology functions are more easily adaptable than others. For instance, technologies related to vertical integration, energy efficiency, and traceability (such as sensors with MES and ERP systems) are often considered first-stage maturity functions. Factory virtualisation and automation only come at a later stage, indicating the need for advanced readiness. Finally, flexibility is the hardest to achieve. As compared to the other two clusters, more companies deploy flexible technologies.

Smart Manufacturing remains a crucial component of Industry 4.0 and is closely linked to the development of smart products (Frank et al., 2019), through foundational technologies.

Digitalisation in Smart Manufacturing

Big Data

Big Data has played a significant role in the development of Industry 4.0, affecting our daily lives and transforming production. Big data in manufacturing has led to the proliferation of disciplines such as statistical mathematics, econometrics, modelling, and optimisation, and the corresponding need for their guidance. The implementation of advanced Industry 4.0 technologies is leading to the generation of vast amounts of data. One of the most illustrative examples of a big data generation case relates to the sensing and recording of any event. Exploring the field of Big Data further provides additional opportunities to extract valuable insights that can support decision-making in manufacturing. Handling big data is essential for addressing complex manufacturing issues, such as product quality, production costs, manufacturing processes, and meeting customer needs. It is clear that big data will have a major impact on manufacturing, and it is no longer sufficient to use conventional tools to collect, manage and analyse this data.

When it comes to dealing with Big Data, it raises complex questions that require careful consideration. Defining Big Data is a complicated task due to the issue of data diversity. Data generated in production can be in various formats such as image, numerical, and textual. Additionally, the data collected can describe different content, such as production line sensors, product composition, and infrastructure parameters. Furthermore, the quality and completeness of the collected data may also vary, which can create difficulties (Wuest et al., 2016).

So, what is big data and when does it become big data?

The Big Data definitions have been evolving, becoming more and more detailed. Whereas in the past it was content to be limited to its size, the definition is now based on the 7V (volume, variety, velocity, veracity, variability, volatility, and value) concept. Big Data is best described by the following seven parameters or qualities (Belhadi et al., 2019):

- volume (the size/scale of the data);
- variety (the form/format of the data);
- velocity (the rate of the data being produced);
- veracity (the uncertainty/reliability of the data);

- variability (this means that data from different sources have to be combined, matched, cleaned and transformed because they are complex and have strong differences between them);
- volatility (the capacity to store and maintain data for convenient and safe access while preserving its speed); and
- value (the value of big data lies in its ability to be processed and analysed, which is lost when the data is only aggregated).

There are two main ways in which Big Data in production is emerging. One stream of Big Data corresponds to necessary production processes (e.g. process operation, control computers and information systems) and the other provides monitoring functions (internet of things, censoring, videos, and indirect measurement technologies).

Data Mining and Big Data Analytics in Manufacturing

Data mining is one of the first, and probably one of the most important, stages of knowledge discovery, based on database operations to search for hidden information when we accumulate Big Data. Data mining gained significance with the advent of IT's capacity to handle large databases, prompting the emergence of specialised techniques in the field. Database technology serves as the foundation for storing, organising, and performing other data mining operations.

Data mining is realised through a distinction between two main functions: The descriptive functions and the predictive functions (Cheng et al., 2018).

Descriptive functions of data mining are used to explore the potential properties and relationships that already exist in the data. These functions include generalisation, association, and clustering. By using these functions, it is possible to identify critical parameters or states of a phenomenon and then create effective solutions that can be used to predict targets.

- *Generalisation* is a method of summarising information that describes the characteristics of a particular type of data. It involves creating a single or higher-level form of knowledge summarisation. Detailed data is usually stored in a database, but this can make it difficult to interpret. Generalisation helps to find a single or higher-level description of knowledge. This knowledge is often used for classification and prediction.
- *The association* reflects the dependence between an event and other events. Based on the different rules of association, it can be divided into simple association, temporal association, causal association, etc. Association rules are quite convenient to use in business practice, including customer behaviour analysis, market segmentation, and target customer selection.
- *Sequential pattern* mining involves identifying patterns in a set of organised data sequences. Unlike association, its goal is prediction. It is commonly used to predict sales of goods, anticipate peak electricity demand, study common gene combinations in a group of patients, and more.
- *Clustering* is a technique used to group data into similar categories by identifying the smallest possible differences between individuals in the same group and

the largest possible differences between individuals in different groups. A commonly used criterion for measuring similarity is the geometric distance. Nowadays, neural networks are being used more frequently for clustering purposes.

The prediction function is designed to analyse relevant trends in the data to predict future states. It performs tasks such as classification, prediction, and time series analysis. The forecasting function uses historical data to make inferences about future attributes and predict future values and trends.

- *Classification* process involves creating a model, known as a classifier, that can assign objects in a dataset to known classes based on their characteristics. This process involves two main steps: first, building the model through training, and then using the classifier to predict the class of new objects. Classification is commonly used in areas such as guest models, image pattern recognition, and target market positioning.
- *Prediction* is a method of predicting future trends by analysing historical and current data. It is used to identify missing or unknown data values and to develop continuous curves. This technique is commonly used in various fields such as medical diagnostics and performance forecasting, especially in manufacturing.
- *Time series analysis* is a method of analysing data that has time as its most distinguishing feature. It is widely used to describe financial phenomena and to make both short-term and long-term predictions. A time series is a particular type of data that assumes that sequential values have an impact on future values. This type of data is often useful for observational purposes, as it captures changes in a phenomenon over time.
- *Knowledge extraction* is a technique used in data mining to identify anomalies and analyse deviations. This method is focussed on detecting behaviour in the data that is different from the expected behaviour of objects. Knowledge extraction is commonly used to trace banking transactions and other instances of fraud.

Ontology

Most studies on data mining use traditional methods with some modifications to emphasise the importance of newly added mechanisms. The Ontological Data Mining Agent (OntoDMA) (Lin & Yang, 2019) can select new data mining options based on web content. Ontological analysis highlights that self-developed data mining techniques are more suitable for specific application domains and are not limited by applicable tools. Ontology-based selection rules for data mining are based on the conversion of the semantic distance of search terms and the conversion of hypernyms, hyponyms, synonyms, and antonyms of corresponding search terms. This makes ontology-based analysis more productive. The location of ontology data must consider the data mining architecture and be complemented by a data warehouse structure. Ontology-pure indexing should intervene upstream of the prediction rule detection link, thus naturally integrating these rules into the overall criterion prediction task.

Big Data Analytics

By leveraging data mining techniques along with established methods, the objective is to devise approaches that can be easily adapted to manufacturing data's peculiarities while satisfying manufacturing's needs at the least possible expense. In order to achieve specific production objectives that are better suited for data usage and created within knowledge management systems, another concept, big data analytics, is to be introduced, combining Big Data management with data mining techniques.

Big Data analytics in manufacturing processes gained significance following the digitalisation process from 2014 onwards (Belhadi et al., 2019). It is evident that there is an increasing interest in using Big Data Analytics (BDA) to solve manufacturing problems. Let's take a look at the most popular analytical tools used in manufacturing.

- *Descriptive analytics*: Backwards-looking analytics show what has already happened or what could happen. The most commonly used tools for this purpose are regular and adhoc reports, as well as alerts such as scoreboards, dashboards, and periodic summaries of indicators, which provide monitoring capabilities. This is one of the most direct ways to promote data-driven decision-making.
- *Inquisitive analytics*: This analysis is based on the results of descriptive analytics, which involves using data to describe a phenomenon. The aim of this analysis is to identify the causes of the phenomenon, which can include examining dependence, similarity, correlation, generalisation, associations, sequential patterns, and clustering.
- *Predictive analytics*. It involves predicting the future trends based on historical and current data. This is achieved using forecasting and statistical modelling techniques, which include supervised, unsupervised, and semi-supervised learning models. There are two main approaches distinguished for prescriptive analysis performance. The first is statistical analytics-oriented, which uses mathematical models to analyse existing data and predict unknown situations. This includes polynomial logit models, regression methods, and Bayesian methods. The second approach is trend discovery, which does not require any prior assumptions. This approach involves machine learning methods such as neural networks, multiple backpropagation, self-organising map, fuzzy set, genetic algorithm, and association rule.
- *Prescriptive analytics*. Optimisation of process models is achieved through normative decision theory, which relies on predictive analytical models to find causal links for selecting actions.

The logic between Big Data Analytics approaches and particularly methods, the goal of analyses is presented in Fig. 2.4.

Belhadi et al. (2019) identified three levels of conceptualisation when assessing the prevalence of Big Data Analytics in manufacturing.

The first level: The initial layer of BDA capabilities deals with the challenges linked to production processes that usually create significant value.

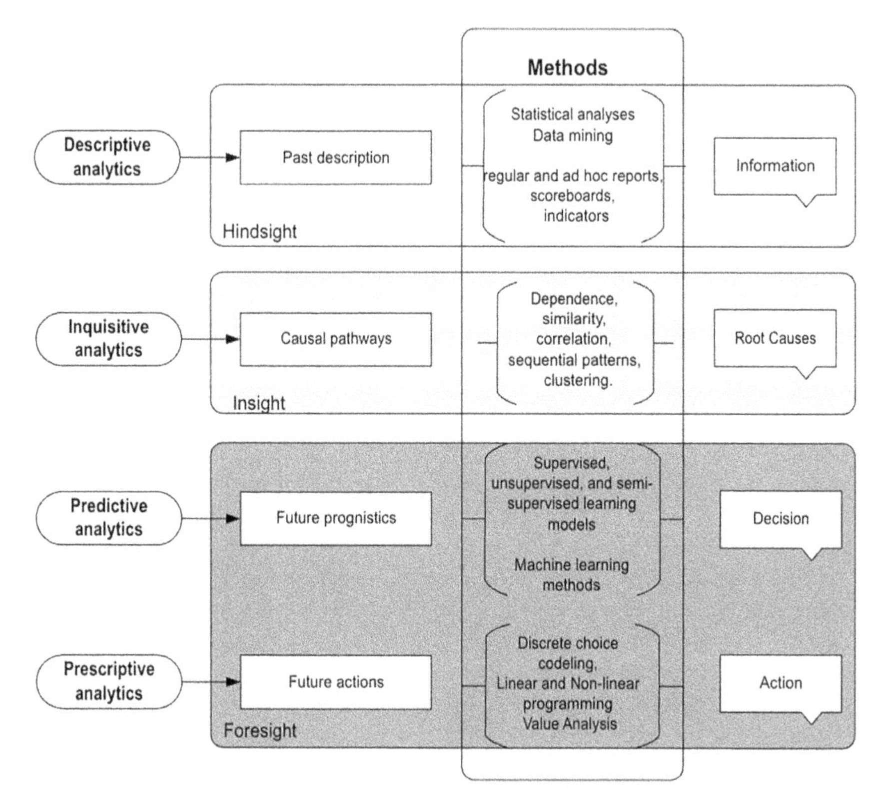

Fig. 2.4. Classification of Big Data Analytics and Used Methods for Manufacturing (Belhadi et al., 2019).

These challenges encompass quality and process control, energy and environmental efficiency, proactive diagnostics and maintenance, and safety and risk analysis. Naturally, these activities generate the necessary data flow that is compliant with Big Data standards.

The second level: The second level is connected to the first level, and it pertains to the necessary capacity and infrastructure that are needed to support the flow of data produced by the process. This includes four components: Data Warehouse, Data Aggregation and Integration, Data Analytics and Modelling, and Data Governance Culture.

The third level: The third level refers to the value that the BDA must bring to the production process. Only when this value is present can the BDA be truly integrated into the production process. This value is about increasing process transparency, improving performance, supporting decision-making, and enhancing knowledge.

In order to link data mining techniques to the needs of analytics in the production process, it is necessary to identify the most important points in the manufacturing process where analytics are most needed. In terms of production processes, advanced planning and scheduling (APS) (new rules and optimisation), quality improvement (parameter optimisation), defect analysis (classification), and fault diagnosis (time and state of the defect) are some essential points to consider (Cheng et al., 2018).

Let's take a look at the available approaches for manufacturing regarding data mining techniques.

- *Advanced planning and scheduling in manufacturing*: Advace planning and scheduling can be seen as an umbrella technology with multiple functions, covering both the time factor, but also the level at which the planning needs to be done, whether it is strategic, tactical or operational planning (Van Amstel, 2003).
  - *Extracting new scheduling rules*: There are two approaches to identifying new planning rules. The first approach, called SA (statistical analyses)-oriented data mining techniques, uses local clustering to modify a predefined schedule and calculate a measure of forecasting quality. The second approach, known as knowledge discovery-oriented data maining technology-based applications, uses data mining classification and optimisation to extract new planning rules from historical planning data. Research shows that the extracted planning rules can produce better planning schemes than the historical scheduling themselves.
  - *Dynamically choosing the optimal scheduling rules*: A possible approach to handling knowledge is through the real-time integration of modelling, data mining, and statistical process control. The suggested planning system periodically chooses a dispatch rule from a decision tree, created by extracting knowledge from production data. The decision tree can adjust to changes by being dynamically updated, with the possibility of using artificial networks for this purpose.
- *Applications in quality improvement for manufacturing*:
  - *Parameter optimisation tasks*: It involves determining the ideal level of a process to achieve a desired level of quality based on learned high-quality characteristics. *SA-oriented* Data mining technology-based applications utilise either Monte Carlo or Bayesian nonlinear models to estimate ideal accuracy for given massive datasets, or set control limits to build a causal sampling diagram. These applications also incorporate human-machine co-intelligence techniques to optimise parameters, typically used to identify quality deviations.
  - *Quality classification tasks*: The process of procedural contentiousness involves the use of predefined clustering algorithms to group together similar sub-assembly products. This method can lower the cost of non-standard sub-assembly products by separating them into groups of similar, usable products and defective products.
  - *Process monitoring tasks*: It is crucial to monitor manufacturing processes to detect the need for remanufacturing as early as possible, as well as to identify any unusual patterns. In some cases, it may be necessary to quickly identify the source of an uncontrolled signal or a point of quality change.

Some applications use Bayesian Posterior Update Sequentially and Hierarchically to detect anomalies in the system. This is achieved by monitoring relevant measurement specific and assessing how they affect the shape of posterior cumulative distribution functions. The proposed method not only allows for fast and reliable anomaly detection through real-time data processing, but also outperforms previous methods in terms of having a low computational cost and a low number of false alarms.

  - *Products/processes description:* To describe or characterise products and/or processes using statistical data mining, it is necessary to identify the attributes or variables that have a significant impact on quality. These attributes or variables should be ranked in order of importance. This can be done by assigning products to categories such as low, medium or high quality.

- *Defect analysis in manufacturing:* Defect analysis aims to identify and categorise different types of defects. This can be achieved by either clustering defects with varying degrees of accuracy, which may result in new defect types, or by classifying defects into known categories. The ultimate goal of defect analysis is to gain a better understanding of the types of defects that occur and to develop effective strategies for preventing them in the future. The proposed method aims to provide a more accurate and efficient way of analysing general defect patterns. It involves a multilevel defect analysis approach that uses the nearest neighbour noise removal method and two cluster analysis methods to obtain clustering results with varying degrees of accuracy. This method estimates the number of defect clusters and their spatial patterns, and it outperforms existing methods in terms of both detection accuracy and computational speed.
- *Fault diagnosis*: The problems refer to identifying and predicting the type of faults in machines or parts by analysing their operating states. Such analysis can be used to predict maintenance durations and provide optimal maintenance solutions. Currently, most of the available applications rely on knowledge development. This type of data mining involves the use of artificial intelligence networks to predict selected variables. Decision tree algorithms are also used to build classification models.
- *Pricing and customer orientation*: Data mining has a variety of applications, including working with customers. By using data mining, businesses can reach more of the right customers and increase their market share and profits. Managers can gain a better understanding of their customer groups and use this information to develop optimal offers and products tailored to their individual needs. Additionally, data mining can help businesses differentiate their pricing strategies, moving beyond traditional cost models, and pricing according to new paradigms (Han et al., 2012).

Applications of Artificial Intelligence

Digitalisation has revolutionised manufacturing, offering new opportunities for optimisation, automation, and customised production. Digitisation has made mathematics and computer science results easily accessible and usable. The existence of easy-to-use and often free software tools has opened up immense potential for the sustainable transformation of manufacturing and the perception of increased data repositories in the manufacturing industry. The increasing demand for Big Data Analytics has opened doors for analysing production data in real-time and dynamically (Belhadi et al., 2019). However, this has also created significant challenges in terms of the ability to identify faults, defects and some other unusual situations in real-time and to link these analytical results to timely decision-making. Various artificial intelligent techniques have made these developments even more conclusive. In the following paragraphs, we will explore the methods provided by artificial intelligent and some of the most popular general applications in manufacturing.

Machine Learning (ML)

Machine learning has become a major breakthrough in the digitalisation process, where data mining, artificial intelligence and knowledge discovery from databases have been made available for a wide range of industrial applications, including manufacturing. However, while machine learning has enabled much deeper intelligence solutions, the high complexity and paradigm-shifting nature of machine learning still hinders the exploitation of the Big Data generated by the manufacturing sector.

Definition. According to Samuel, machine learning is the mechanism allowing computers to solve problems without being specifically programmed to do so (Samuel, 1959).

Machine learning is a subset of artificial intelligence (AI) that involves the use of data analysis to create computer models that can learn from data in order to make decisions or predictions.

One of the most important features of machine learning that is valuable in manufacturing is the ability to find highly complex and, most importantly, nonlinear trends in data of different types and sources, and to turn the raw data into meaningful data by giving it relevant attributes. The basic analysis of such data is further applied to forecasting, discrepancy detection, classification and regression. There are numerous methods utilised in machine learning designed to analyse large amounts of data, particularly high-dimensional data exceeding 1,000 dimensions (Wuest et al., 2016).

Machine Learning Pipeline and Quality Attributes

ML pipeline (Aloisio et al., 2022) is an important tool for data scientists, practitioners, and researchers in various fields. These pipelines allow for the creation of multiple machine learning models, covering the entire analytical process, from raw datasets to the final model (Bodendorf & Franke, 2021). Over time, many solutions have been developed that automate the development of machine learning pipelines. These solutions have mostly focussed on the semantics and characteristics of the input dataset.

The pipeline is crucial for machine learning applications. It takes raw data as input, pre-processes it, and calculates necessary changes. First, the data is

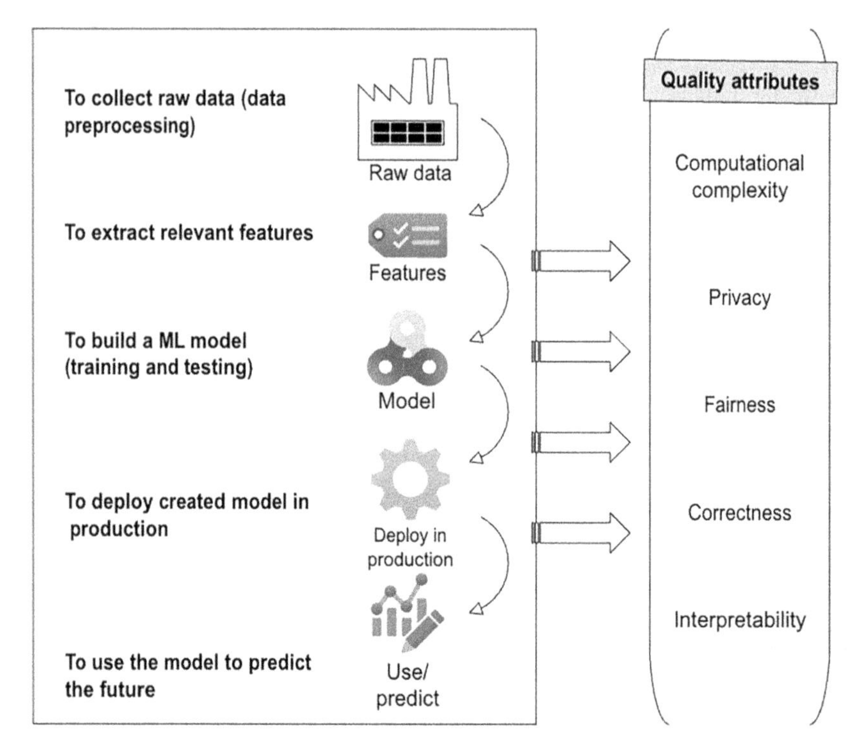

Fig. 2.5. Generic Machine Learning Pipeline with Quality Attributes.

analysed and important features are extracted to build a machine learning model. The model is then trained and tested. It can be evaluated further with human intervention, and once it's implemented, its performance is continuously monitored (see Fig. 2.5). The machine learning pipeline itself is an iterative process, which means that most of the steps depicted can revert back to previous steps to achieve a qualitative performance. It's important to note that these steps aren't mandatory and some can be skipped if unnecessary.

When discussing the quality criteria, it is worth taking into account the specific steps in the pipeline. To attain high-quality machine learning systems, it is crucial to identify and define the key quality attributes of such systems by selecting the most significant ones and formalising them in the context of pipeline development.

There are three main components that are agreed upon for the identification of quality attributes in machine learning systems (Aloisio et al., 2022):

- *Training data*: The quality of training data is usually measured in terms of features such as privacy, bias, number of missing values, and past pressure.
- *Machine learning models*: When referring to a machine learning model, the authors refer to the trained model used by the system. The quality of this component is usually assessed in terms of fairness, explain ability, clarity, interpretability, and security.

- *Machine learning platforms*: The machine learning platform is a system that is heavily impacted by security and computational complexity. The quality of the platform can be measured by various factors such as data quality, ethics, privacy, fairness, performance of machine learning models, and more. Additionally, security and explain ability can also be considered as important attributes.

Data privacy and integrity are particularly relevant in the pre-processing stage, where data manipulation takes place. In the second phase of feature engineering, the aim is to select the best features from the dataset to be used to train the ML model, and the objective is to provide good predictions, where an important quality element is the correctness of the prediction, which is dependent on the computational complexity, as an incorrect selection of the features can make the computations more difficult. Subsequently, the training of the ML model depends on the correctness of the prediction, the computational complexity and the fairness. Here it is possible to return to the feature engineering stage for verification (see Fig. 2.5).

Machine Learning Models (Techniques)

Machine learning involves the use of sophisticated algorithms, which are also known as models or techniques. There are four main types of machine learning algorithms, which include *unsupervised, semi-supervised, supervised, and reinforcement* learning (Sarker, 2021) (see Fig. 2.6). These algorithms can be used to intelligently analyse large sets of production data, and they can also be used to make realistic predictions based on the data. The most important algorithms can be divided into four group broad groups, which are implemented using a number of specific techniques (regression, classification, clustering, and association). In the following section, we will present the most characteristic algorithms based on the literature (Sarker, 2021; Shalev-Shwartz & Ben-David, 2014). We have selected the most common and production-related algorithms.

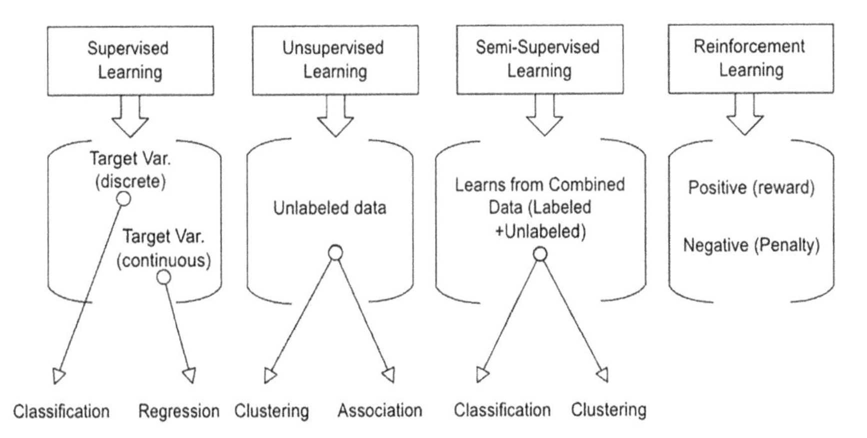

Fig. 2.6. Machine Learning Models (Techniques) (Based on Sarker, 2021).

Regression

Logistic regression is a widely used statistical model in machine learning for solving classification problems. It estimates probabilities using a logistic function and can handle high-dimensional datasets. It can be utilised to determine a causal relationship between independent and dependent variables. Simple regression involves one relationship between one independent variable and the dependent variable. Regression is a type of supervised learning that is used to model continuous variables and make predictions. However, it is limited by the assumption of linearity between dependent and independent variables. Logistic regression can be used for both classification and regression problems, but it is more commonly used for classification.

Classification

Classification is a type of supervised learning method used in the field of machine learning to solve prediction problems and create models. Mathematically, it involves creating a function that maps input variables to output variables based on a set of target labels or categories. This algorithm can be used on both structured and unstructured data.

A decision tree is a well-known non-parametric supervised learning method used for both classification and regression problems. These algorithms are efficient for consumer behaviour analysis and cyber security analysis. The classification of instances is done by sorting down the tree from the root to some of the leaf nodes. Cases are classified by checking the attribute defined by that node, starting from the root at a node in the tree, and then moving down the branches of the tree corresponding to the attribute value.

The Random Forest (RF) classifier is a well-known machine learning and data science technique used for various applications, including price prediction tasks. It is an ensemble classification method which combines multiple decision trees. The method identifies subsets of the dataset by majority vote or by averages, leading to increased accuracy of the prediction and easier parameter control. Thus, the RF learning method is generally more accurate than a model based on a single decision tree. To create a series of decision trees with a control variable, a combination of special aggregation and random feature selection is used. The technique can be adapted for both classification and regression problems (see Fig. 2.7).

Clustering

Cluster analysis, which is also known as clustering, is a machine learning technique that helps to identify and group related data points in large datasets without predicting the outcome. It performs a clustering of a set of objects so that objects of the same category, called cluster, are more similar to each other than objects in other clusters. This method allows the discovery of interesting trends or patterns in the data and is considered an unsupervised learning method.

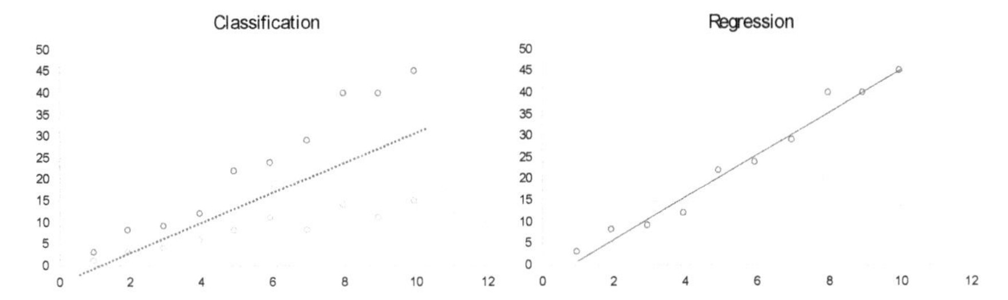

Fig. 2.7. In Classification, Two Classes are Distinguished by A Dashed Line Representing the Separation, While in Regression, the Model is Used to Represent the Axis of the Relationship Between Variables.

Mean shift clustering: Mean shift clustering is a non-parametric clustering method that does not require prior knowledge of the number of clusters or the shape of the cluster boundaries. It aims to detect 'blobs' in a uniform distribution or density of samples. To form the final set of centroids, these candidates are dispersed during the processing stage to remove near-duplicate objects. Mean shift has the disadvantages of being computationally expensive and having limited accuracy when the number of clusters changes suddenly.

Association

Association rule learning is a machine learning technique used to find relationships between variables in large data sets. Association rules have a wide range of applications.

Artificial Neural Networks (ANN)

Deep learning is a type of machine learning that is based on ANN and representation learning. It uses a computational architecture that combines multiple processing layers, including input, hidden, and output layers, to learn from data. The main advantage of deep learning over traditional machine learning approaches is that it can learn from large datasets more effectively, resulting in improved performance. The realisation of machine learning is often achieved using Artificial Neural Networks. Artificial Neural Networks are mathematical structures inspired by the human brain that perform computations or signal processing. They are used to solve problems related to classification, prediction or regression, and have several key properties. ANNs can learn and generalise knowledge, adapt to changing conditions, and have low sensitivity to input errors. There are many successful applications of ANNs in various scientific fields, including manufacturing. ANNs are useful for estimating and costing, particularly in regression problems (Juszczyk, 2017). Their learning algorithms can uncover hidden relationships in data without prior assumptions about the regression function, which makes ANN-based cost estimation non-parametric. ANNs are especially appropriate for early or conceptual cost estimates, as they allow for the development of a full cost estimate for a specific project or facility type. Their application in manufacturing offers significant benefits for businesses and researchers alike.

Machine Learning in Manufacturing

In the manufacturing industry, machine learning is already being used for a variety of purposes including predicting material properties, distortions, and failures, as well as for smart manufacturing, natural language processing, and object recognition. As Big Data continues to grow, there is already a lot of information available regarding the prevalence and benefits of ML. Studies have shown that companies that have adopted smart manufacturing techniques using ML have achieved higher levels of efficiency (82% of companies report this achievement) and increased customer satisfaction (45% of companies report such achievements) (Shaikh et al., 2022).

The uniqueness of machine learning lies in its ability to operate on existing data sets. Manufacturing can use machine learning to ask and answer complex questions, allowing for not just the description of the current situation, but also the prediction of changes in parameters. This is the case in the manufacturing sector, where machine learning both uses existing data collected from the regular manufacturing process and encourages the creation of new data sets, providing for new reference points (for sensors or automation). According to Wuest et al. (2016), the most advanced applications of data and data analytics in manufacturing are as follows:

- quality improvement initiatives;
- estimating production costs;
- process optimisation; and
- better understanding of customer requirements.

We can observe significant progress in both engineering and management processes achieved by data science. Let's now explore the specific engineering processes where data science has produced substantial outcomes. In the following, we will outline the primary methods of manufacturing and the key elements of machine learning applications, described by Shaikh et al. (2022).

Welding

Welding is a manufacturing process that involves joining two or more metals using heat. ML can be used in welding for various purposes such as predicting the temperature of the welding bath, detecting welding defects, analysing the welding bath, and determining process parameters. Various ML methods are applicable to all types of welding: arc welding, laser welding, gas welding, ultrasonic welding and friction stir welding.

A system has been developed that uses both a classification model and regression to create welding images and predict the depth of welding. This system has achieved an accuracy of up to 95%, with errors of less than 1 mm. Additionally, neural networks are being utilised to predict the quality of laser welding, which is crucial for ensuring the quality of electricity in batteries.

Moulding

Moulding is a process that shapes a liquid or flexible material using a rigid frame called a mould.

Machine learning is currently being widely used to analyse various parameters and predict defects in processes such as injection moulding, liquid composite moulding, and blow moulding. Popular techniques such as ANN and decision trees have proven to be highly effective in optimising and monitoring the injection moulding process in real-time. There are advanced applications available, such as an algorithm for controlling the speed of cylinder injection in a casting machine using a feedback signal configuration. This is achieved by using an iterative learning algorithm. Additionally, a model approach based on a neural network is used to estimate the impact of mould temperature and flow rate on the diameter and thickness of bulges. These methods have been described in the literature.

Machinery

Machine learning methods are already being extensively used in the machinery industry. In machining processes, they aid in boosting productivity rates, monitoring proper system maintenance, improving product quality, and optimising design and process parameters. Artificial intelligence guarantees that new products have better quality and lower costs.

A machine learning algorithm based on support vector machines has been developed for a tool fracture detection system. The system captures multiple force and energy applications using sensors to monitor tool wear. The algorithm enables tool users to automatically adjust process parameters to maintain a stable process. In turning operations, machine learning techniques can be useful for selecting the cutting tool, choosing cutting parameters (such as cutting speed, feed, and depth of cut), and determining the optimal cooling environment. The machining performance can be improved by using artificial intelligence neural networks and genetic algorithms. These predictive elements are used to extend tool productivity, lapping time, side wear, surface roughness, drill hole cleaning, and drag reduction. To predict machining performance, various machine learning methods such as polynomial regression, support vector regression, and Gaussian process regression are used. Among these methods, Artificial Neural Networks is the most popular method.

Forming

Forming is a critical manufacturing process that involves plastically deforming components to give them the desired shape and size. The process includes incremental forming, which is difficult to predict under compression, torsion, and tensile stresses. Machine learning has the ability to transform a complex and unpredictable process into a precisely controlled or even precisely predictable one.

When using machine learning in roll forming, it's crucial to carefully select the controllable parameters. In this process, the rolling force and rolling moment, rolling speed, contact angle, and power demand are all important parameters. To create accurate models for rolling, artificial neural networks are often employed. These models are trained using a variable learning rate to optimise performance. Machine learning is a useful tool for producing objects with a consistent cross-sectional profile. One of the key benefits of machine learning is its ability to predict surface roughness during additive manufacturing. There are six commonly used ML algorithms for modelling: Random forest, adaboost, classification and regression trees, support vector regression, ridge regression, and random vector feature network, and random vector feature function network.

Digital Platforms for Manufacturing

As demand for customisation grows and actual customised production volumes expand, manufacturers are facing increasing pressure to automate their ordering processes in response to the need to quickly meet customer demands. This is resulting in the emergence of 'performance-based contracting through product and service systems (PSS)' (Rodríguez et al., 2020), where a PSS is a system in which products and services are linked together to respond to the customer's needs for a specific set of functions.

Industrial Platforms

The Industrial Internet platform is built on advanced technologies like artificial intelligence, Big Data, cloud computing, and virtual simulation. Primarily the platform is designed to efficiently collect, analyse, and process industrial data, connect people, machines, objects, information within the systems. It acts as an integration tool to achieve efficient results.

An industrial internet platform typically includes five layers: the resource layer, edge layer, infrastructure as a service, the platform as a user interface, and the software layer (Li et al., 2022):

- The resource layer contains all types of data generated at any stage of production, including the full-cycle history data.
- The edge layer comprises various data collection tools that ensure the collected data is accessible to all devices, external systems, and product specifications. It is a set of protocols for integrating complex data.
- The infrastructure layer ensures data security and confidentiality.
- The User Interface/Platform Interface connects industrial data with industrial machinery. Its primary function is to transform data into knowledge and ensure the utilisation of industrial data. It is an application development environment.
- The software layer provides smart applications and service products that perform specific tasks on the data and is designed to model specific production scenarios.

Components of industrial platforms

To achieve optimal results, it's critical to collect and analyse data related to the entire life cycle of a sub-object present on the equipment and its history. When creating an analytical system for personalised production, it is crucial to provide *seamless access* to data and data repositories. It is also necessary to integrate the most appropriate *data mining techniques*. The user interface module can have various functionalities and modules. Some of these functionalities can be directed towards data specialists who generate queries for data mining. Other functionalities can be directed towards production specialist users who do not need to modify the data mining techniques but can manipulate the data output to generate a specific query for the system. Each platform includes a data mining system, the functionality of which is ensured by the architecture of the system.

A generally accepted standard *data mining* architecture for data mining systems includes the following mandatory elements (see Fig. 2.8):

- *Databases*: A database, whether it's a database or repository, World Wide Web, or an information repository, requires data cleansing and integration activities.

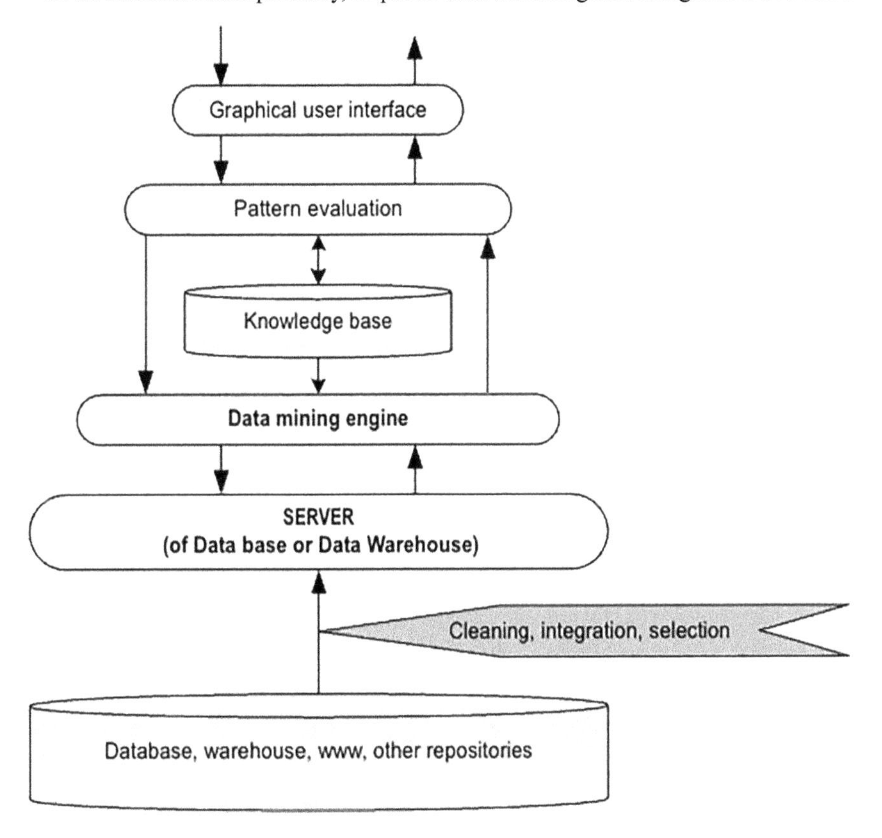

Fig. 2.8. Architecture of a Typical Data Mining System (Han & Kamber, 2006).

- *Database or repository servers*: Retrieve relevant data in response to user queries.
- *Knowledge base*: This is the specific domain knowledge that is used to search or assessing the validity of the resulting models. Such knowledge may include conceptual hierarchies used to map attributes or attribute values to different levels abstractions.
- *Data mining engine*: Ideally, the system should consist of functional modules that are designed to address specific tasks by selecting data mining methods such as characterisation, association and correlation analysis, classification, prediction, and others.
- *Model evaluation module*: This component usually uses curiosity tools and interacts with data mining modules to focus search to target interesting patterns. This module is particularly recommended to be included as deep as possible in the mining process.
- *User interface*: This module enables users to communicate with the data mining system by specifying a query, creating a two-way communication relationship.

Warehouse architecture. The production business data warehouse is a crucial component of the industrial digital platform and is distinct from other platforms. A data warehouse architecture model should reflect how data is stored and organised for efficient business use. In the context of the blast furnace ironmaking applications (Li et al., 2022), collecting and storing complex field data involves multiple systems and sources, resulting in intricate reports. As a result, independent data warehouse architectures are emerging.

The overall design of warehouse architecture according to Li et al. (2022) is as following five layers:

- *A buffer layer* is needed for storing primary raw data (storing up to 1 day without changing the primary data type).
- *An operational data* store used for storing data after the dimensions have been standardised. This module integrates, reorganises, and stores all the data.
- *The application layer* stores data that are already personalised according to the data procedures of the individual analytical applications.
- *An indication layer* contains the key indicators of the production process, allowing the most convenient extraction of indicator data and the most efficient analysis and application of the underlying data.

The quality of a platform's data warehouse is inversely related to its functionality. A high-quality data warehouse enables the platform to provide rich and fast visual analysis functions that can distinguish between the types of data analysed by different workers. This is done through the aggregation function calculation, statistical dimension unification, and drag-and-drop merging of other data.

Interface of communication. Every platform needs a distinct interface to interact with data and smart applications. This interface is a graphical representation of essential information that helps management professionals to understand the data with ease, generate queries and reports, summarise information, and identify patterns in the data on their own. Consumers often require indicator-based information that describes the condition of equipment or provides indicators of the overall state of production. Engineers, on the other hand, tend to focus more on design, general condition, raw material, and equipment condition indicators. Meanwhile, managers tend to focus on economic, technical, project management, and design indicators.

Small and Medium Enterprises' Readiness to Embrace Industry 4.0 Possibilities

There is a lot of debate about the significant potential of Industry 4.0 for businesses. However, leveraging the latest technological opportunities is not a simple process. It presents many adaptive challenges, some of which may not even be addressed at the individual company level. Moreover, when considering the small and medium-sized enterprise segment, which is crucial for national economies, there are even more challenges. As SMEs make up a large part of the business environment, it is crucial for them to keep up with the latest innovations.

The role of SMEs in the economy is recognised both in terms of the number of employees and the impact on the economy. The SME sector employs a large number of workers. Two-thirds of nonfinancial sector workers are employed in SMEs, while 29.1% of workers are employed in micro-enterprises with fewer than 10 employees, 20.2% of workers are employed in small enterprises with fewer than 49 employees, and 17.1% of workers are employed in medium-sized enterprises with fewer than 250 employees. SMEs of all three sizes contribute almost equally to value added in the 28 EU Member States. In addition, some countries are particularly dependent on SME welfare, with the majority of their GDP being generated in the SME sector. In addition, the SME sector continues to grow: for example, 13.8% by 2017 (Matt et al., 2020).

Given the significant impact of small and medium-sized enterprises on the global economy, it is crucial to promptly comprehend their challenges.

Among all the challenges that small and medium-sized enterprises will face in order to take advantage of new technological solutions, are those that are recognised as necessary adaptive changes for any manufacturing company. These include:

- financial challenges (raising or distributing new, sometimes very large, and funds);
- organisational challenges (adapting the business model to new technological processes);
- digitalisation challenges (investing in data management applications);
- quality management challenges (developing and maintaining new standards); and
- human resources challenges (attracting specialists with the right skills).

In addition to these issues of adapting to the new environment that each business is concerned with, there are also issues related to the smallness of the SME.

First of all, when looking at the field of opportunities for SMEs, it should be borne in mind that SMEs are technically the weakest. They have limited or no resources to ensure transformation. Research consistently shows that large corporations, from the innovation field as a whole, are more likely to invest in process and product innovation, and are therefore quite receptive to Industry 4.0 inventions. This trend cannot be replicated in the SME sector because while large companies can raise substantial amounts of money and invest in technological infrastructure, this is not always feasible for small businesses.

It is important to recognise that the smaller the SME, the greater the risk that it will not be able to capitalise on this revolution (Matt et al., 2020).

Factors That Are Linked with the Success of Industry 4.0

Let's first discuss the most critical success factors for businesses to adopt new technologies before highlighting the most significant Industry 4.0 challenges for SMEs.

Data Driven Culture

It seems that a data-driven culture may act as one of the harder brakes on the expansion of analytical capabilities (Pech & Vrchota, 2022). In addition to the social context in manufacturing companies, the openness to disruption and the short-term objectives of companies in a competitive market act as a disincentive to innovation. The need to transform data into action is a challenge that requires additional effort and high-level thinking skills.

Big Data Analytics and Regionalisation

If we consider the Big Data Analytics as an indicator of how prepared countries or companies are to benefit from the opportunities of Industry 4.0 through digitisation, it becomes evident that this issue is not universally relevant across all countries. It is understandable that certain countries are more engaged in the manufacturing of Big Data Analytics than others. However, it is worth analysing the current trends. The USA and China are the top contributors, followed by Germany, South Korea, and Sweden (Belhadi et al., 2019). Regional issues arise when adapting to Industry 4.0 opportunities, as other countries may have achieved success but cannot claim leadership.

Economical Implication

The economic potential of Industry 4.0 is closely tied to the technological and economic progress of a country. While the manufacturing sector spans across multiple countries, the success of Industry 4.0 is contingent on the development of a particular country's economic and technological development. This is particularly true as studies have indicated that the first successful results of Industry 4.0 integration have been achieved in countries where wages are relatively high (Kagermann et al., 2013).

Complexity

Production processes are complex and sophisticated. Industry 4.0 can facilitate these processes and reflect their complexity. However, this does not happen by itself, as it must be borne in mind that companies (regardless of their size) still tend to deploy most of the front-end technologies first, rather than just some of the core technologies (Frank et al., 2019). The use of front-end technologies is crucial in the enterprise value chain. Without these technologies, businesses would not be able to provide a seamless and efficient user experience. Therefore, it is essential to invest in front-end technologies to ensure the success and growth of your enterprise. Advanced manufacturing technologies, such as high-precision machining, reconfigurable production facilities, additive manufacturing, and others, are transforming not only production strategies but also processes and entire production systems (Matt et al., 2020). It is possible to gradually introduce Industry 4.0 technologies, starting from the front end to reduce complexity, and highlight the core purpose.

SME-specific Maturity and Readiness Models

Smart Manufacturing Maturity Models

The Smart Manufacturing Maturity Model is a good theoretical construct to understand the steps that still need to be taken in a specific case to get closer to the practical application of Industry 4.0 innovations. This branch of the literature is quite well developed and there are a number of components of the model that can quantify the theoretical maturity constructs.

One of the characteristics of maturity models is the *maturity levels* (Vance et al., 2023) with specific scale of values 'not explicitly defined', 'defined in terms of the organisation's performance', 'defined in terms of the technology', which are often on a scale of 0 to 10 points.

Another element of the model is the maturity *dimensions* (about 10 dimensions are currently being defined). Here is the list of common categories of dimensions: information technology and cyber-physical system and data; strategy and organisation; products and services; supply chain and logistics, culture and employees; technology and capabilities; customer and market; cybersecurity and risk; leadership and management; governance and compliance (Vance et al., 2023). A grouping of dimensions is common as well with specific focus on vertical integration and horizontal integration.

SME Specific Maturity Models

According to a systematic review of SME maturity models, there were identified six key dimensions of a maturity model: (1) Strategy, (2) products/services, (3) technology, (4) people and culture, (5) governance, and (6) processes (Vance et al., 2023).

The unique characteristics of SMEs are such as resource constraints, low production volume, high flexibility, and high specialisation, make it challenging to apply traditional SME maturity models and roadmaps. These models are not easily adaptable to SMEs with lower technological, infrastructural, human, and financial capacities. Additionally, SMEs often lack the financial resources to seek advice on digital transformation. Unlike large companies, which can start a smart manufacturing transformation from a theoretical level 1, SMEs typically start from level 0 with limited information and resources. Therefore, government agencies and industry associations are crucial in supporting SMEs in their digital transformation process (Ghobakhloo, 2018).

Maturity models tailored to suit the requirements of small and medium-sized enterprises include the Maturity Model towards Industry 4.0 for SMEs and the Maturity Model of Digitisation for SME (Vance et al., 2023). Therefore, it's important to evaluate the suitability of all existing maturity models for SMEs, taking into account more detailed dimension categories. The researchers are critical of business, without singling out SMEs for not seizing the opportunities of Industry 4 or for being too fragmented in their deployment of existing innovations. However, significant barriers have to be acknowledged, especially for sectors close to the frugal innovation wing.

Barriers to the Implementation of Industry 4.0

As we have already pointed out, digital transformation is a complex and deliberative process, especially for manufacturing companies, notably SMEs. There are various technical and technological barriers that hinder digital transformation and servitisation in some cases. Let's review the most evident barriers (Peillon & Dubruc, 2019):

Technological Barriers

> The main drivers of the digital transformation are complex and diverse, including internet of things, big data and analytics, cloud computing, cybersecurity, mixed and augmented reality, intelligent manufacturing solutions, additive manufacturing, connected machine modelling, and artificial intelligence. SMEs are financially constrained and cannot quickly build up contingency funds for technology deployment, especially when it comes to previously used technology infrastructures that require significant first-generation costs.

Organisational Barriers

> Organisational barriers also exist, as new technologies only bring tangible benefits when they are immediately integrated into all business and production processes. Integration requires a strategic posture that creates the preconditions for business model transformations, including changes in business processes, product development phases, organisational structure, and decision flows. This also raises the need for entirely new competences, resources, and

cooperation, changing many decision paradigms such as innovation management, powerful customer service functions, and knowledge management systems. However, there is usually human resistance to change and cultural barriers to action.

Human Resource-related Barriers

Digitalisation is facing a major challenge due to the shortage of skilled workers. The problem is particularly acute in sectors such as furniture manufacturing, where IT-related skills have traditionally been irrelevant. These skills are usually developed by specialists who work exclusively in the IT field, while manufacturing focusses on its own product development issues. However, digitalisation has significantly increased the complexity of the entire production process, requiring every production employee to possess these skills. Furthermore, service integration links the beginning of production to its end through the customer interface, further increasing the complexity of the required competences.

Customer-related Barriers

Digitalisation can bring customers closer to the production process, making it difficult to demonstrate new value. There is a need to better understand the customer, which raises new cost issues. It is important to get closer to the customer without compromising their security or the security of the production system's know-how.

Continuity-related Barriers

Studies have shown that companies that utilise advanced technologies and continuously adapt to them are more likely to integrate new technologies into their existing systems rather than completely replace them (Frank et al., 2019). This is because completely changing technology requires a complicated adaptation process, takes longer, and is met with greater resistance from the environment. As a result, Industry 4.0 progress involves an increasing number of additional technologies that are added to existing ones, ensuring a smoother growth process. It is therefore essential to understand the operating environment of small and medium-sized enterprises and develop support systems in a timely manner to facilitate their transformation.

We have reviewed the challenges that companies encounter while integrating Industry 4.0 capabilities, as documented in the literature. However, these obstacles may not affect all companies equally in every region. Therefore, researchers are exploring the regional context to gain more insights. Let's take a look at some of them. France for instance.

France case (Peillon & Dubruc, 2019). A 2019 review of French small and medium-sized enterprises shows that companies, particularly in traditional industries, are not yet considering digitisation as a strategic issue and risk delaying its implementation. On the other hand, successful attempts have already been made, especially among larger companies.

The French manufacturing companies that were surveyed are in the early stages of digitisation and digital service provision. This means that they may have to overcome organisational and cultural barriers first. When focussing on SMEs that have started to digitise, they face two main challenges. The first challenge is internal, which means that they are often confronted with the practical organisation of the tools to be implemented or the digital services to be offered (teleservices). The second challenge is external and customer-related: their customers are often hesitant to give them access to their data due to privacy and security reasons. Technical/technological and human resource barriers were not significant in the cases studied. This reflects the early stage of digitalisation of these companies, which may not be sufficiently advanced to face this type of problem.

As a completely different case of regional specificity, let's look at the Romanian example.

Romania case (Türkes et al., 2019). The case of Romania highlights the challenges involved in transitioning from traditional manufacturing to Industry 4.0 processes. A survey of managers revealed that most SME managers are familiar with the fourth industrial revolution. They understand that digitalisation in the industry requires determination, significant capital investment, and a well-trained workforce. Therefore, the few managers who see great potential in Industry 4.0 are seeking partnerships to implement new technologies, and 26.1% of SMEs have already signed contracts with specialists in this field.

Regarding relevant technologies, the most popular ones are 'Autonomous robots' (35.2%), 'Horizontal and vertical integration of systems' (27.8%), 'Big data and analytics' (21.6%), 'Internet of Things' (21.6%), and 'Cybersecurity' (17.6%). It is worth noting that Romania identifies itself as an Industry at Level 2. However, none of the 176 SMEs studied have reached levels 4 and 5, indicating that they are not yet ready to make the leap from Industry 2.0 to Industry 4.0 (Türkes et al., 2019). In an attempt to gauge the readiness of businesses to adopt current technology, it has become apparent that there is a general lack of knowledge of Industry 4.0, a lack of standards, a focus on activities at the expense of enterprise development, and insufficient human resources. Radical developments in companies require strategically committed units and expert groups, which, unfortunately, is not the case in small and medium-sized enterprises.

Barriers in the Furniture Industry

The perception of value in the engineering industry is hindering the progress of Industry 4.0. A study on the barriers that affect value engineering in the furniture industry has identified several obstacles, including a lack of knowledge about value engineering, product quality issues, reluctance to support vendors, distributors, and retailers, and uncontrolled production losses due to material wastage. However, as the value chain becomes more manageable, digitisation and robotics will be integrated more quickly, resulting in significant benefits (Sharma, 2016).

The engineering value chain has more potential in furniture product design solutions, which are highly influenced by the demand for customised production. Therefore, by utilising innovative design and industrial conceptual solutions, it is possible to increase the value added.

Design solutions are often considered a weak driving force in the furniture industry, along with insufficient materials, complex inventory, and resistance to change. However, they are actually sensitive elements that are influenced by other constraints. Therefore, in order to implement innovative computing solutions provided by Industry 4.0 and make design modifications, it is necessary to reduce other barriers in the furniture industry. On the other hand, the usual challenges of furniture production, such as product quality problems, reluctance to support dealers, distributors, and retailers, and uncontrolled production losses due to material wastage, have a strong driving force and a significant impact on the industry (Sharma, 2016).

References

Abiri, J. A., & Kundur, D. (2020). Fundamentals of cyber-physical systems. In C. Anumba & N. Roofigari-Esfahan (Eds.), *Cyber-physical systems in the built environment* (p. 359). Springer.

Aloisio, G., Marco, A. Di, & Stilo, G. (2022). Modeling quality and machine learning pipelines through extended feature models. In *Proceedings of ACM conference (Conference'17)* (Vol. 1). Association for Computing Machinery.

Aquino, S., Rapaccini, M., Adrodegari, F., & Pezzotta, G. (2023). Augmented reality for industrial services provision: The factors influencing a successful adoption in manufacturing companies. *Journal of Manufacturing Technology Management*, *34*(4), 601–620. https://doi.org/10.1108/JMTM-02-2022-0077

Belhadi, A., Zkik, K., Cherrafi, A., Sha, Y. M., & El, S. (2019). Understanding the capabilities of big data analytics for manufacturing process: Insights from literature review and multiple case study. *Computers & Industrial Engineering*, *137*, 106099. https://doi.org/10.1016/j.cie.2019.106099

Bodendorf, F., & Franke, J. (2021). A machine learning approach to estimate product costs in the early product design phase: A use case from the automotive industry. *Procedia CIRP*, *100*, 643–648. https://doi.org/10.1016/j.procir.2021.05.137

Caputo, A., Marzi, G., & Pellegrini, M. M. (2016). The Internet of Things in manufacturing innovation processes of a conceptual framework. *Business Process Management Journal*, *22*(2), 383–402. https://doi.org/10.1108/BPMJ-05-2015-0072

Chen, Y. (2016). Industrial information integration—A literature review 2006–2015. *Journal of Industrial Information Integration*, *2*, 30–64. https://doi.org/10.1016/j.jii.2016.04.004

Cheng, Y., Chen, K., Sun, H., Zhang, Y., & Tao, F. (2018). Data and knowledge mining with big data towards smart production. *Journal of Industrial Information Integration*, *9*, 1–13. https://doi.org/10.1016/j.jii.2017.08.001

De Backer, K., & De Stefano, T. (2021). Robotics and the Global Organisation of Production. In J. von Braun (Ed.), *Robotics, AI, and humanity* (pp. 72–84). https://doi.org/doi.org/10.1007/978-3-030-54173-6_671

Dosso, M. (2020). Technological readiness in Europe EU policy perspectives on Industry 4. 0. In L. De Propris & D. Bailey (Eds.), *Industry 4.0 and Regional Transformations* (p. 276). Routledge.
Davies, R. (2015). *Industry 4.0: Digitalisation for productivity and growth*, EPRS: European Parliamentary Research Service. Belgium.
European Commission, Directorate-General for Internal Market, Industry, Entrepreneurship and SMEs, Simons, M., Van de Velde, E., Copani, G. et al. (2016). *An analysis of drivers, barriers and readiness factors of EU companies for adopting advanced manufacturing products and technologies*. Publications Office, https://data.europa.eu/doi/10.2873/715340
Ford, H., & Crowther, S. (1922). My *life and work*. https://books.google.lt/books?id=4K82efXzn10C
Frank, A. G., Dalenogare, L. S., & Ayala, N. F. (2019). Industry 4. 0 technologies: Implementation patterns in manufacturing companies. *International Journal of Production Economics*, *210*, 15–26. https://doi.org/10.1016/j.ijpe.2019.01.004
Ghobakhloo, M. (2018). The future of manufacturing industry: A strategic roadmap toward Industry 4.0. *Journal of Manufacturing Technology Management*, *29*(6), 910–936.
Han, J., & Kamber, M. (2006). *Data mining: Concepts and techniques* (2nd ed.). Morgan Kaufmann Publishers.
Han, J., Kamber, M., & Pei, J. (2012). *Data mining: Concepts and techniques* (3rd ed.). Morgan Kaufmann.
Juszczyk, M. (2017). The challenges of nonparametric cost estimation of construction works with the use of artificial intelligence tools. *Procedia Engineering*, *196*(June), 415–422. https://doi.org/10.1016/j.proeng.2017.07.218
Kagermann, H., Wahlster, W., & Helbig, J. (2013). *Recommendations for implementing the strategic initiative INDUSTRIE 4.0. Final report of the Industrie 4.0 Working Group* (p. 678). Acatech—National Academy of Science and Engineering.
Li, H., Li, X., Liu, X., Bu, X., Li, H., Lyu, Q., & Li, H. (2022). Industrial internet platforms: Applications in BF ironmaking. *Ironmaking & Steelmaking*, *49*(9), 905–916. https://doi.org/10.1080/03019233.2022.2069990
Lin, H., & Yang, S. (2019). A smart cloud-based energy data mining agent using big data analysis technology. *Smart Science*, *7*(3), 175–183. https://doi.org/10.1080/23080477.2019.1600112
Mouël, F. Le, & Carrillo, O. (2023). PIS : IoT & Industry 4 . 0. In M. K. Pinheiro, C. Souveyet, P. Roose, & L. A. Steffenel (Eds.), *The Evolution of Pervasive Information Systems* (pp. 123–155). Springer International Publishing.
Matt, D. T., Modrák, V., & Zsifkovits, H. (2020). *Industry 4.0 for SMEs challenges, opportunities and requirements*. Palgrave Macmillan.
Nishioka, Y. (2015). *Industrial value chain initiative for smart manufacturing*. Tokyo, Japan.
Park, J. (2015). *Korea smart factory program*. Tokyo, Japan.
Pech, M., & Vrchota, J. (2022). The product customization process in relation to industry 4.0 and digitalization. *Processes*, *10*(539), 1–30. https://doi.org/10.3390/pr10030539
Peillon, S., & Dubruc, N. (2019). Barriers to digital servitization in French manufacturing SMEs. *Procedia CIRP*, *83*(1), 146–150. https://doi.org/10.1016/j.procir.2019.04.008
Ploos van Amstel, W. (2003). Advanced planning and scheduling (APS). In A. R. Van Goor, W. Ploos van Amstel, & M. Ploos van Amstel (Eds.), *European distribution and supply chain logistics* (p. 526). https://doi.org/10.4324/9781003021841-20
Rodríguez, A. E., Pezzotta, G., Pinto, R., & Romero, D. (2020). A comprehensive description of the product-service systems' cost estimation process: An integrative review. *International Journal of Production Economics*, *221*, 1074–1081. https://doi.org/10.1016/j.ijpe.2019.09.002

Rosin, F., Forget, P., Lamouri, S., & Pellerin, R. (2020). Impacts of Industry 4. 0 technologies on Lean principles. *International Journal of Production Research*, *58*(6), 1644–1661. https://doi.org/10.1080/00207543.2019.1672902

Samuel, A. (1959). Some studies in machine learning using the game of checkers. *IBM Journal*, *3*, 210–229. http://dx.doi.org/10.1147/rd.33.0210

Sarker, I. H. (2021). Machine learning: Algorithms, real-world applications and research directions. *SN Computer Science*, *2*(3), 1–21. https://doi.org/10.1007/s42979-021-00592-x

Shaikh, A., Shinde, S., Rondhe, M., Chinchanikar, S. (2023). Machine Learning Techniques for Smart Manufacturing: A Comprehensive Review. In A. Chakrabarti, S. Suwas, M. Arora (Eds.), *Industry 4.0 and Advanced Manufacturing. Lecture Notes in Mechanical Engineering*. Springer. https://doi.org/10.1007/978-981-19-0561-2_12

Shalev-Shwartz, S., & Ben-David, S. (2014). *Understanding Machine Learning : From Theory to Algorithms*. Cambridge.

Sharma, V. (2016). Analysis of furniture industry through value engineering by eliminating the barriers. *International Journal of Science Technology & Engineering*, *3*(4), 51–58.

Smart Manufacturing Leadership Coalition. (2011). *Implementing 21st Century Smart Manufacturing: Workshop Summary Report*. Washington D.C.

Tay, S. I., Lee, T. C., Hamid, N. Z. A., & Ahmad, A. N. A. (2018). An overview of Industry 4.0: Definition, components, and government initiatives. *Journal of Advanced Research in Dynamical and Control Systems*, *10*(14), 1379–1387.

Türkes, M. C., Oncioiu, I., Aslam, H. D., Marin-Pantelescu, A., Topor, D. I., & Capus, S. (2019). Drivers and barriers in using Industry 4.0: A perspective of SMEs in Romania. *Processes*, *7*, 1–20. https://doi.org/10.3390/pr7030153

Vance, D., Price, C., Nimbalkar, S. U., & Wenning, T. (2023). Smart manufacturing maturity models and their applicability: A review models. *Journal of Manufacturing Technology Management*, *34*(5), 735–770. https://doi.org/10.1108/JMTM-03-2022-0103

Witkowski, K. (2017). Internet of things, big data, Industry 4.0 – Innovative solutions in logistics and supply chains management. *Procedia Engineering*, *182*(1), 763–769.

Wuest, T., Weimer, D., Irgens, C., & Thoben, K. D. (2016). Machine learning in manufacturing: Advantages, challenges, and applications. *Production & Manufacturing Research*, *4*(1), 1–23. https://doi.org/10.1080/21693277.2016.1192517

Yang, F., & Gu, S. (2021). Industry 4. 0, a revolution that requires technology and national strategies. *Complex & Intelligent Systems*, *7*(3), 1311–1325. https://doi.org/10.1007/s40747-020-00267-9

Chapter 3

Furniture Industries: Challenges of Regionalisation, Customisation and New Paradigm of Pricing

Birutė Mockevičienė and Tomas Vedlūga

Mykolas Romeris University, Lithuania

Abstract

The chapter is designed to discuss the preconditions for the competitiveness of the furniture industry, global networks and regional perspectives, as well as the competitive advantages of different regions such as the USA, Europe and the East. The challenges created by customisation and the needs of consumers for individual products are also discussed. As consumers become more and more focussed on furniture designed exclusively for them, the furniture business has to reorient its production and has to deal with a number of management issues. It is necessary to reconsider not only how to involve consumers but also how to keep prices competitive because even for an individual order, the customer is less and less willing to pay more. The issue of new product development is also discussed. It delves into the management of furniture companies, the characteristic organisational structures, and management models that could ensure the sustainability of the business. Particular attention is paid to the digital issues of furniture manufacturing and enterprise resource planning (ERP) in particular. An examination of how the furniture sector evaluates prices and costs, which are the most popular methods and which can be used for forecasting, looks at the most important global trends. Such cost estimation methods as cost-based, competition-based, analogous-based, and expert-based are discussed, highlighting the limits of their applications. Then discusses current trends and the current IT supply, which unfortunately does not fully meet the needs of customised furniture production, and digitisation

Participation Based Intelligent Manufacturing:
Customisation, Costs, and Engagement, 47–100

doi:10.1108/978-1-83797-362-020241003

within a small company becomes more difficult. So, companies have to recognise the limits of digitisation.

Keywords: Frugal innovation; sustainability; furniture manufacturing; ERP; expert judgement; customised manufacturing

Sustainability Trends in the Furniture Industry

More and more consumers are setting sustainability expectations for the furniture industry. Pursuing a more environmentally friendly business model for the goods we produce isn't just the conscientious thing to do for our planet, it's now a prerequisite, particularly for Millennials and Gen Xers entering their prime spending years. According to a 2021 study by the World Wildlife Fund, web searches for sustainable products increased by an astounding 71% in just 5 years (Stewart, 2021). Consumers at mass now expect their favourite brands – from furniture and fashion to food production – to be taking sustainable business actions. Customers make purchasing decisions based on these efforts and are willing to pay a premium for products that are sourced, manufactured, and packaged more sustainably.

Customers of furniture companies want environmentally friendly sources for furniture design and production. Today's customers, that is, millennials and Generation Xers, are much more likely to look at a catalogue of custom furniture fabrics and finishes and are more likely to consider not only the appearance of the furniture but also its durability and sustainability. From wood and metals to fabrics and fittings, making more environmentally friendly choices starts in the design process. Furniture companies usually source their wood from sawmills and timber warehouses that may have certain sustainability certificates. A small number of furniture companies own sawmills. For example, the certificates certify and promote environmentally sound, socially beneficial and economically viable management of the world's forests. In other words, they set the gold standard for the ethical use of timber, ensuring that what is taken from nature is replanted for future generations. As a consumer, the Sustainability Certification Label lets you know whether the company you buy wood from is sourcing sustainably. Fabrics can also be chosen specifically for their low impact on the environment. Cotton and linen, for example, are both considered eco-friendly and sustainable. Upcycled materials can also be used in furniture pieces, like thread made from recycled water bottles.

Consumers are also interested in how companies dispose of their waste. On the surface, you may be thinking, 'Why would anyone care what we do with our trash?' However, the initiative to go green started with recycling decades ago, so naturally, the trend today is to creatively repurpose your discarded manufacturing materials. For example, in the furniture industry: sawdust and wood scraps can be taken to local facilities to be used for heat and power, foam scraps can be ground up and used for carpet padding, cardboard waste can be recycled, and fabric scraps can be sent off for use in handbags, accessories, etc. Modern furniture companies aim to think sustainably – to become zero-waste factories (Yaolin et al., 2023).

Consumers are more inclined to pay a premium for furniture pieces and home goods if their sustainability and quality expectations are met (Laroche et al., 2001). Buying new pieces year after year from fast furniture stores has lost its lustre for many who seek to reduce their carbon footprint and pass down high-quality furniture to children and grandchildren. Many opt to invest in high-quality pieces periodically to grow their heirloom collection gradually (Laroche et al., 2001). When sustainability is kept in mind from the design stage to manufacturing and delivery, quality is a by-product of these business decisions. Therefore, the rejection of short-lived items and fast-moving furniture is one of the four sustainability trends in the furniture industry.

Another way furniture companies can reduce their environmental impact is by sourcing local or even indigenous raw materials whenever possible. We have briefly discussed which materials have a lower environmental impact, such as sustainability-certified wood, cotton and linen fabrics, as well as recycled materials, but the distance travelled by materials is just as important when it comes to carbon dioxide emissions. Furniture companies looking to reduce their carbon footprint can focus on sourcing wood from forests close to the production site. This reduces the company's carbon footprint associated with international maritime transport and outsourcing (Cariou et al., 2019). Furniture that meets the expectations of today's consumers is the furniture they buy and demands accountability, reliability and transparency. Modern furniture companies that do not proactively adopt more sustainable solutions will find that this will affect their bottom line, while companies that continue to move forward on a greener path will find greater customer loyalty (Wijekoon et al., 2021).

As a culture, we've become used to getting what we want when we want it – and that includes furniture. If we decide to buy a new sofa or dining room table, we surf the internet for one that suits our style, or we drive to the nearest big box store for the lowest price point. Both options generally deliver within a matter of days. Most furniture available in the market today is not made to last. These substandard pieces are made in bulk out of cheap materials and sold for the lowest dollar possible. As a result, consumers consider these short-term home goods completely disposable – too expensive to repair and not worth the hassle of selling – so they hit the curb.

Because most fast-moving furniture is made of low-quality composites and toxic adhesives, it is almost impossible to properly separate and recycle it, and more than 80% of it ends up in landfills (Wang et al., 2023). Choosing furniture made from reusable resources such as solid wood, metals and adhesives that comply with California's safety laws makes it easier to recycle high-quality furniture when the time comes. While remnants of fast furniture are frequently found in landfills after a couple of years of use, quality pieces fill homes for generations. Consumers who invest in high-quality pieces pass their collection down to children and grandchildren, leaving an inheritance of long-lasting, beautiful home goods that stand the test of time. But it's not just about whether whole pieces of furniture find their way to the trash within a season or adorn living spaces for a lifetime. Furniture companies that prioritise sustainability look for innovative ways to dispose of their manufacturing waste responsibly. For example, polyurethane foam waste (foam scraps) can be repurposed for carpet padding or recycled plastics, rather than decomposing in a landfill for 30–50 years (Kibria et al., 2023). Even sawdust can be used for power generation or chemically converted to a hydrocarbon for use in gasoline. Furniture manufacturers with a green mindset are forming partnerships that proliferate a circular economy – make, use, reuse, remake, recycle, etc. One company's waste is another's treasure! When selecting pieces for a client's home or choosing pieces to fill your showroom floor, ask the furniture manufacturer what they're doing to properly dispose of their waste materials. The Fig. 3.1 illustrates the sustainability breakpoints in the furniture industry.

Another very important aspect is deforestation due to the irresponsible sourcing of wood for furniture production. Forests are the lungs of the earth and provide essential ingredients for our ecosystem: oxygen, clean air and water, timber, homes for wildlife, etc. According to the scientific journal Nature, around 15 billion trees are cut down every year (Ehrenberg, 2015). Most deforestation takes place in tropical areas. So, if you have the option of buying from a US producer who sources the wood locally, you are usually buying furniture made from wood sourced from a certified sustainable forest. As a designer, buyer or consumer, you will know whether the company you are buying from sources its wood sustainably, through the Sustainability Certification Seal of Approval, which ensures that what is taken from nature is restored for future generations. You can also find out where the wood comes from and what efforts the furniture manufacturer is making to protect the Earth's natural resources.

In America or Europe, we know that most of what we buy is made in China and transported in containers. But did you know that China used most of its

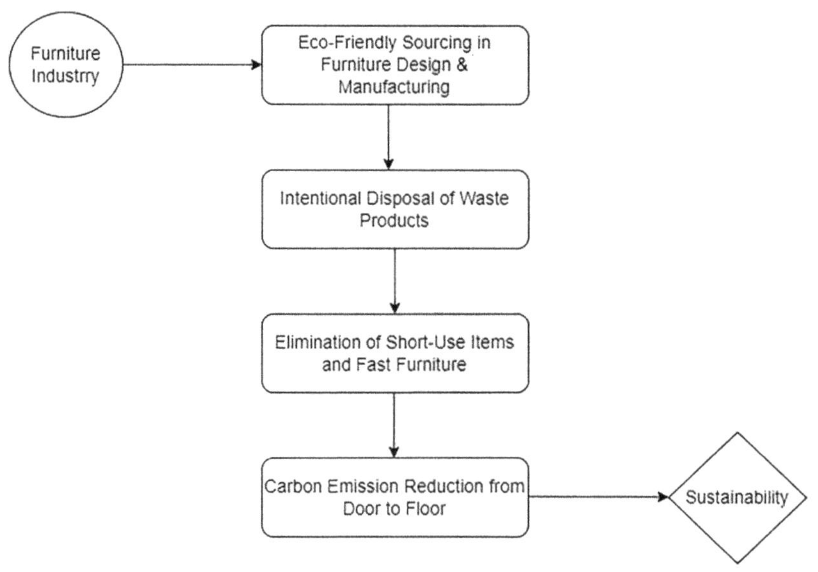

Fig. 3.1. Furniture Industry Sustainability.

forests to produce goods in the 1990s? For this reason, the demand for fast furniture in the USA and Europe is leading not only to massive deforestation in unregulated forests all over the world, but also to huge transport emissions as the wood is transported from Africa to China for production and then to the USA or Europe for sale. Figures show that these container ships consume more than 110 tonnes of fuel oil per day and can take more than two weeks to reach port (Berg, 2016). That's more than 1,500 tonnes of fuel per ship going one way! In addition, ship fuel still contains sulphur, which is a major pollutant and is estimated to account for 17% of all human carbon dioxide emissions by 2050 (Berg, 2015). In spite of this potential, shipping hasn't been prioritised in any of the international agreements coordinated through the UN Framework Convention on Climate Change, and the latest agreement coming out of the COP 21 talks in Paris does not include stipulations on shipping or the high emissions caused by air freight (Paris Agreement, 2024). Still, progress is underway. From technological improvements such as retrofitted rudders and propellers to enhanced weather routing, shipping companies are eyeing many ways to improve their efficiency. For instance, newer ships have been designed to carry more without a proportional increase in fuel use. The biggest ship today is capable of transporting close to 20,000 of the type of containers typically carried by a semi-trailer on the highway, a huge jump from the roughly 2,500 that the first purpose-built containerised ships could carry in the 1970s (Berg, 2016). And as this capacity has grown, ports have adapted to handle the influx. Big ships can use more than 100 metric tons (110 tons) of fuel oil per day and can take two weeks or more to traverse oceans (Berg, 2015). Shipping's international nature makes it tricky to control; measures such as fuel regulations and emissions standards have long implementation periods and are slow to achieve greenhouse gas reductions and environmental goals. Standards vary inside and outside so-called 'emissions control areas' established by the International Maritime Organisation, a United Nations agency focussed on shipping.

The era of huge container ships has led to the development of logistics hubs, with rail yards, truck bays, and massive warehouses that receive, sort, and redistribute all these goods. Transporting freight by rail is more energy-efficient than transporting it by truck, says Asaf Ashair, an emeritus research professor with the University of New Orleans's National Port&Waterways Initiative. But while it makes sense energy-wise to transport freight on rail for most mid- and long-range hauls in the USA, for example, the flexibility of trucking and the wide geographic spread of the country means that most stuff is eventually moved to its point for sale o ruse by truck. According to the American Trucking Association, trucks carry about 70% of the tonnage of stuff moving throughout the USA annually, requiring 3 million trucks and more than 37 billion gallons (140 billion litres) of diesel fuel (Berg, 2015).

Ideas for addressing the congestion and energy requirements of the so-called 'last-mile' issue range from centralised delivery boxes to cargo bicycles. Big companies such as FedEx are investing in hybrid or all-electric delivery vehicles. Amazon is famously investigating the potential of delivery by battery-powered drones, which could reduce its reliance on traditional vehicles and their emissions.

But many experts say the idea is just speculation at this point. With the rise of 3D printing, some technologists are looking at the potential of distributed manufacturing – factories interspersed throughout urban areas where machines can print whatever part or product a consumer could want or need, eliminating the need to ship a part across an ocean or put it in a box in the back of a delivery truck. Such fabrication labs may serve a niche audience, but they're unlikely to be able to compete economically with the large-scale manufacturing system already in place. As the economic efficiency of shipping increases on sea and land, it no longer will make sense to concentrate huge factories in places such as China. It is predicted that there will be more factories in more places, with parts and raw materials moving between them at lower costs and more efficiently than today. Efficiency gains and developments in automation may have the biggest influence on how the environmental footprint of our global system of goods movement evolved in the coming years. And even if self-driving trucks and delivery drones eventually revolutionise the movement of stuff over land, almost all of that stuff will still start its long journey on a boat.

The carbon footprint of transporting raw materials and finished furniture is one of the main reasons why sustainability in furniture production is the best way to tackle this huge carbon footprint – buy from furniture companies that make furniture and source materials locally. Not only will you help the planet, but you'll also be rewarded with shorter lead times, a superior product, and the knowledge that you contributed to the local economy.

Regionalisation and Frugal Innovation in Manufacturing

The furniture industry is feeling the pressure to become more and more sustainable. Businesses are looking for alternatives to the conventional production methods promoted by the global market. While Industry 4.0 has helped somewhat, it has not fully met the high standards required for sustainable manufacturing. Let us now examine the ways that have already been discussed in the literature to address sustainability issues through the reinforcement of consumer interest. The solution can be found within regional integration and the approach of frugal innovation (FI).

Regionalisation

The increasing pressure on national governments is leading to a reduction in public sector spending, so the innovation sphere is becoming more and more financially squeezed. This is also the case in emerging markets and low- to middle-income countries.

Therefore, with budgets under considerable pressure, the innovation sector is left with few opportunities for development. This situation can be observed both in certain markets and regions (developing countries), as well as in certain sectors (low innovation and technology) and a certain segment of the market – small and medium-sized enterprises (SMEs).

The practice of regionalisation is gaining more attention as consumers are becoming more region-specific. The regional dimension is starting to impact the speed of innovation, production of new products, innovation penetration, and its expression (Mourtzis, 2018). There is a growing recognition of the value of regional markets and a shift towards adapting furniture mass production to better suit the needs of these regions. This change is being driven by customer-focussed planning and control methods within global production networks. We are no longer just talking about fragmented regionalisation of production (supply of individual components and global supply chains), but the regionalisation of the entire system, bringing the entire production system physically closer to the consumer (Zheng et al., 2023). The above situations necessitate specific solutions, which can be effectively addressed by applying the concept of FI, in the hope of fostering a more sustainable development including the elements of a circular economy.

The Concept of FI

The concept of FI has circulated since 2010 and has become popular and proven its value in a variety of sectors worldwide such as manufacturing, automotive, food, energy, or healthcare (Zeschky et al., 2014).

The FI paradigm emerged as a response to the challenges of regionalisation. It is a mindset and approach focussed on finding efficient and inventive ways to improve product solutions in the simplest and most cost-effective manner. This need became apparent when companies, particularly SMEs, encountered difficulties aligning their business processes and product requirements, particularly in terms of price and affordability.

Frugal as a concept is understood as functional, robust, user-friendly, affordable and local and can be found in many industries (Mourtzis, 2018). FI can also be seen as a low-cost and more affordable temporary solution to a problem.

FI involves using a resource-smart approach to product design. This means developing highly functional products tailored to specific, non-differentiated markets where consumers differ in terms of their purchasing power and needs. This approach ensures that products can be developed with optimal price and quality (Winterhoff et al., 2014).

FI is even more challenging than good enough innovation. FIs are similar to good enough innovations, but they involve developing new applications specifically for resource-constrained environments. This creates a completely new value proposition for the product. The value proposition of a FI is so unique to the emerging market that it cannot be easily transferred to other markets. Interestingly, such innovations often create entirely new markets in the Western world (Zeschky et al., 2014), not just in emerging markets. These innovation efforts often focus on reaching new customers in underserved areas, and companies have to learn how to develop new products with entirely new parameters in mind.

Frugal's innovation concept also comes with its own business models (Mamasioulas et al., 2020) that would allow for the reduction of complexity and help reduce the overall life cycle costs of a product while at the same time providing high-value and affordable solutions to consumers in emerging and developing markets. To a large extent, FI is attached to emerging market economies.

The concept of FI becomes valuable when it is challenging to incorporate customer feedback into production. Customers' unique requirements shape the entire product design and influence the design of the frugal supplier network. There is evidence of a 17% reduction in time to market (Mourtzis, 2018) due to the frugal criterion, making it important to gather extensive customer information. They should be given the opportunity to take on different roles such as consumer, country or regional manager. Considering the knowledge and experience level of each role, as well as the regional needs, allows production to be oriented towards smoother information processing. This involvement of the consumer strengthens their perception of the product and makes them an active part of the production network, as their preferences influence the design of the production network.

The concept of FI involves active participation in processes without excessive spending. In contrast, clusters that include partners from diverse backgrounds often create complexities and may only give the illusion of collaboration (Holm et al., 2019). FI, on the other hand, emphasises user participation within familiar environments (recognises the participation of the user locally, where the user resides, in an environment the user understands and knows).

Summarising, we can identify three key criteria that are unique in distinguishing an innovative species as frugal: (1) significant cost reduction, (2) focus on core functions, and (3) optimised productivity/performance levels (Hindocha et al., 2021).

Despite these three important criteria, their content is well revealed by key characteristics and associated keywords:

- Affordability (unnecessary costs; low cost; cost-conscious; cost-effective; minimum-cost; and nonessential costs).
- Adaptability (addressing specific needs; flexible; conforming to needs; and suffice the needs).
- Accessibility (availability; convenience; ease of access; and in any segment of the economic pyramid).
- Sustainability (eco-friendly; ecosystem; low environmental impact; repurposing; and low environmental intervention).
- Simplicity (easy to use; user-friendly; user-friendliness; and no previous training required).
- Emerging market (low purchasing power, resource constrained; resource scarcity; unserved customers; and underserved consumers).
- Developed market (top of the pyramid; high- income market; developed countries; and developed world).
- Functionality (essential; appropriate; minimalistic; and minimalism).
- Performance (approach; procedure; and moving counterparts).
- Process (approach; procedure; and moving counterparts).
- Outcome (result; conclusion; and consequence).
- Mindset (psychological; cognitive; behaviour; lifestyle; and mentality).

- Good-enough (sufficient; functional; acceptable; and appropriate).
- High value (potential; profitable; scalable; attractive; and novelty).

Despite the well-formulated problem of customised production addressed by the FI concept, its definition remains ambiguous. Its constituent elements, evaluation criteria, and uniqueness are disputed. However, its appeal has also given rise to a multitude of definitions, mostly based on case studies and lacking a sound conceptual or theoretical basis. It is difficult to call FI a new concept or strategy in innovation management. It is more of an adopted and flexible approach to balanced consumption and smart use of resources (Hindocha et al., 2021). This lack of coherence is further explained by the fact that the concept overlaps with other types of innovation, which are also less characteristic and less unambiguous (Hindocha et al., 2021).

FI may seem like a result of technological progress, but it needs external incentives to emerge and sustain in certain areas. According to research (Mourtzis, 2018), government funding, policies, legislation, and the capacity to design processes play a crucial role in maintaining lean innovation. Then we can discuss the determinants of the FI environment, mentioning 10 of them and merging various support and business subsystems (Shibin et al., 2018):

- Government funding.
- Government policies and regulations.
- Process design capability.
- Supply chain talent.
- International rules and regulations.
- Infrastructure quality and connectivity.
- Technology adoption.
- Competition.
- Environmental awareness and knowledge.
- Social values and ethics.

The 10 above mentioned elements that are crucial for regionalisation and the adoption of FI are not equally important. Studies indicate (Shibin et al., 2018) that government funding, government policies and regulations, and process design capacity are key drivers. These are only partially supported by international rules and regulations and environmental awareness and knowledge. Other important elements include the role of technology, infrastructure quality and connectivity, competition, social value and ethics, and the role of talent in the supply chain. These elements are essential mediators to ensure the coherence of the support system.

Impact of FI on Customised Furniture Manufacturing

The concept of FI is still being put into practice. There are ongoing attempts in various markets to develop a frugal-based system for decision-making (Mourtzis, 2018) or to establish a model that connects sustainable supply chains with FIs (Shibin et al., 2018).

In customised production, being frugal becomes a self-handicapping criterion that influences the final decision. Being frugal is associated with reliability, efficient lead times, and local sourcing. This is how suppliers are chosen and how networks of frugal suppliers are established, often located in close proximity or locality (Mourtzis, 2018). For some regions, FI presents an opportunity to enhance their product markets, as it focusses on economic efficiency and optimising prices downwards. The price reduction aimed at aligning with consumer budgets and preferences, as well as being geographically accessible, is challenged in the frugal paradigm due to the focus on sustainable resource consumption. On the other hand, some may view FI innovation as a cheaper and less effective solution to a short-term problem, especially as the drive for 'bigger and better' solutions continues to expand.

FI is likely to become a method for companies to address economic, social, and environmental challenges, as it has the potential to maintain their markets and uphold quality. As a result, it is becoming a strategic area of activity for certain sectors, bridging the gap between the market and the producer. This paradigm is highly acceptable to SMEs wishing to develop affordable products and services.

Customised Manufacturing: Challenges and Opportunities

Basic Principles of Customised Manufacturing

Frugal's innovation concept aligns with customised production and consumer-driven product design. Let's explore how we can make our products more personalised. So, let's take a deep insight into how customised manufacturing is emerging and how it fits in with Industry 4.0 opportunities.

Flexible processes, automation, robotics and AI, e-commerce, 3D printing and flexible manufacturing enable meeting customer needs and customising products on the same production line. This completely transforms production and revolutionises the approach to product creation, production, and customer interaction. The growing individual needs of customers can no longer be met by traditional mass production, purely because working on a traditional basis may no longer be able to keep costs within reasonable limits. Technology and new process methods have, therefore, become vital in manufacturing, making Industry 4.0 an important instrument for the customisation and personalisation of consumer needs. More and more opportunities are emerging to improve customisation manufacturing by introducing automation, data management and fourth-generation digitisation. Until customisation options were formed, mass production and mass customisation were strengthened and exploited (Duguay et al., 1997).

In today's global market, competition is intense, and companies are constantly seeking new innovative ways to bring high-quality products to the market in a timely manner. They are also taking into account the diverse needs of their customers (Kamrani et al., 2012). Companies that are able to offer a wide range of new products to cater to the personalised needs of their customers are the ones that can stay competitive in the market. This trend towards catering to personalised customer needs started to emerge as early as 2003 (Ma et al., 2003) and since 2010, with the concept of Industry 4.0, they have acquired a certain tangibility and a basis for being realised through technology.

When we refer to customisation, we use the additional new concept of 'smart', which means that consumer products are equipped with smart user toolkits for co-designing to personalise a newly developed product. Smart management allows the creation of intelligent management rules that are adaptive, contextual, dynamic and personalised, using new technologies, digitisation, and artificial intelligence.

The orientation towards customisation triggers the entire production organisation, from the organisation's management to internal communication processes or the development of new supporting services. In addition to the main changes in production, the focus on customisation also offers the possibility of expanding the service area in search of added value for the customer (servitisation) (Pech & Vrchota, 2022). From a customer value perspective, servitisation is seen as a process whereby customers are offered compensatory services to facilitate the sale of the product (maintenance and financing), adaptive services (customisation of the product based on knowledge sharing) and substitute services. Many well-known companies, such as Dell and Harley-Davidson, are now producing customised orders and organising their production accordingly.

Definition of Customised Manufacturing

Product customisation refers to the process of manufacturing personalised products according to the specific requirements and preferences of individual customers (Pech & Vrchota, 2022). Such production is characterised by low volumes and long delivery times (Zennaro et al., 2019). This represents a significant shift in the way production operates. Not all personalisation results in personalised production. Customisation refers to the extent to which customers can access and alter product features in order to meet their specific needs and interests (Anshari et al., 2019).

Production is evolving to accommodate both mass customisation and full individual customisation production, allowing for more personalised and tailored products to be produced at scale. Examples of mass customisation are growing. Some of the brighter examples have come from large companies, but likewise, small companies are increasingly moving towards customised production as an essential production approach needed by the market. Car manufacturers provide customers with the option to select the engine model along with the interior and exterior colours of their choice. Nissan, for instance, credits up to 25% of its sales to the practice of customisation. On the other hand, shoe manufacturers, particularly sports shoe manufacturers, offer the service of personalising a pair of shoes according to the customer's design, as illustrated by examples from Nike's literature (Pech & Vrchota, 2022). Customisation is increasingly becoming popular in the furniture industry. The main reason behind choosing customised furniture is that it allows consumers to have a specialised piece of furniture that fits their requirements. Custom-made furniture can have the right size of drawers, the desired number of drawers, as well as additional functions that are not necessarily specific to that piece of furniture. Moreover, it can be combined or disconnected as required, repositioned, and made to fit specific spaces. This specific need can be common in certain business segments, such as book and paper conservation laboratories (Zachary & Boal, 2012), where custom-made furniture may even be cheaper than those available in the market.

In order to determine the location of customised manufacturing across the entire production typology, a customer order cut-off point can be utilised. Therefore, in addition to mass, serial, or continuous production, we will continue to rely on the manufacturing typology based on four modes (Sackett et al., 1997):

(1) manufacturing from stock (MTS);
(2) manufacturing to order (MTO);
(3) assembly to order (ATO); and
(4) engineer to order (ETO).

Manufacturing Strategies for Customisation

Since the decision to move to customised production is not an easy one, as it requires significant changes in performance, certain external pressure elements are needed which are typical of business logic. When considering manufacturing strategies for choosing customisation manufacturing, a wide range of factors must count. These include rational reasoning, timing to benefit, cost-effectiveness, desired volume, and potential manufacturing challenges (Spring & Dalrymple, 2000). The rationality of customised strategies is frequently linked to the potential *to enter the market* with the goal of altering the customer portfolio.

Customisation can serve as a tool *for learning* and improvement. Beneficiaries are required to make customisation decisions, which involves planning for appropriate design engineering capabilities and a suitable cost accounting system.

Customisation Process

Customisation impacts every stage of the production value chain, including product design, manufacturing, assembly, and distribution. Individual process customisation is also possible.

The customised production process consists of many typical production steps, but a few specific steps have a significant impact on the final product and production overall. Below, we describe the standardised process of customised production according (Pokojski et al., 2018).

- *Order request*: When a user makes a request for a product, it's important to clarify the requirements and possible functionalities. However, it's common for the requirements, which must be resolved in and has a significant impact on all subsequent design phases. Selecting and describing a list of requirements is a complex task that often leads to communication problems between the customer and manufacturer. Any error or misunderstanding during this stage can result in significant costs during the subsequent stages. Be contradictor next step. This stage is crucial for the whole project's success.

- *Design*: The process of designing begins with a request for a proposal. In order to provide a quotation, a preliminary conceptual design must be prepared. At the beginning of the project, the design is formulated and then refined and elaborated later on, with several sub-steps identified. This design usually requires a high level of accuracy in customised production, so detailed design elements must be selected at this stage. The design process involves a kind of compromise where engineers ideally draw on the company's knowledge resources and even external advice.
- *Prototype* production is crucial in customised production as it acts as a safeguard against uncertainties that arise in the client's specifications.
- *Specification*: Once the prototype has been tested, refine the technical information and formulate the product specification. This detail is crucial for the later stages of production.
- *Pricing*: It's important to complete the project specification before pricing can be determined. It's recommended to review the project details that were already clarified in the prototyping phase, as part of the decision-making process.
- *The production* area is more or less a classical manufacturing activity, although it is equally complex and depends on process planning and management. Due to customisation, processes may be non-standard and take an unforeseen amount of time.
- *The sales and distribution* start when the product is ready to be transferred to the customer and include activities such as delivery and after-sales service.

Manufacturing Challenges

Customised production is a challenging process that involves the entire product lifecycle and production process. It's not just about the benefits of Industry 4.0 or the customer-manufacturing interface; customisation also affects many inline issues. Customisation comes with the challenges of low cost and short delivery times.

When it comes to personalisation, it's important to distinguish between full personalisation and customisation of only some components or parts. Components can be categorised into two groups: standard components, which are generally common, with few spare parts and are less expensive; and highly customised components, which are voluminous, complex, and expensive (Zennaro et al., 2019). The production difficulties are mainly due to the uniqueness of the project, which requires new control tools and highly complex models.

In terms of the major issues faced by manufacturing companies with customised product orientation, they can be categorised into some groups that encompass all levels of management (Zennaro et al., 2019):

- *Application of Industry 4.0 paradigms and tools*:
  The implementation of Industry 4.0 involves planning the production environment and sourcing materials, utilising integration tools to optimise

customisation and production processes. This encourages the exploration of digitisation opportunities, automation of production processes, and integration of data into business systems. The concept of the Internet of Things and the automatic identification system are gradually being utilised. This is expected to help customised manufacturing businesses connect with customers more effectively and take advantage of opportunities to establish a real-time link between customers and the production process, where their input is most needed.

- *Parts management and feeding and space constraints consideration*:
  Customised production may involve not only non-standard and first-time production parts, but also non-standard items for which storage and handling space may not be available. There is a clear lack of attention to all aspects of the manufacture of such large and complex products. This can lead to a drop in productivity and delays in the production process. Therefore, solutions are needed for customised production in terms of the optimum configuration of the production system layout and the best material handling and storage strategies.
- *Human resource diversity assessment and consideration*:
  The challenge lies in the availability of human resources because customised production requires a large group of workers at every stage of the production process. The workers' skills, physical attributes, abilities, competencies, and limitations all vary. Customised production would better accommodate the personal characteristics of the workers, leading to improved ergonomics in manual processes and increased efficiency.

In summary, it is crucial to consider how the benefits of customised production for consumers and producers through new products are reflected in the technologies and technical capabilities that Industry 4.0 can offer. The benefits of customisation intersect with the challenges that have arisen, and Industry 4.0 offers practical solutions to address them are presented in Fig. 3.2.

This means that consumers receive a customised product, ensuring their satisfaction as they are involved in the development and detailing of their needs. While the process may seem complex due to increased consumer involvement, producers also benefit significantly. Motivated consumers are willing to pay more and quality mistakes are minimised during the prototyping phase, as the consumer continually refines their input. This ultimately reduces quality errors and shortens the development phase, eliminating the need for repeated error correction in the sales process.

But even if the benefits are already realised, the challenges are much considerable.

Achieving increased profitability often means dealing with greater complexity in production, management, and sales. The prototyping phase remains challenging and elusive, in line with the demands of new product development. Pricing custom furniture during the development phase, particularly in the early stages, proves to be difficult, as does implementing life cycle management. As a result, the organisation naturally evolves into a learning entity, grappling with new

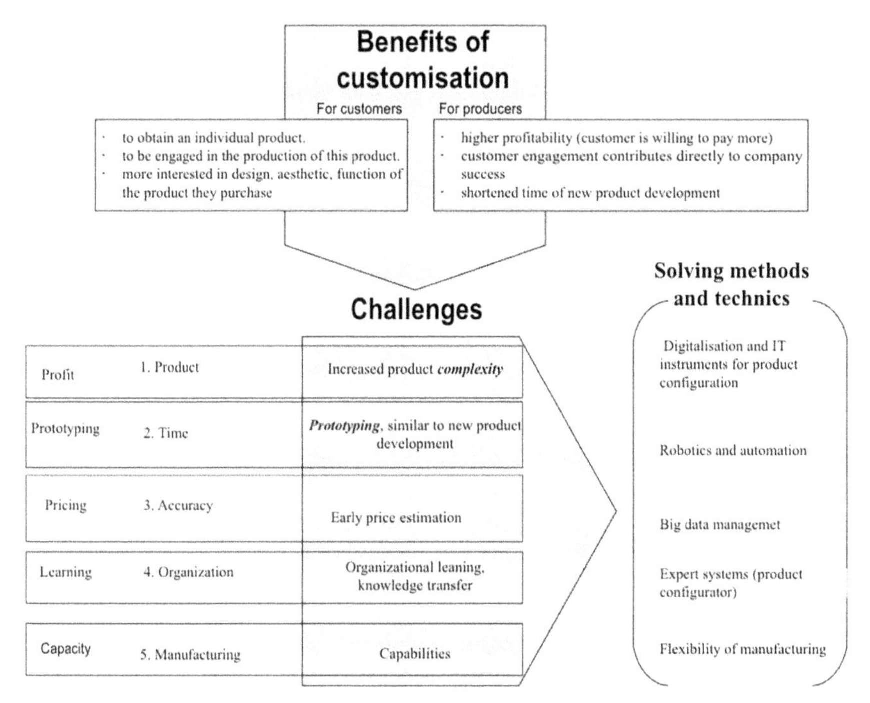

Fig. 3.2. The Benefits of Customisation Intersect with the Challenges That Have Arisen, and Industry 4.0 Offers Practical Solutions to Address Them.

capability issues, and ensuring the flexibility of employee development cycles and the interdisciplinarity of competencies.

Fortunately, the challenges outlined above can be mitigated through the innovative solutions offered by Industry 4.0. These solutions include more extensive digitisation, the integration of robotics and sensors, the use of data mining and big data analysis methods, and the incorporation of expert systems (in the form of user participation), among others.

Customer Integration

Customer Attitudes Towards Customisation.

Competition among companies is getting tougher as consumers' needs increase. Therefore, companies should focus on improving customer experience and personalisation to stand out. In 2020, the American Customer Satisfaction Index (ACSI) has shown a decline in customer satisfaction across the retail sector (Pech & Vrchota, 2022). To overcome this challenge, companies need to develop new business models that revolve around the customer's needs and also deploy smart and flexible manufacturing technologies to meet individual customer preferences while keeping costs manageable. Nowadays, even mass production must respond to the individual needs of the consumers.

Customisation is no longer just an expensive option; nowadays, consumers are willing to spend extra time and money to personalise the products they purchase. This is a positive development for businesses that want to adapt to new production possibilities. While customisation does inevitably lead to higher costs, consumers are not put off by this fact. Instead, they are more than happy to pay a premium to get a product that has been customised to their specific needs. In fact, consumers are even more likely to support the development of pricier customised products. According to Deloitte's research from 2015 to 2019 (Deloitte, 2019), customers are willing to participate in the personalisation process of expensive products such as furniture and holidays. They also want to contribute in areas where they can express themselves, like clothing and fashion accessories. However, customers prefer less involvement in areas where personalisation is limited, such as flights, hotels, and electrical goods.

The emergence of big data, the Internet of Things and cloud computing, data storage and sharing in the cloud has created more opportunities to know the emotions and needs of consumers, thus bridging the gap between the unmet needs of the consumer and the ability of the manufacturing company to meet those needs. In addition to the qualities that have long been the hallmarks of a new product, the importance of design, reliability and usability must be emphasised in the customised manufacturing process.

While the cost of producing customised products continues to rise, consumers are still willing to pay a price for the value they expect to receive from these products. Consumers are also willing to contribute their knowledge and time to product development. Therefore, while customisation depends largely on consumer input, the challenge for a long time has been how to involve consumers and find the most appropriate production stages for their involvement.

Classification of Customer Involvement

If we consider customer involvement to be based on co-design principles, we can distinguish between three process-oriented modes of development: co-design, co-development and co-production (Pech & Vrchota, 2022). Additionally, co-configuration also is valuable to have in mind.

The degree of consumer involvement is another way to measure customised production. As far as the degree of customisation is concerned, customisation can be very easy when adapting a product name. But it can also be fully customised, where the consumer not only builds their own product, but also takes part in its design and production. It is therefore worth creating customisation categories for clarity. The simplest categorisation to understand the company-based manufacturing is (Pallant et al., 2020):

- collaborative customisation (customising products for customers after articulating their needs);
- adaptive customisation (allowing customisable designs that can be altered later based on one standard product);

- cosmetic customisation (presenting a standard product differently to different customers); and
- transparent customisation (providing unique goods or services without customers explicitly knowing about the customisation).

Customer Engagement Management

Creating digital environments in manufacturing and managing their resources requires a unique understanding of how digitisation and customisation create benefits and added value for different customer references (Pech & Vrchota, 2022). In addition to developing new skills to work in the digital space, a company must have the capacity to adapt its supply chains in order to develop IT-based customer engagement tools to ensure that customers are retained on platforms where customers have only a satisfactory experience when working on a customised product and are willing to continue to work with the company (Pech & Vrchota, 2022).

The challenges for the implementation of customisation are to coordinate the communication of information flows, to ensure that channels are set up for the exchange and sharing of strategic information, and to ensure that the different functional units are involved in the creation of a customised product, sharing information in a timely manner and accordance with the user's needs. Customisation is also made effective by additional external integration, that is, when companies and their suppliers and customers have a way of working together. Another extension of integration is when other market players or the data they generate are recognised as suppliers. In this case, customisation benefits from processes where market actors can share valuable data about customers or their preferences. Often, companies developing a new product do not have the capacity (time and experience) to get to know the customer's needs as well as they might need to for a customised product. In such cases, customer needs can be understood through shared data sets or third-party services. Another alternative is to collaboratively develop customer data repositories that can be used for future analytical tasks or machine learning products.

Whatever the level of customer involvement and at what stage we engage consumers, we will have to talk about the customer involvement process based on marketing strategies. To implement the principles of customised production, it is necessary to integrate customer-related processes into the production processes, which are essentially based on the following principles, which are already widely used:

- segmenting the customer;
- managing the relationship with the customer; and
- emphasising the added value from the customer's point of view.

Customer collaboration can be achieved through two methods: customer-centric collaboration, which involves dyadic customer-manufacturer relationships as resources and customisation opportunities; and customer-centric integration, which involves the co-development of products in a networked relationship cantered on product design (Pech & Vrchota, 2022).

The process of designing activities begins with identifying user requirements. The success of customised production largely depends on the designers having a correct and complete understanding of the user's needs. Therefore, the primary focus is on communicating with the user. This is achieved through methods such as (Zheng et al., 2023):

- interviews (a natural way of communicating with users, with the designer leading the conversation);
- user story telling (where the narrator leads the conversation); and
- scenarios (a narrative structure describing events, used to refine requirements by detailing product use scenarios) consumer.

Another alternative method for customer requirements collection is the knowledge-based engineering (KBE) approach, which refers to 'a design method that captures and reuses engineering knowledge in a convenient and maintainable way'. It helps designers avoid repetitive development work, increasing efficiency and reducing lead times and costs. First, users and designers are defined, and then a specific semantic relationship is established between functional requirements and modules to identify the modules that meet the user requirements. Next, using the two compatibility rules specified in the ontological knowledge model, semantic inference and querying technology can be used to find all valid combinations in which the components are compatible. In the final step, a multi-criteria analysis is used to select the appropriate architecture based on the user's non-functional requirements. The model guarantees a continuous iterative process to match user needs. More about KBE will discuss further, embracing more digitalisation instruments useful for customised manufacturing management.

Knowledge-based Engineering for Customised Product Lifecycle Management

The ability to customise and adapt products became possible when changes were made not only to the product's functions but also to the management of the product's entire life cycle. Managing and controlling the parameters of the life cycle itself became a challenge in manufacturing. Customised manufacturing can only be considered of high quality if the entire product life cycle is taken into account. To achieve such a manufacturing process, knowledge capture tools are necessary.

KBE and knowledge management systems go through several stages of development before they can provide the necessary tools for customised production (Pokojski et al., 2018). The first phase of knowledge modelling involves identifying the states that meet the current needs. These states are subject to change as the needs evolve. The second phase involves developing the KBE application version. The remaining phases are dedicated to the iterative development of the application.

Based on Pokojski and co-workers' work (Pokojski et al., 2018), it is possible to formalise the concept of the KBE system. This system enables the design of customised production within an overall knowledge management system. The lifecycle of a knowledge model in such a system includes the following phases: Pattern creation, pattern use, pattern reuse, evolution, and adaptation. The KBE system stores knowledge about production and product design, allowing for the possibility of using and reusing patterns to create new concepts or modify existing ones.

In their study, Pokojski et al. (2018) explored the importance of the needs-based adaptation model for achieving knowledge management goals. They outlined four individual purposes (creation, evolution, direct reuse, and non-direct reuse model) that are involved in recognising the knowledge pattern based on the production requirements (see Fig. 3.3).

Manufacturing companies rely on several product knowledge management systems. These systems may either compete with each other or become highly specialised and unique in the entire customised manufacturing process. One noteworthy management tool is called Product Lifecycle Management (PLM) (Mourtzis, 2018). The primary function of PLM is to gather customer feedback and analyse it to identify new product configurations and improve existing ones. PLM solutions serve as central repositories that enable manufacturing companies to manage their products and services throughout their life cycle. Many companies currently use these systems to meet their engineering data management needs and facilitate collaboration between design departments by allowing core software tools to exchange data with each other. PLM systems are often deployed in manufacturing companies as a tool for developing modular products that form the basis for customised production.

Currently, PLM research is focussed on ensuring seamless interoperability of different systems within an enterprise at every stage of the product lifecycle (Liao et al., 2014). Interoperability, which refers to the ability of multiple systems or components to exchange information, is often dependent on the involvement of stakeholders.

Different ontological models of knowledge capture are being developed to create interoperability and make knowledge understandable to diverse stakeholders

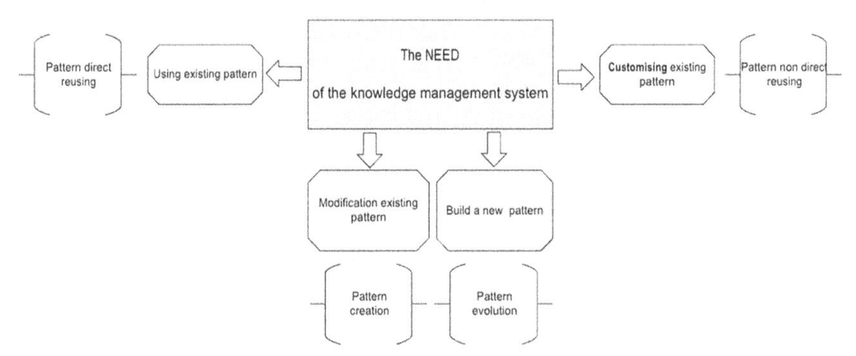

Fig. 3.3. Lifecycle of Knowledge Pattern. *Source*: Adopted According to Pokojski et al. (2018).

with varying levels of experience, competencies, and interests. These models are based on the idea that production can be described as a sequence of processes forming a linear product life cycle, facilitating information sharing. This approach to knowledge management enables the exchange of accurate and unbiased information.

The main product lifecycle processes that are the objective for semantic knowledge annotation are the following:

- Analysis of requirement.
- Product engineering.
- Manufacturing engineering.
- Production planning.
- Manufacturing.
- Sales and distribution.
- After sales service.
- Recycling.

These are the points in a product's lifecycle where stakeholders collaborate through their individual systems and essential information accumulates. This forms the Knowledge Creation and Management Package. Information from production processes is traced to a knowledge repository via semantic links created by a Semantic Annotation and Processing Agent (SAPA) model.

ERP and Customisation

ERP stands for enterprise resource planning. It is a type of software and integrated suite of applications that organisations use to manage and streamline their business processes. ERP systems are designed to facilitate the flow of information and data between various departments and functions within an organisation, such as finance, human resources, manufacturing, supply chain management, customer relationship management (CRM), and more. ERP systems help the enterprise to share and transfer data and information across all functions units inside and outside the enterprise (Elmonem et al., 2016). Sharing data and information between enterprise departments helps in many aspects and aims to achieve different objectives (Elmonem et al., 2016). It helps break down information between inventory, production, planning, materials, engineering, finance, Human resources, sales, marketing, operation, and all other departments in the enterprise (Elmonem et al., 2016). Popular ERP vendors include SAP, Oracle, Microsoft Dynamics, and Infor, among others. The choice of an ERP system depends on the organisation's size, industry, and specific needs. Implementing an ERP system is a significant undertaking and often requires careful planning, customisation, and training for employees. When successfully implemented, ERP systems can improve efficiency, reduce costs, and enhance overall business performance.

The main functions of the ERP are illustrated in Fig. 3.4 and will be discussed below.

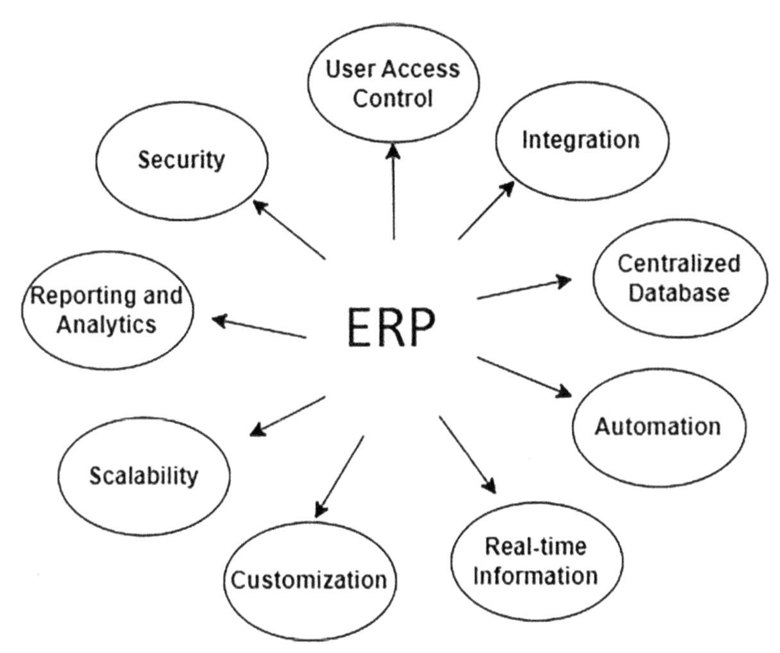

Fig. 3.4. Key Features and Aspects of ERP Systems.

Integration: ERP systems integrate data and processes across different departments and functions. This integration helps eliminate data duplication, improve data accuracy, and provide a unified view of business operations. ERP integration refers to the process of connecting an ERP system with other software applications or systems within an organisation or with external systems, such as suppliers, customers, or partners. The goal of ERP integration is to streamline and automate data exchange and business processes, ensuring that information flows seamlessly between different systems. ERP Integration is Important because of Data Consistency: Integration ensures that data is consistent across all systems, reducing the risk of errors and data discrepancies. Also, it automates data transfer and reduces manual data entry, saving time and effort, which increases efficiency. Integration enables real-time data sharing, which is crucial for making timely and informed decisions. Different departments can work more collaboratively when they have access to the same data which improves relationships with customers and suppliers: integration with external systems can improve interaction with customers, suppliers, and partners.

Types of ERP integration:

- *Internal integration*: This involves connecting various modules or components within the same ERP system to ensure seamless data flow. For example, integrating the finance module with the inventory module to update inventory levels when a sale is made.

- *External integration*: This involves connecting the ERP system with external systems or third-party applications. Examples include integrating the ERP system with e-commerce platforms, CRM systems, or supply chain management software.
- *Cloud integration*: As more organisations move to cloud-based ERP solutions, cloud integration tools are used to connect cloud ERP systems with other cloud or on-premises applications.
- *B2B integration*: This type of integration is common in supply chain management, where ERP systems connect with those of suppliers, distributors, or logistics providers to automate order processing and inventory management.

When integrating ERP, consider how well the integration solution can scale as the organisation grows or changes, because regular maintenance and updates are necessary to ensure that integrations continue to function correctly as systems evolve. Integration projects can be costly, so it's important to weigh the benefits against the expenses. Additionally, consider the long-term benefits of integration, such as improved efficiency and productivity, when evaluating the return on investment (ROI) for your integration project.

Centralised Database: ERP systems typically use a centralised database to store and manage data, making it easier for employees to access the information they need. ERP, as standardised software package employed to integrate business functions through a single database (Klaus et al., 2000). With the development of cloud computing technology, the cloud has been a popular option to host an ERP system instead of a traditional on-premises hosting (Paulsson & Johansson, 2023). In the UK, Accenture report indicates that nearly 70% of firms are employing variations of cloud computing technology for their ERP solutions (Paulsson & Johansson, 2023; Peng & Gala, 2014). It's worth noting that while centralised databases offer many advantages, they also come with challenges, such as the need for robust data governance, data access controls, and the risk of a single point of failure. Organisations need to carefully plan their database architecture and ensure that it aligns with their ERP and overall business strategies. Many ERP systems are designed with centralised databases as a core component to support seamless data integration and access across different modules and functions. This centralisation plays a crucial role in the ERP's ability to provide a unified view of an organisation's operations. Key aspects and benefits of a centralised database in ERP:

- *Data consistency*: In a centralised database, all data is stored in one location, ensuring consistency and reducing the risk of data duplication and inconsistencies. The database should have a transaction table that should include invoices, transaction items, furniture components, and so on (Farhat & Owayjan, 2017). This consistency is critical for accurate reporting and decision-making.
- *Efficient data retrieval*: Centralised databases make it easier and faster to access data because it's all in one place. This efficiency is crucial for ERP systems, where users need to retrieve information from various modules and departments.
- *Streamlined data updates*: When data is centralised, updates and changes can be made more efficiently. A change in one part of the organisation can instantly

affect other parts, ensuring that everyone is working with the most up-to-date information.

- *Simplified data management*: Managing data security, backup, and maintenance is more straightforward in a centralised database. Security measures, such as access controls and data encryption, can be implemented consistently.
- *Easier reporting and analytics*: Centralised data makes it easier to create comprehensive reports and perform analytics since all the required data is readily available. Users can generate reports that span multiple aspects of the business.
- *Reduced data redundancy*: With data stored centrally, there is no need to duplicate data across different systems or departments. This reduces data redundancy, which can save storage space and simplify data maintenance.
- *Data integrity*: Centralised databases often have mechanisms in place to ensure data integrity, such as referential integrity constraints, which prevent data anomalies and errors.
- *Scalability*: Centralised databases can typically be scaled as an organisation grows. You can add more storage capacity or processing power as needed to accommodate increasing data volumes.
- *Backup and disaster recovery*: Centralised data makes it easier to implement consistent backup and disaster recovery plans. Organisations can ensure that critical data is backed up regularly and can be restored in case of data loss or system failure.
- *Security*: Security measures can be implemented consistently across all data stored in a centralised database. This is especially important for sensitive or confidential information.
- *Data governance*: Centralised databases facilitate data governance practices, including data quality control, data stewardship, and compliance with data regulations.

Automation: ERP systems automate various business processes, reducing manual data entry and the risk of errors. The ERP system creates and improves the automated production planning process according to the company's conditions and adapts the production planning process automation methodology to the company's conditions (Syreyshchikova et al., 2020), especially for furniture production. This can lead to increased efficiency and productivity. Implementing ERP automation requires careful planning, customisation, and integration with existing systems. It often involves a combination of software tools, workflows, and business process reengineering. While there are upfront costs associated with automation, the long-term benefits in terms of efficiency, accuracy, and cost savings can be substantial, making it a valuable investment for many organisations. ERP automation refers to the use of technology and software to streamline and simplify various business processes within an ERP system. Key aspects and benefits of ERP automation:

- *Data entry and processing*: Automation eliminates manual data entry by automatically capturing and processing data from various sources. This reduces data entry errors and frees up employees for more value-added tasks.

- *Workflow automation*: ERP systems can automate workflow processes by routing tasks and approvals to the appropriate personnel based on predefined rules and conditions. This ensures that processes are executed consistently and efficiently.
- *Inventory management*: Automation in ERP systems can optimise inventory levels by automatically tracking stock levels, generating reorder alerts, and even placing purchase orders when inventory falls below a certain threshold.
- *Order processing*: Automated order processing can accelerate the fulfilment process by automatically processing customer orders, generating invoices, and updating inventory levels in real-time.
- *Financial processes*: ERP automation can handle financial tasks such as accounts payable and receivable, invoice generation, and reconciliation. This reduces the risk of errors and ensures accurate financial records.
- *Reporting and analytics*: Automated reporting tools within ERP systems can generate customised reports and dashboards, providing real-time insights into key performance indicators (KPIs) without manual data compilation.
- *Document management*: ERP automation includes document management features, allowing organisations to digitise, store, and retrieve documents easily. This reduces the need for paper-based processes and manual filing.
- *Human resources*: ERP automation can streamline HR processes, including employee onboarding, payroll processing, leave management, and performance evaluations.
- *Supply chain management*: Automated supply chain processes can optimise the procurement, production, and distribution of goods, ensuring timely delivery and cost savings.
- *Customer relationship management (CRM)*: Automation in CRM modules of ERP systems can automate marketing campaigns, lead tracking, and customer support processes, enhancing customer engagement and satisfaction.
- *Maintenance and alerts*: ERP systems can automatically generate maintenance alerts for equipment and assets, helping organisations schedule preventive maintenance and reduce downtime.
- *Compliance and reporting*: Automation ensures that organisations can meet regulatory compliance requirements by generating necessary reports and maintaining accurate records.
- *Integration with external systems*: ERP automation extends beyond internal processes and can automate interactions with external systems, such as suppliers, customers, and e-commerce platforms.
- *Notifications and alerts*: ERP systems can send automated notifications and alerts to users when specific events or thresholds are met, ensuring timely actions.

Real-time Information: Users can access real-time data, allowing for better decision-making. The direct benefits are the simplification of operations and agility in the decision-making process (Aboabdo et al., 2019; Chopra et al., 2022; Guisea et al., 2023). Managers can monitor KPIs and make informed decisions based on up-to-date information. This real-time information is critical for making timely and informed decisions, as it allows users to monitor and respond to changes as they occur. To achieve real-time information in ERP, organisations often invest in advanced technology infrastructure, including fast and reliable data networks, databases, and ERP systems that can handle real-time data processing. Additionally, ERP systems may incorporate features like dashboards, reports, and analytics tools that provide users with immediate access to relevant real-time data. Here are some key aspects and benefits of real-time information in ERP:

- *Timely decision-making*: Real-time information enables decision-makers to react promptly to changing conditions, market trends, or operational issues. This agility can give an organisation a competitive advantage.
- *Inventory management*: Real-time inventory data helps organisations track stock levels, monitor demand, and avoid stockouts or overstock situations. This is particularly important for businesses with variable demand.
- *Order processing*: Real-time order processing allows organisations to respond quickly to customer orders and inquiries, improving customer satisfaction and order fulfilment
- *Production planning*: Manufacturing organisations can adjust production schedules in real-time based on demand fluctuations, ensuring efficient resource utilisation and minimising lead times.
- *Supply chain visibility*: Real-time information provides visibility into the entire supply chain, allowing organisations to track shipments, monitor supplier performance, and address any issues as they arise.
- *Financial insights*: Real-time financial data helps finance teams manage cash flow, monitor revenue, and make informed financial decisions.
- *Sales and customer service*: Real-time CRM data enables sales and customer service teams to provide timely responses to customer inquiries and track sales opportunities.
- *Employee productivity*: Managers can monitor employee productivity and performance in real-time, addressing any concerns promptly and optimising workforce management.
- *Quality control*: Real-time data can be used to monitor quality control processes in manufacturing or service delivery, ensuring that products or services meet predefined quality standards.
- *Compliance monitoring*: Organisations can use real-time data to monitor compliance with industry regulations and internal policies, reducing the risk of non-compliance.
- *Alerts and notifications*: ERP systems can generate automated alerts and notifications when predefined thresholds or conditions are met, enabling proactive actions.
- *Data analytics*: Real-time data can be analysed to identify trends, patterns, and anomalies, leading to data-driven insights and improved decision-making.
- *Customer insights*: Real-time customer data allows organisations to personalise marketing and sales efforts, enhancing the customer experience.
- *Production efficiency*: Real-time production data helps optimise equipment utilisation and minimise downtime through predictive maintenance.

- *Data accuracy*: Real-time information reduces the reliance on outdated or manually-entered data, improving data accuracy and reliability.

The ability to access and act upon real-time information is becoming increasingly important in today's fast-paced business environment, allowing organisations to adapt and thrive in rapidly changing markets.

Customisation: Many ERP systems can be customised to meet the specific needs of an organisation. This involves configuring the software to match the organisation's business processes. While ERP systems offer a wide range of standard features and functionalities, every business is unique, and customisation allows organisations to adapt the ERP system to align with their particular processes, industries, and goals. Tailoring might be an important procedure during ERP implementations, in which the ERP customisation takes place in order to ensure the compliance with the organisations' critical business processes and requirements (Hustad et al., 2016). Customisation ensures that the ERP system aligns perfectly with the organisation's unique workflows and processes, improving efficiency and productivity. Different industries, especially in furniture manufacturing companies, have specific compliance standards and operational needs, so customisation allows the ERP to address these industry-specific requirements, can provide a competitive advantage by enabling unique features or workflows that competitors may not have, have a higher user adoption rate since they are designed to fit the way employees work.

Types of Customisation:

- *Configuration*: This involves setting various parameters and options within the ERP system to match specific requirements. It often doesn't require programming and can be done through the system's administrative tools.
- *Custom development*: When standard configuration options are insufficient, custom development involves writing code to create new functionalities or modify existing ones. This is typically more complex and expensive than configuration.
- *User interface (UI) customisation*: Organisations can customise the ERP's user interface to make it more user-friendly and intuitive for their employees.
- *Reporting and dashboards*: Customised reports and dashboards can be created to provide specific insights and data visualisation tailored to the organisation's needs.
- *Integration*: Customisation can involve integrating the ERP system with other software applications and databases to enable seamless data exchange.

ERP customisation can be a powerful tool for organisations seeking to maximise the value of their ERP investment. However, it should be approached strategically, with a focus on addressing specific business needs and ensuring long-term sustainability. In particular, the implementation of an effective ERP system can lead to the development of sustainable organisations (Anaya & Qutaishat, 2022; Pohludka et al., 2018). ERP systems can enhance the organisations ability to capture sustainability data and can provide increasingly favourable disclosures, resulting in better sustainability assessment scores by rating agencies and indices

(Anaya & Qutaishat, 2022; Simmonds et al., 2018). Must be taken into account long-term needs and scalability when customising ERP systems to avoid frequent modifications.

Scalability: ERP systems are often scalable, which means they can grow with the organisation. New modules and features can be added as needed. Scalability in the context of ERP refers to the system's ability to handle increased workload, data volumes, users, and transactional complexity as an organisation grows. ERP scalability is a crucial consideration for businesses, as it ensures that the ERP system remains effective and responsive even as the company's needs evolve.

Here are some key aspects and considerations related to ERP scalability:

- *Vertical scalability*: Vertical scalability, also known as scaling up, involves upgrading the hardware and infrastructure components of the ERP system to handle increased demands. This can include adding more powerful servers, increasing memory or storage capacity, or enhancing processing capabilities. Vertical scalability is often suitable for smaller to medium-sized businesses or when an organisation experiences moderate growth.
- *Horizontal scalability*: Horizontal scalability, also known as scaling out, involves adding more servers or nodes to distribute the workload and accommodate growing demands. This approach is more common in large enterprises with high transaction volumes. Load balancing and clustering technologies are often used to achieve horizontal scalability, ensuring that requests are evenly distributed across multiple servers.

Scalability should be considered as part of the long-term ERP strategy. Organisations should plan for future growth and ensure that their ERP system can adapt to evolving business needs. Scalability is a critical aspect of ERP system design and management, as it ensures that the system remains an asset rather than a limitation as an organisation expands and evolves. Proper planning, architecture, and infrastructure choices are key to achieving ERP scalability effectively.

Reporting and Analytics: ERP systems typically include reporting and analytics tools that help organisations analyse their data and gain insights into their operations. They provide organisations with the ability to gather, analyse, and present data from various parts of the business, enabling informed decision-making and strategic insights. These tools often offer predefined templates and customisable options to tailor reports to specific business needs. But also, custom reporting capabilities enable organisations to create reports that address their unique requirements. Users can define report layouts, select data fields, and apply filters to extract relevant information. Custom reports are valuable for tracking KPIs and performance metrics specific to the organisation's goals. Some ERP systems offer self-service reporting capabilities, allowing non-technical users to create ad-hoc reports without IT assistance. This empowers business users to access the data they need when they need it. Advanced analytics capabilities within ERP systems enable organisations to perform deeper data analysis, including trend analysis, predictive analytics, and what-if scenarios. These analytics can uncover patterns, identify opportunities, and support strategic decision-making.

Also, ERP systems can be integrated with external data sources, such as customer databases, market data, or social media, to provide a broader perspective for decision-making. Modern ERP systems are becoming more and more integrated, more efficient and more strongly interconnected, but they also offer multiple options to ensure or increase the safety and security of data because we live in the era of Cyber Insecurity where there are so many threats to data (Chirvase & Zamfir, 2023; McCrevan, 2020).

The reporting and analytics capabilities of the ERP system should be scalable to accommodate increasing data volumes and user demands as the organisation grows. Effective reporting and analytics within an ERP system play a pivotal role in driving organisational efficiency, identifying opportunities for improvement, and supporting strategic planning. It's essential for organisations to invest in training and user adoption to ensure that employees can fully leverage these capabilities to make data-driven decisions.

Security: Security is a critical aspect of ERP systems. ERP system is becoming the system with high vulnerability and high confidentiality in which security is critical for it to operate (She & Thuraisingham, 2007). They often have robust security features to protect sensitive business data. ERP systems contain sensitive and valuable data, making them a prime target for cyberattacks and unauthorised access. Effective ERP security measures are necessary to protect the confidentiality, integrity, and availability of data and ensure that the system operates securely. Here are key aspects related to ERP security:

- Implement role-based access control (RBAC) to ensure that users have appropriate permissions and access only the data and functionality required for their roles. Enforce strong password policies and multi-factor authentication (MFA) to enhance user authentication.
- Encrypt data both in transit (when transmitted between systems or over networks) and at rest (when stored in databases or files). Encryption safeguards data from unauthorised access or interception.
- Keep the ERP system and all related software components up to date with security patches and updates. Regularly apply security updates to address vulnerabilities.
- Deploy firewalls to monitor and filter network traffic to and from the ERP system. Intrusion detection systems (IDS) can help detect and respond to suspicious network activity.
- Implement regular data backup procedures to ensure data can be restored in case of data loss or system failure. Verify the effectiveness of backups through periodic testing.
- Enable auditing features within the ERP system to track user activities and changes to data. Audit trails provide a record of who did what and when, aiding in security investigations.
- Classify data based on its sensitivity and importance. Segment the network and isolate critical systems and data from less critical ones to limit the potential impact of security breaches.

- *Third-party integrations*: Assess the security of third-party integrations and ensure they do not introduce vulnerabilities into the ERP system. Vet third-party providers for security compliance.
- Ensure that the ERP system complies with relevant industry regulations and data protection laws, such as GDPR, HIPAA, or SOX, depending on the organisation's industry and location.
- Conduct regular vulnerability scans and penetration tests to identify and remediate potential weaknesses in the ERP system's security defences.
- Implement continuous security monitoring to detect and respond to security threats in real-time. Use security information and event management (SIEM) tools for this purpose.

ERP security is an ongoing process that requires vigilance and a proactive approach. By implementing robust security measures and staying up to date with security best practices, organisations can minimise the risks associated with ERP systems and protect their critical data and operations. User access control in an ERP system is a fundamental aspect of ensuring data security, protecting sensitive information, and maintaining the integrity of the ERP environment.

User Access Control: ERP systems allow organisations to control user access to different parts of the system, ensuring that employees only have access to the information and functions relevant to their roles. It involves defining and managing who can access the ERP system, what they can do within the system, and under what conditions they can perform these actions. Roles are typically associated with predefined sets of permissions, simplifying the management of user access what data and functionalities users can access within the ERP system. Permissions should be granted on a need-to-know basis. Here are key aspects related to ERP security:

- Classify data within the ERP system based on its sensitivity and confidentiality. Segment data to restrict access to specific user groups or roles. Isolate and protect highly sensitive data, such as financial or personal information.
- Establish a formal process for users to request access to specific parts of the ERP system. These requests should be reviewed and approved by authorised personnel before granting access. Periodically review and audit user access rights to ensure they align with job roles and responsibilities.
- Apply the principle of least privilege, which means that users should be granted the minimum level of access necessary to perform their job tasks. Avoid granting excessive permissions that could lead to misuse or data breaches.
- Enable user activity monitoring and logging within the ERP system. Log and review user actions to detect and respond to suspicious or unauthorised activities.

User access control is a crucial component of ERP security, and organisations should establish robust policies and procedures to manage user access effectively. By implementing these measures, organisations can reduce the risk of data breaches, unauthorised access, and other security incidents within their ERP environment.

Empowering People in Manufacturing Industries: Participation-Based Management

The success of an organisation, including a manufacturing organisation, depends on its employees and their motivation to contribute their skills and efforts to common goals. Therefore, the improvement of skills and responsiveness of employees to new changes in the environment is a necessity in modern organisations. Extensive research has shown that such growth is driven by a combination of determinants such as flexibility, well-being, or mastery-enhancing preconditions, including participation (Sharma et al., 2022).

Today, there is no longer any doubt about the positive impact of participation on organisational culture and performance. By participating, people can understand the underlying principles that led to the development of a solution or the introduction of a new technological process. Participation becomes more realistic if a sense of ownership is created: understanding the underlying principles, the value of the new technology, the expected outcomes, and the opportunities to influence the features of the technology to better meet their needs (Ito et al., 2021).

The concept of employee participation has become an important management research framework in a variety of social research contexts, both in political science (Sarzynski, 2015) and stakeholder management (Ninan et al., 2019). Participation is most often identified with a person-centred orientation, recognising both oral and written participation. It is generally considered that participation is divided into: direct communication; upward decision-making; representative participation; and financial participation (Wilkinson et al., 2010).

While employee involvement is primarily associated with the performance of employees' job responsibilities, the concept has now been considerably broadened and employee involvement is measured at the level of the whole organisation, regardless of the forms of involvement applied (Wikhamn et al., 2022).

Employee participation is associated with both organisational-level and individual-level outcomes in the following areas: organisational innovation, organisational performance, individual job satisfaction and psychological ownership (Wikhamn et al., 2022).

Participation is aligned with participation in decision-making and the capacity to be part of the processes that take place in the organisation. Therefore, participation is therefore a shared responsibility between employees and managers, which brings members together to solve problems and make decisions by working in a team (Alsughayir, 2016). In this context, teams are understood much more broadly than just formal work teams within a hierarchical structure. Employee participation is associated with both organisational-level and individual-level outcomes in the following areas: organisational innovation, organisational performance, individual job satisfaction and psychological ownership (Wikhamn et al., 2022).

When it comes to employee participation, decision-making is usually taken into account (Ito et al., 2021). This approach to performance management helps to smooth the implementation of change in organisations and achieve employee adoption. Employee participation is becoming a prerequisite for success, which needs to be ensured as well as a continuous movement towards greater optimisation of production processes. According to research, employee participation leads to increased opportunities to develop their specific skills and competences (Sharma et al., 2022).

There is an increasing amount of evidence to suggest that involving middle managers and shop floor workers in production can reduce the operating costs of projects, provided that enough time is taken to develop the involvement techniques and make them a part of the culture. This approach is promoted in many countries, particularly Germany, where the unique characteristics of the national manufacturing sector make this approach a particularly good fit (Wagner et al., 2015).

Total Quality Management (TQM) is often successful when employees participate in continuous improvement activities. Participative management is seen as a key component in enabling employees to take the lead in such activities. This not only benefits the organisational objectives, but also enhances perceived accountability, job satisfaction, and productivity. Therefore, every company that aims to produce high-quality products should strive to maximise employee involvement, awareness, and the possibility of having channels for feedback. Involvement becomes a crucial factor in the structure of a quality management system (Huang & Gong, 2019).

There is an increasing amount of evidence to suggest that involving middle managers and shop floor workers in production can reduce the operating costs of projects, provided that enough time is taken to develop the involvement techniques and make them a part of the culture. This approach is promoted in many countries, particularly Germany, where the unique characteristics of the national manufacturing sector make this approach a particularly good fit (Wagner et al., 2015).

Stages of Participation

Participation can be fostered at different stages of change: planning (deciding to apply the technology), prototyping (developing the technology), and implementation (implementing or even updating the technology). When exploring ways to improve employee engagement, a commonly used approach involves a dynamic three-step process. This process comprises of (Oikonomou, 2018):

- an analysis phase,
- a formulation phase, and
- an implementation phase.

It is important to understand that different practices may be used by senior management in each of these stages to foster employee engagement. However, not all practices are suitable for every phase.

There are various alternative approaches that managers can utilise to integrate employee participation with the overall strategy. These include (a) consultative participation, (b) employee ownership, (c) representative participation, (d) informal participation, (e) employee participation, (f) employee participation groups, (g) works councils and employee representation at board level, and finally, (h) social media (Oikonomou, 2018).

Instrumental Approach Towards Participation

From a managerial perspective, we are always on the lookout for ways to involve our employees in a more meaningful and productive manner. We understand that modern society offers numerous options for achieving this goal, including changes to the participative culture in a live organisational setting, as well as the use of digital means to get closer to each employee on an individual basis. The following are the primary forms of engagement that we consider to be of the utmost importance:

- Online platforms that allow employees to be involved in all company processes, from observing and being informed to actively suggesting and operating together.
- Decision support systems (often also called expert systems), which are mainly directed at helping employees to make decisions under conditions of uncertainty, are another form of engagement.
- Total quality management (TQM). Often the involvement is already built into existing quality management systems or LEAN, where initiating feedback involves employees in the consideration of non-conformities, while TQS involves employees in periodic quality reviews.
- Industry Peer Networks (IPNs) bring together non-competing small businesses to work together to improve their skills and stay competitive in the market. These networks consist of peer firms that belong to the same industry segment, and provide similar goods or services to different customer groups in non-competing geographic areas. IPNs require a higher level of participation and involvement from its members.

Higher Scale of Participation and Engagement

Industry Peer Networks (IPNs) are a higher form of participation and should be discussed in more detail.

Alongside organisational participation and the involvement of individuals, professional networks are beginning to grow, which are however essentially distinguishable from other forms of self-organisation, such as associations, which are more concerned with the representation of interests. The IPNs are more concerned with sharing and strengthening their skills (Leung et al., 2019).

IPNs a participatory initiative, bring together non-competing small businesses in a collaborative framework to improve their skills to remain competitive in local and regional markets. IPNs are a unique form of 'parallel peer enterprises', where member firms belong to the same segment of a given industry and use similar resources to provide similar goods or services to different customer groups in non-competing geographical areas. Such a peer network requires a higher level of participation and involvement. A key feature of IPN is the peer dimension. It is like a network of experts who meet regularly and participate in activities in small groups. For such a network to work, a high level of confidence is essential. In terms of the content of the peer dialogue, it covers knowledge of management and marketing, industry trends both within and outside one's sector, discussing issues related to the company's performance and constructively criticising peer companies. However, so far, such networks are more common in the retail and service sectors, and performance improvement is the main motivation for a member firm to join an IPN (Leung et al., 2019). But their explicit value-added character may also be relevant for Industry 4.0 sensitive sectors working in the FI segment, as the Industry 4.0 context has triggered a different type of learning. Whereas previously collaborative learning (e.g. information sharing) dominated in dyadic relationships between collaborating organisations, the new context and the spread of technology are now promoting competitive learning (e.g. ranking) between parallel peer organisations (Leung et al., 2019).

Employee Participation in the Era of Industry 4.0

Engagement is crucial for manufacturing companies, just as it is for any other organisation. Firstly, it is beneficial for the reasons mentioned above, such as increased efficiency, reduced costs, and better control over the final product. Additionally, it can make a significant contribution to the individualisation of production processes, as motivated workers are better equipped to handle complexity and constant uncertainty. The phenomenon of inclusion, particularly in the manufacturing sector, is so universal that it is independent of education and other demographic characteristics (employee age, gender, education qualifications, and length of employment) (Chan et al., 2016). The only thing it depends on is the effectiveness and acceptability of the engagement process.

As Industry 4.0 comes with major changes not only in performance but also in thinking, with new paradigms changing established practices, including the set of competences required in manufacturing, and the willingness of employees to acquire new competences, to explore and use new technology, becomes a crucial factor for the penetration of Industry 4.0.

As the vast majority of industrial engineers would prefer to use a technology that they have developed themselves or on which they have been consulted during the implementation process (Ito et al., 2021) new Industry 4.0 innovations are nevertheless delayed at the application stage, as they are difficult to grasp and to be adopted into the paradigm shift. Participation is explored as a means to reduce resistance to Industry 4.0 innovation (Ito et al., 2021). A great number of studies have focussed on this, looking at what enables these transitions to be facilitated. It has been found that when employees are involved in working with new tools, new technologies, and new channels to develop a participative organisational culture, then transitions are more smoothly experienced since participation creates a place in the organisation to share knowledge (Sharma et al., 2022).

In manufacturing SMEs, employee involvement becomes an even more sensitive organisational tool, as the limited pool of employees means that the range of skills required must be covered and that flexible adaptability to change must be allowed to ensure that performance is not undermined (Wikhamn et al., 2022).

Assuming that small enterprises are even more affected by the benefits of employee participation, some authors have identified three important aspects of participation and areas where the contribution of SME employees is particularly important. These are work-role participation, human resources management participation and strategic participation (Wikhamn et al., 2022):

- Work role participation is a commonly used form of participation where the employee is involved in issues directly related to his/her job function and responsibilities.
- Participation in human resources management refers to the involvement of the employee in the consideration of recruitment issues. It is argued that in small companies, that cannot afford to have an HR department or a dedicated specialist, it makes sense to involve all employees in team-building issues to ensure a coherent HR policy. Such involvement is positively perceived by the employees themselves.
- Strategic involvement refers to participation in business decisions such as collaboration, goal setting, and investment in assets, business models, products, services, processes and working methods.

Based on the current knowledge of employee participation, it is possible to formulate a concept of prerequisites for employee participation specifically for furniture customised manufacturing companies, which will typically be small or medium sized enterprises, and which will be oriented towards a significant change in manufacturing principles as a response to the new technologies offered by Industry 4.0. A visualisation of the concept is given in Fig. 3.5.

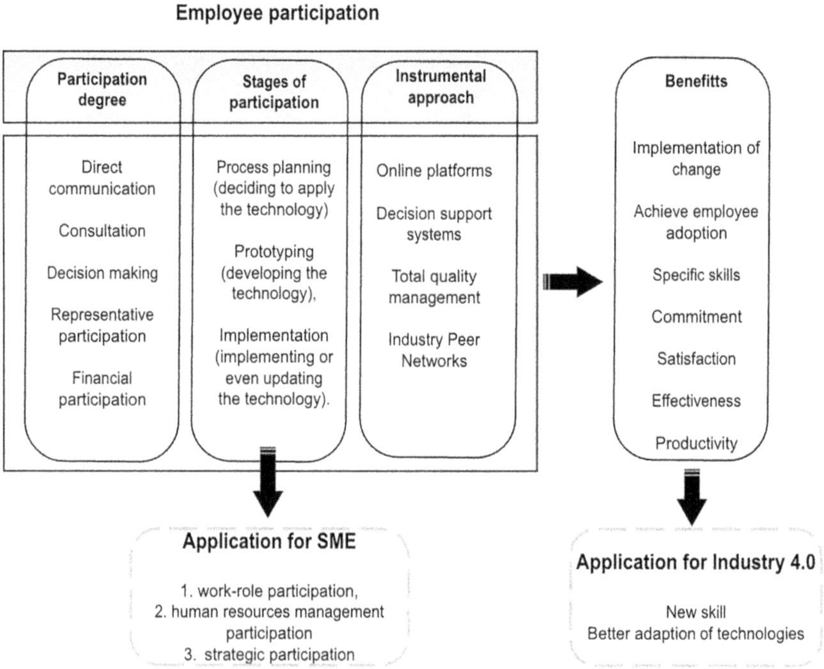

Employee participation

Participation degree
Direct communication
Consultation
Decision making
Representative participation
Financial participation

Stages of participation
Process planning (deciding to apply the technology)
Prototyping (developing the technology),
Implementation (implementing or even updating the technology).

Instrumental approach
Online platforms
Decision support systems
Total quality management
Industry Peer Networks

Benefitts
Implementation of change
Achieve employee adoption
Specific skills
Commitment
Satisfaction
Effectiveness
Productivity

Application for SME
1. work-role participation,
2. human resources management participation
3. strategic participation

Application for Industry 4.0
New skill
Better adaption of technologies

Fig. 3.5. The Concept of Employee Participation.

The idea of employee participation is based on three dimensions:

- the level of participation, which includes both direct communication and representative participation;
- the stage of participation, which corresponds to the most common stages of customised furniture production; and
- the instrumental approach required.

Considering the specific situation of SMEs, the three dimensions of participation can be refined by emphasising the importance of involving employees in the allocation of work roles, building teams through HR tools, and making strategic decisions. If a company can implement the elements of this concept successfully, it is highly probable that it will enjoy various benefits. These benefits may include increased efficiency, reduced costs, and the ability to adapt to new changes. In the context of Industry 4.0, these benefits may manifest as improved adaptability to new production technologies and the development of new employee capabilities that can absorb paradigm-shifting production changes.

Cost Estimation Approaches

After discussing the importance of customised production facilitated by digital business processing tools such as ERP and the role of employee involvement, it is now crucial to address the pricing of such products. This is particularly important in the case of new products, which represent a paradigm shift in production practices.

Pricing is a critical factor in the production of customised products. Customers want to buy at a reasonable price. Customised products are typically produced in small batches and tailored to meet specific customer needs, which requires a significant amount of design and manufacturing work, hence the higher price compared to standard products.

Traditional Costing Methods as Life Cycle Costing

Manufacturing costs are the core activity of every manufacturing business. Estimating the cost of a product is a crucial and important task for any manufacturing company in building its long-term competitiveness. High-tech manufacturers seem to be more concerned about these problems (Chou et al., 2010), but it has to be acknowledged that all businesses, whether high-tech or low-tech, are faced with a need to address this problem with a similar urgency. Although costing seems to be primarily related to the engineering content of the production of a product, it is a key issue for sales managers when they are working on the front line with customers and offering them a product at a price that is acceptable to them.

In most cases, the objective of determining the final cost of production is to allocate the raw materials used, labour and overheads to the finished products. Extensive research has shown that, in all industries, the cost of production accounts for between 60% and 70% of the final selling price (Quesada-Pineda,

2010). In order to optimise production costs, it is important to have efficient cost allocation systems. However, such a simple task is not so trivial, especially when dealing with new product development (Chwastyk & Kołosowski, 2014). Customised production is a situation where every product on the factory line is also a completely new product in the production cycle, making the issue of cost even more complex and sensitive for the whole production process.

Modern manufacturing aims to ensure the life cycle of the product, by linking the costing process to it. This has led to the development of the life-cycle costing approach, which is a cost analysis method that collects cost information. This helps designers or decision-makers to analyse project trade-offs, by producing a cost model that represents the total cost of a product over its entire lifetime. The results of this analysis are then used in design processes (Darla, 2017). Let's review the pathway of frameworks used for cost evaluation.

When developing costing systems, it is important to have knowledge of furniture manufacturing processes, which, although common to any other manufacturing industry, are essential to the development of a new product in the furniture industry. The process of developing a new product starts with a design solution, which involves working with the user, building a prototype and testing it. The design phase is then revisited to improve it before production begins.

Design. The design phase is essential to create a new product. It is the design phase that is the most sensitive focal point for unforeseen costs. It has been estimated that 70–80% of the total cost of a product is expected at the design stage, while only 10–15% of the cost will actually be incurred during the design stage (Darla, 2017; Quesada-Pineda, 2010). As most product decisions are made in the early stages of product development, it is important to estimate costs at that stage. Most quality mistakes and inaccuracies are caused by errors of omission particularly in the design phase.

Prototype. Prototype is the testing of a new product design before the production phase. This phase contributes significantly to reducing costs, ensuring the desired quality of the product and shortening the product development time (Vestad & Steinert, 2023). In addition to classical prototype testing, iterative prototyping through design (Vestad & Steinert, 2023), build and test cycles are becoming more and more popular, which is important for selecting the optimal product design, but clearly raises the cost of production, while minimising the caste associated with quality failures in the production itself.

Production. During the production phase, it is relatively easy to determine the actual cost, because, at least theoretically, it is tangible. However, several issues arise at this stage, including the complexity of calculating the fixed costs or the variable costs involved in producing multiple ingredients or products. Nevertheless, at this stage, the type of cost can be easily predicted, planned, and measured.

In practice, different ways of classifying expenditure are used, primarily to meet the requirements of the accounting frameworks. In the furniture industry, as in product-oriented manufacturing, it is important to have detailed information on how much each product requires in terms of costs (Quesada-Pineda, 2010). In the furniture industry, we can identify the following cost components: raw materials, human labour, processes, equipment, and other costs resulting from complexity. However, the allocation of some overheads or indirect costs to final products requires the creation of allocation or cost management bases.

Common cost categories for manufacturing are the following:

- *Direct materials (DM):*

  Materials that are consumed in the manufacturing process and physically incorporated in the finished product.

  Materials whose cost is sufficiently large to justify the record-keeping expenses necessary to trace the costs to individual products.

- *Direct labour (productivity, utilisation):*

  Labour time that is physically traceable to the products being manufactured.

  Labour time whose cost is sufficiently large to justify the record-keeping expenses necessary to trace the costs to individual products.

- *Manufacturing overhead (MOH):*

  All of the costs of manufacturing excluding direct materials and direct labour.

  Indirect materials (IM) – materials, used in the manufacturing of products, which are difficult to trace to particular products in an economical way.

  Indirect labour (IL) – labour, used in the manufacturing of products, which is difficult to trace to particular products in an economical way.

Machinery Utilisation

All other types of manufacturing overhead (depreciation on machinery, depreciation on factory building, factory insurance, and utilities for factory).

- *Non-manufacturing Costs:*

  Marketing or selling costs – costs incurred in securing orders from customers and providing customers with the finished product.

  Administrative costs – executive, organisational, and clerical costs that are not related to manufacturing or marketing.

Classification of Traditional Cost Methods

However, even if the essential categories of manufacturing are specified, accounting precision does not bring genuine precision. In manufacturing companies, including furniture companies, the most common cost approach is to record the cost of goods sold as inventories in the income statement, balance sheet and cash flow statement. The calculation of the cost of sales is a reporting requirement under government financial accounting regulations. This calculation helps an enterprise to control the information on how much raw materials, work in progress and finished goods it holds in inventory.

The traditional direct method is very popular because of its simplicity and low maintenance costs. However, they can also cause many problems. For example, a single input indicator can hide critical cost problems that can undermine the company's performance.

So, we have more than one approach to cost accounting. The most popular approaches to costing are the following:

- *Traceable costs*: In the furniture industry, a company therefore aims to know exactly how much raw material is used to produce a good. This determines that traceability becomes a dimension of cost evaluation. Unfortunately, while some raw materials can be accurately traced (direct materials), others are difficult to quantify and even more difficult to allocate correctly (added or indirect materials). For example, it is relatively easy to determine how many wood panels have been used, but it is rather difficult to determine the exact amount of glue needed to produce a given piece of furniture (Quesada-Pineda, 2010).
- *Fixed or variable cost*. Other difficulties arise in determining the fixed or variable part of costs. 'Fixed' costs are critical to a company's profitability because they remain stable even when production is disrupted. Other costs become variable depending on the volume of production (e.g. wages) (Novák & Popesko, 2014).
- *Value-added*. This classification encourages a focus on only those activities that are necessary for the production of services. The costs of these activities will be recovered on the basis of cost drivers and the inputs will be transferred to services. This approach encourages a focus on meaningful activities rather than on costs. For example, in the production of a product, the transfer of parts from one workshop to another has proved to be too costly, so that, the activity needs to be analysed and cost optimisation sought as a non-value-creating activity.

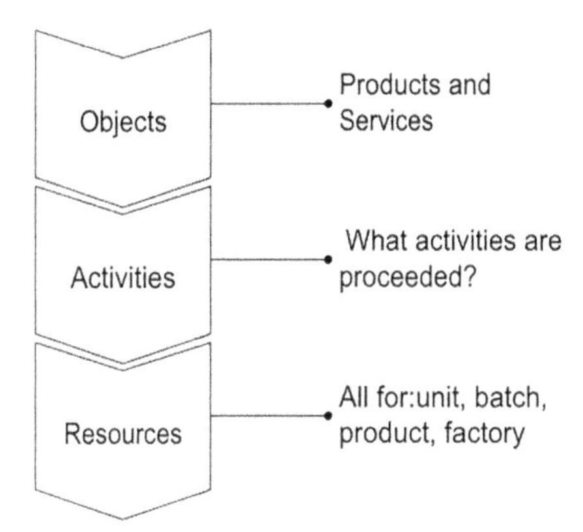

Fig. 3.6. The Concept of the Activity-based Costing Approach.

- *Financing reporting*. In manufacturing companies, costs are generally reported on the basis of the cost of the product. Under this reporting format, costs need to be broken down into direct materials, direct labour and overhead/indirect costs.
- *The activity-based costing method* (Fig. 3.6). In the ABC method, costs are allocated to specific activities and these activities are related to cost objects by activity factors (Ievtushenko & Hodge, 2012). Subsequently, each activity will be related to the resources of the organisation by resource factors. Cost drivers can be defined as the cause and effect relationship between cost objects and activities. These factors are also considered as an assessment of cost behaviour and help to find critical cost drivers. It is agreed to identify cost drivers at four different levels: unit, batch, product and factory levels.

Managerial Approach to Costing

Uncertainties Regarding New Product Development

There is overwhelming evidence that the issue of cost estimation is a critical element in the production of possible inaccuracies or errors due to hard-to-predict inputs. Manufacturers of new products attribute failures to inappropriate contractual decisions (in terms of pricing), underestimated costs and overly optimistic promises by manufacturers (Rodríguez et al., 2020). This problem is further exacerbated when service and product systems are integrated and the complexity of the new designed product system increases, creating new difficulties in assessing the behaviour of the system. This is very common at the ordering stage of a new product, when the customer is trying to define the desired functionality and manufacturing is trying to capture it as precisely as possible.

There is a growing debate that traditional cost assessment methods (e.g. Life Cycle Costing) no longer provide an adequate methodological basis to capture the dynamics inherent in the service-product interaction and therefore urges the development of a new paradigm of cost assessment. Addressing the content of the new cost/price paradigm leads to a deeper underlying understanding of the issues involved. Challenges associated with the price evaluation paradigm in the modern manufacturing environment include (Rodríguez et al., 2020):

- identifying relevant sources of uncertainty;
- identifying appropriate cost approaches;
- identifying cost drivers;
- incorporating the measurement of uncertainty into the evaluation process;
- exploring behavioural factors associated with decision-making under cost uncertainty; and
- proposing a cost assessment framework for a specific application area.

The cost estimate is commonly referred to as a baseline for determining an approximate price, while both top-down and bottom-up approaches to costing are still used. The top-down approach is used when insufficient data is available on the product being produced. In this case, the assessment is often done on the basis of the results of previously produced products, where future costs are predicted on the basis of mathematical relationships derived from the similarities and differences with previous projects. The bottom-up approach is based on the collection of highly disaggregated cost data, usually in a deterministic manner. The nature of the data also determines the character of the estimate itself, if the data is well defined then deterministic methods are used, if the data is probabilistic then stochastic methods are more common. To begin with an analysis of costs, data from all stages of the life cycle are required. The main difficulty is the fact that in most cases real data is only available late in the life cycle, but at that stage, the costs are already high.

From Costing to Pricing Strategies

The success of manufacturing companies is largely dependent on their strategy. The overall engineering approach and strategy chosen are influenced by the results of cost analysis. Managers are concerned about the final price of the product. The principle used to determine the final price of a product is that it should be what the consumer is willing to pay for a product that is tailored to their needs. This means that the final cost is not the only factor that determines the price of the product. This is especially true since pricing strategies are heavily influenced by competition and it's impossible to know the final cost of customising the product when the conversation with the consumer about the customised product begins. New approaches to pricing are emerging alongside the conventional and traditional approaches to costs. Two such practices are value-based pricing (VBP), which measures the value created for consumers by a product or service, and pricing based on the average market valuation of the product, guided by competition (taking into account costs and expected profit margins) (Aydede & Turkoglu, 2017). Recognising how heterogeneous the task of pricing strategy is, the managerial orientation of pricing requires the efforts of many specialists: engineers, accountants, sales managers, researchers when it comes to assessing product costs, competition in the segment, and the product portfolio.

The constraints that limit the pricing straightforward setting are defined as the following.

- *Uncertainty*: These arise from information limitations (incompleteness of data and information, non-measurability of customer value and inconsistency of competition.
- *Organisational structure*: The organisation's extent of formalisation reflects its members' perceived decision-making and self-monitoring abilities. Which price estimation strategy an organisation chooses depends on the form of the organisation. The formalisation of the organisation implies detailed procedures, standardised rules, and hierarchical distribution of responsibilities. In larger companies, pricing specialists who pool their competencies and draw on the necessary expertise from other departments emerge. Hierarchical management and centralisation are considered to be a key aspect of the structure of the organisation, with mixed effects on specific decisions such as pricing. Hierarchical centralisation is defined as the concentration of decision-making power at the top of the hierarchy. In most cases, high centralisation is justified in situations that operate under full certainty, which means that rules can be agreed unambiguously. Unfortunately, in situations that are less defined (e.g. in the case of customisation), where the participatory practice of employees is used to reduce uncertainty, decentralisation is preferred. The choice of a pricing strategy in the face of high uncertainty, therefore, poses considerable challenges for the organisational structure in combining existing centralised structures with distributed responsibility structures. In addition, in such cases the pricing strategy is associated with specific competences and is often localised to relatively high levels of centralisation.

Manufacturing companies often face pricing challenges, and they need to find effective solutions to overcome them. In addition to the traditional cost-based pricing strategy, companies also consider pricing according to their competitors' prices. This is because if the price of a manufacturer's product is significantly higher than similar products available on the market, it becomes necessary to find rational solutions to adjust the price or change production methods to match competitors' prices. By adapting production and investing more in the consumer experience, manufacturers can add more value to their products beyond the basic function they serve. If consumers are willing to pay a higher price for a product that provides them with a high level of value beyond the basic function, then a VBP strategy becomes more appropriate.

The challenges faced in the pricing process have resulted in management attitudes towards pricing and three-fold strategies could be defined:

- *Cost-based pricing*: Cost-based pricing is the usual classical approach to costing, where the price is evaluated through defined costs according to the principles described above.
- *Competition-based pricing*: Competition-based pricing is a popular strategy among companies. It suggests adjusting prices based on the average market

price. Pricing below the market average increases sales, while pricing above it decreases sales. Different strategies include stay-out pricing, bundled pricing, and penetration pricing. All are related to competition-based pricing based on the overall average market price.

- *Value-based pricing* (VBP): VBP works if the organisation has a culture of mutual trust, promotes member engagement and increases motivation to embrace change. VAP only works if it enables experimentation and learning by observing experiences. According to most marketing scholars, this approach is considered advantageous. However, studies report that this method is the one which is most rarely used and causes all sorts of difficulties (a meta-analysis for 1983–2006 showed a prevalence of 17%) (Liozu et al., 2011).

Whichever approach is used, it is likely that four steps will be required to determine price: information acquisition and processing, pricing objectives, policies and beliefs, organisational decision processing and organisational response.

Pricing strategies have been a topic of debate for many years. Research indicates that different types of companies tend to avoid certain strategies. For instance, SMEs usually use competition-based pricing approaches (Liozu & Hinterhuber, 2012). This is not surprising since VBP is quite challenging to implement and integrate into daily production activities. The difficulties associated with VBP are fundamental in nature and are rooted in basic organisational design and behavioural characteristics. This was demonstrated by Liozu's (2013) development of a conceptual model of integration based on three pillars: (1) organisational confidence, which encompasses team spirit, corporate resilience and collective capabilities to meet success; (2) championing behaviour, which is based on transformational leadership and ownership; and (3) organisational change, which is related to the centralisation of expertise in pricing, the clarification of responsibilities and pricing research. These characteristics are closely connected and interdependent, forming a 'circle' of internationalisation and transformation of VBP. The complexity of VBP lies in the significant organisational changes that are altering the company's identity and its stakeholder's identity. This transformation is a gradual shift from a focus on cost and competition towards a preference for customer value. To achieve this shift, there must be a profound change in the established structures and culture, followed by changes in processes and systems.

New Paradigm of Cost Evaluation

Today, with Industry 4.0, the reduction of labour in companies, and the increase in automation, overheads have started to rise steadily and the old methods of cost estimation no longer serve their purpose. The furniture industry has been hit hard when Computer Numerical Control (CNC) equipment has become an essential part of the equipment and has generated a huge price rise in costs. Traditional costing methods, including the direct method with traditional accounting performance indicators such as direct labour productivity and utilisation, machinery utilisation and others no longer reflect the actual situation in production and therefore lead to inappropriate strategic decisions. For example, there is no information on how many ingredients will be in a product, what is the lead time and price of a specific material. In addition, such a cost approach does not satisfy innovative and entirely new product development. Customisation has further complicated the issue and has come with a new approach to cost issues in the industry.

While attempts are made to operate with deterministic quantities, a new cost paradigm is inevitably coming because the uncertainties have not been resolved at the engineering level. For some time, attempts have been made to manage cost uncertainty within the framework of stochastic definition. However, only about half of the cases reported in the literature recognise the stochastic nature of costs and therefore quantify the uncertainty associated with them (Rodríguez et al., 2020). Others attempt to handle epistemic uncertainty using other, more modern, approaches such as fuzzy set theory, neural networks, possibility theory and evidence theory (Rodríguez et al., 2020).

The new paradigm comes with the realisation that there is a non-linear relationship between cost and factors of influence. The paradigm shift is illustrated by the new classification of cost estimation methods, highlighting five groups of methods: intuitive, analogue, parametric and analytical methods (Niazi & Dai, 2006), and incorporating the capabilities of artificial intelligence.

- *Intuitive* methods encompass case-based methods and decision support methods, including rule-based, fuzzy logic and expert system methods.
- *Analogical* methods refer to regression analysis models and back-propagation neural networks. Analogy-based methods use the costs of the analogous or previously produced product and incorporate any new adjustments known to be present. This approach is based on the assumption that historical costs are a good predictor of future costs (Rodríguez et al., 2020). Parameters such as production length, product size, cost, and design complexity become important considerations in the search for analogues. This approach becomes a useful technique when costs need to be estimated at an early stage of the project life cycle, when naturally little information is available about the new product. It is then important to have sufficient knowledge and expertise on analogues. Due to the detailed nature of the method, the cost estimation steps are more time-consuming than other methods, but also less accurate. Situations can result in a cost estimation process where some parameters may appear to be the same, but another parameter, such as the use of technology, can lead to a significant change in the circumstances in which the estimates are derived.
- *Parametric* models are based on statistical methodologies (e.g. regression models and distribution fitting) and express costs as a function of their constituent variables. Parametric costing establishes statistical relationships between historical costs and parameters. They are based on the assumption that factors that have affected costs in the past will continue to affect them in the future.
- *Analytical* approaches (Activity based costing) include the operations-based approach, the decomposition approach, tolerance-based cost models, attribute-based costing and activity-based costing. The essence of the methods is to

be able to link product inputs to a process and to decompose it into manageable tasks, operations or activities.

- *Artificial intelligence* (AI) (artificial neural network, case based methods). Case based methods use the nearest neighbour algorithm and is based on the characterisation of projects according to a number of key attributes.

A Machine Learning Approach to Estimate Early Costs of New Product

While AI plays an emerging role in cost estimation, its capabilities are increasingly being exploited and there are already many good applications of AI for cost estimation in the manufacturing sectors, computer and software engineering, process industry, parallel engineering, construction industry, R&D management and product development (Chou et al., 2010). Machine learning has become increasingly popular for price evaluation. There are several useful studies demonstrating the great potential of machine learning. It has been widely used in highly competitive sectors, such as the automotive industry, to address quality issues in production and pricing vetting (Zhang et al., 2020). There are several useful studies demonstrating the great potential of machine learning. For example, Bodendorf and Franke (2021) conducted a study in the automotive industry and found that machine learning models have high accuracy and precision in cost estimation. However, on average, the models tend to underestimate total costs. Nonetheless, the findings are extremely valuable in supporting cost engineers in making informed decisions.

Artificial Neural Network

It has been shown that for early cost forecasting, *Artificial neural network (ANN)* in particular can model complex systems with even a minimum amount of data. Specifically, ANNs have been applied to process vessels, the automotive industry, sheet metal parts, TFT-LCD yield, machine-tool selection, software development, product lifecycle cost, vertical high-speed machining centres, and building construction (Chou et al., 2010).

Case-based Reasoning

Case-based reasoning (CBR) uses information from earlier manufactured products to adapt an existing design. These cases are collected in a base of cases and selected according to the attribute most relevant to the new product. The CBR assumes that when assessing the cost of a new product, the modifications foreseen in the new product will also be taken into account. This approach is particularly appropriate where there is no intention to develop a totally new product from scratch, thus saving engineering and pricing costs and time. The CBR approach is well suited to mass customisation, where existing products are developed with minor modifications. CBR is often combined with other well-known methods to improve the quality of cost estimation: feature counting, analytic hierarchy process, multiple regression analysis, decision trees, genetic algorithms, and artificial neural networks (Relich & Pawlewski, 2018).

In the case of the CBR approach, regression analysis models are often used to complement the artificial neural network analysis. The essence of this analysis is to find the relationship between the cost of previously manufactured products and certain characteristic variables that are related to the properties of the product (most often the number of components in the product). For this purpose, historical product data are always used (Relich & Pawlewski, 2018).

This feature can be exploited in the design phase of a new product development, thus saving limited resources and avoiding quality false.

CBR is carried out through a series of systematic procedures to extract relevant knowledge from the experience and then integrate the case into an established knowledge structure. The procedures include case indexing to match a new case with similar cases and to store cases in the knowledge base for further comparisons. The most common procedure of CBR as follows:

1. *Data about new product*. The CBR method starts with data collection about a new product. The aim is to gather as much information as feasible to identify the characteristic features of the new product (e.g. its purpose, complexity, and shape) and translate them in to attributes. An example attribute list (Relich & Pawlewski, 2018): the number of modifications proposed by customers, the number of customer requirements translated into product specification, the number of components in a new product, the number of new components in a new product, the number of project team members.
2. *Attributes weights*. New product attributes are selected and weights are assigned to these attributes based on their impact on new product costs.
   a. *Case similarity assessments*. Case similarity assessments shall be carried out by comparing the attribute values of a new case (new product) and a historical case stored in the case base. The nearest neighbour method (Chou et al., 2010) is used to define the total similarity function.
   b. *Artificial neural networks* can be used to calculate attribute weights: to calculate the error between a selected case (new product) and a tested case from historical data. The basic algorithm is designed on the principle of minimising the distance between instances in the same cluster and maximising the distance between instances in different clusters. The number of inputs to the neural network is the same as the number of attributes.

Flowchart of CBR is presented Fig. 3.7.

Expert Judgement in Price Estimation

Although the engineering-accounting approach to costing is widely recognised, expert opinion on costing is a popular alternative, particularly in estimating the cost of a new product that is still in the early stages of development and lack of

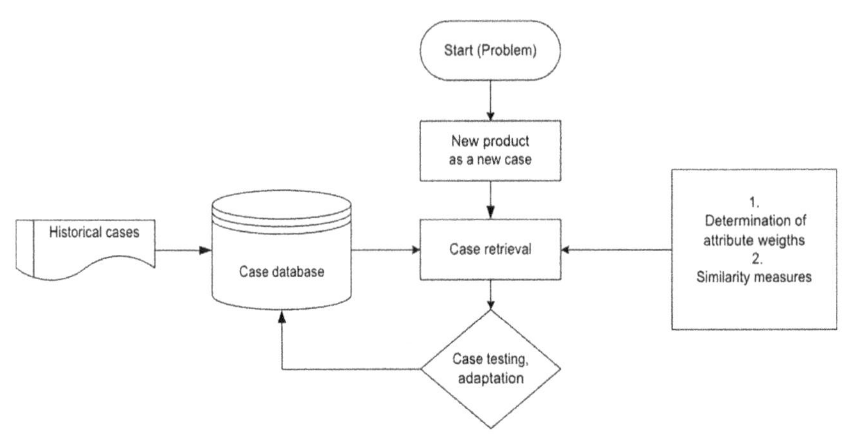

Fig. 3.7. Case-based Reasoning Flowchart. *Source*: Adapted from Chou et al. (2010).

actual production information. During the initial phases of product design, there is usually a scarcity of product information. As a result, the expert's experience becomes a crucial and exclusive source of information. The expert has the ability to consider the entire life cycle expenses of the product, making their opinion practically the only source of information accessible (Darla, 2017).

When expert input is discussed, it's often assumed that it's based on a guess made by an expert using personal knowledge and experience in the field. It is anticipated that cost estimators will utilise a mixture of logic, experience, skills, and judgement while discussing the input of experts in cost estimation. This ensures that the final estimate is appropriate and meaningful. In addition, the process of producing the final estimate involves numerous assumptions, which may be challenging for non-experts to comprehend. Thus, becoming an efficient cost estimator requires a wide range of skills, including economics, accounting, sales, engineering, statistics, and mathematical modelling (Rush & Roy, 2001).

When considering the role of experts in pricing, several issues need to be addressed. Firstly, there should be evidence to show that expert judgement is reliable and valid. Secondly, a valid working methodology needs to be developed to ensure that expert opinions are quantifiable. Lastly, it is important to identify good experts and find ways to connect with them.

Validity of Expert Judgement

Expert judgement is a traditional method used in the early stages of product development and is also prevalent in innovative products or services like software products. It's valuable to use peer review as a complementary approach because it's completely different in nature. Unlike quantitative pricing methods, expert judgement is based on the expert's domain knowledge rather than historical data.

Expert opinion is essential in areas where other types of empirical data on cost cannot be objectively collected or do not exist. For instance, in art history, determining the value of an artistic work relies on expert opinion (Ginsburgh et al., 2019). A similar analogy can be found in customised pricing, where a product is produced for the first and only time. Expert judgement can be very helpful when there are limitations that make it difficult to collect data efficiently and effectively. This approach is important when making decisions based on constraints and strict factors (Chirra & Reza, 2019).

Expert judgement is a widely used approach in many sectors, and empirical studies show that it is a valid method. For instance, studies on air traffic control demonstrate that expert judgements are valid when they are based on specific domain knowledge (Nunes & Kirlik, 2005). In such cases, experts can mitigate the effects of human biases. This mitigation may occur because increasing domain knowledge helps experts to better understand the ecological validity of environmental cues. However, this approach also has its disadvantages. Since the assessment is highly dependent on the expert's domain knowledge and previous experience, it often results in ambiguous cost estimates. As a result, pessimism, optimism, and bias are likely to influence the evaluation process. That's why it is often the case that companies prefer algorithmic and computer models over expert judgement (Rush & Roy, 2001). Nevertheless, by relying on an expert's knowledge and experience, experts are able to provide an honest opinion regarding the best course of action for understanding the impact of a system.

Methodological Implications of Expert Judgement

Expert judgement has long been used in pricing, although it is not considered a criticism, and it is discussed among pricing methods in the same line as mathematically rigid methods such as the parametric/algorithmic method or the analogy method (Darla, 2017).

In general, any methodology for expert evaluation relies on a two-step process: expert evaluation and statistical calibration. During statistical calibration, the model predictions are adjusted according to a defined reference prediction, data collected through questionnaires, or from empirical ontology engineering processes. Various techniques are used to make it easier for experts to assess and researchers to measure the validity of a prediction. One such technique is pairwise comparisons (Peng et al., 2011) where experts' opinions on the relative probabilities of a given failure mode are combined.

There are various systemic methodologies for involving experts in decision-making, assessments and forecasting and we reviewed them further. These include (Arvan et al., 2019):

1. Adjusting quantitative models through evaluation. The process of forecast adjustment typically involves two steps. First, the forecaster determines whether adjustments to the system forecast are necessary and then decides on the direction and magnitude of those adjustments.
2. Forecast combination, which involves combining quantitative forecasts with expert opinions. Different combination methods, such as Theil's optimal linear correction method, mechanical combination, and corrected forecasts

of the first method combined with statistical forecasts have been used in the past. However, studies have shown that combining forecasts does not add any value to the forecasting process when the biases of judgements are already corrected.

3. Decisional bootstrapping, which is used to derive the mean from forecasters' inconsistency. This method aims to understand how forecasters arrive at their forecasts or revisions by using a regression model that regresses the experts' forecasts on various dependent variables.
4. Rule-based forecasting (RBF), which involves developing a set of condition and action statements (i.e. rules) based on the analysis of historical time series and domain knowledge. These rules are then used to combine the forecasts of simple extrapolation methods.
5. Forecasting by analogy (FBA), which is based on the idea of pattern matching. Forecasts are made by finding analogies between the current situation and past situations that have similar characteristics. However, relying on heuristics may lead to basing judgements on recent but not necessarily analogous or unusual cases.
6. Decomposition methods, which involve breaking down the forecasting problem into smaller tasks that require less cognitive ability from the forecaster. This method can reduce the forecaster's dependence on the use of heuristics in the decision-making process.
7. Ensemble forecasting, which can be used in addition to quantitative forecasting. The Delphi method is a popular ensemble forecasting method where anonymous ensemble members independently produce a forecast over one or more iterations. At the end of each iteration, a coordinator announces the experts' forecasts and the reasons for them. This process is repeated until a final forecast is derived as the average of the forecasts of the last iteration.

In order to provide accurate and reliable information, experts more often required to contribute to the development of a model beyond just providing a price as a single estimate. This involvement may include helping to choose variables, specify the model's structure, and set parameters (Arvan et al., 2019). In other words, expert input goes beyond simple reporting and requires a deeper level of involvement in the modelling process.

The Availability of Experts

Although common wisdom suggests that more experienced and respected experts are better predictors, research reveals a different story. Experts tend to rate each other consistently, but their predictions improve when expert panels are formed, with structured question protocols and feedback in place (Burgman et al., 2011). This is the foundation of popular expert vetoing methods, such as SEJ (Hanea et al., 2016).

One way to address the expert availability issue is to use expert recommendation systems, which are web-based systems. These systems aim to find people with extensive knowledge in a particular topic. Historically, experts are individuals who possess knowledge and expertise in a particular field, typically having a business or organisational connection to the subject matter (Peng et al., 2011). They can be sourced from the general public (Kareksela et al., 2020) or even from clients who have become experts themselves (de Graaff et al., 2019). Nowadays, expert recommendation systems are becoming more common. They use information retrieval techniques and search engines to extract information from people's activities and documents related to them (Nikzad-Khasmakhi et al., 2019). An expert recommendation system can help reduce the costs associated with finding and selecting the most suitable experts in knowledge management systems. This system works by following these steps: it accepts a user's query, collects the previous reputation of the experts, ranks their expertise according to a topic classification scheme, and provides a list of the experts whose knowledge is most relevant to the user's query.

These systems provide a list of people ranked according to their level of expertise in a particular topic. This combination of natural language processing systems and search engines can help protect the pool of experts. The popularity of such systems is significant in academic and industrial activities, with platforms like LinkedIn, Quora, Stack Overflow, and Expert Lookup being suitable commercial examples. Expert recommendation platforms can provide data for both general and specific purposes, such as water supply, sewerage services, electricity and communications or apparel (Guan et al., 2016). Expert recommendation systems can be used in software engineering or healthcare. The widespread use of expert platforms already provides a means of strengthening this branch of research by developing web-based expert platforms specialised for furniture manufacturing. This can expand the base of potential experts outside the company and extend the company's capabilities.

References

Aboabdo, S., Aldhoiena, A., & Al-Amrib, H. (2019). Implementing enterprise resource planning ERP system in a large construction company in KSA. *Procedia Computer Science*, *164*, 463–470.

Alsughayir, A. (2016). Employee participation in decision-making (PDM) and firm performance. *International Business Research*, *9*(7), 64. https://doi.org/10.5539/ibr.v9n7p64

Anaya L., & Qutaishat F. (2022). ERP systems drive businesses towards growth and sustainability. *Procedia Computer Science*, *204*, 854–861.

Anshari, M., Nabil, M., Ariff, S., & Al-mudimigh, A. (2019). Customer relationship management and big data enabled: Personalization & customization of services Big Data. *Applied Computing and Informatics*, *15*(2), 94–101. https://doi.org/10.1016/j.aci.2018.05.004

Arvan, M., Fahimnia, B., Reisi, M., & Siemsen, E. (2019). Integrating human judgement into quantitative forecasting methods: A review. *Omega (United Kingdom)*, *86*, 237–252. https://doi.org/10.1016/j.omega.2018.07.012

Berg, N. (2015). The environmental cost of shipping stuff is huge. *Can we fix that?*https://www.vox.com/2015/12/23/10647768/shipping-environmental-cost

Berg, N. (2016). The future of freight: More shipping, less emissions? *GreenBiz*, https://www.greenbiz.com/article/future-freight-more-shipping-less-emissions

Bodendorf, F., & Franke, J. (2021). A machine learning approach to estimate product costs in the early product design phase: A use case from the automotive industry. *Procedia CIRP*, *100*, 643–648. https://doi.org/10.1016/j.procir.2021.05.137

Burgman, M. A., Mcbride, M., Ashton, R., Speirs-Bridge, A., Flander, L., Wintle B., Fidler, F., Rumpff, L., & Twardy, C. (2011). Expert status and performance. *PLoS ONE*, *6*(7), 1–7. https://doi.org/10.1371/journal.pone.0022998

Cariou, P., Parola, F., & Notteboom T. (2019) Towards low carbon global supply chains: A multi-trade analysis of CO_2 emission reductions in container shipping, *208*, 17–28.

Chan, S. W., Omar, A. R., Omar, S. S., & Lim, K. H. (2016). Assessing participation in decision-making among employees in the manufacturing industry. In *2016 International Conference on Industrial Engineering, Management Science and Application (ICIMSA)*, *10*, 1–5. https://doi.org/10.1109/ICIMSA.2016.7503997

Chirra, S. M. R. C., & Reza, H. (2019). A survey on software cost estimation techniques. *Journal of Software Engineering and Applications*, *12*, 226–248. https://doi.org/10.4236/jsea.2019.126014

Chirvase, C. S., & Zamfir, A. (2023). Exploring enterprise resource planning (ERP) development: Challenges, opportunities and how can help companies navigate turbulent contemporary times. *Proceedings of the 17th International Conference on Business Excellence*, 1919–1928. https://doi.org/10.2478/picbe-2023-0169

Chopra, R, Sawant, L, Kodi, D, & Terkar, R. (2022). Utilization of ERP systems in manufacturing industry for productivity improvement. *Materials Today: Proceedings* [Internet], https://linkinghub.elsevier.com/retrieve/pii/S2214785322026955

Chou, J. S., Tai, Y., & Chang, L. J. (2010). Predicting the development cost of TFT-LCD manufacturing equipment with artificial intelligence models. *International Journal of Production Economics*, *128*(1), 339–350. https://doi.org/10.1016/j.ijpe.2010.07.031

Chwastyk, P., & Kołosowski, M. (2014). Estimating the cost of the new product in development process. *Procedia Engineering*, *69*, 351–360. https://doi.org/10.1016/j.proeng.2014.02.243

Darla, S. P. (2017). Product life cycle cost estimation at early design: A review on techniques and applications. *International Journal of Engineering Development and Research (Www.Ijedr.Org)*, *5*(4), 1558–1561.

De Graaff, M. B., Stoopendaal, A., & Leistikow, I. (2019). Transforming clients into experts-by-experience: A pilot in client participation in Dutch long-term elderly care homes inspectorate supervision. *Health Policy*, *123*(3), 275–280. https://doi.org/10.1016/j.healthpol.2018.11.006

Deloitte. (2019). *The Deloitte Consumer Review: Made to Order : The Rise of Mass Personalisation. Deloitte.* Retrieved from https://books.google.lt/books?id=mRe3tQEACAAJ

Duguay, C. R., Landry, S., & Pasin, F. (1997). From mass production to flexible/agile production. *International Journal of Operations & Production Management*, *17*(12), 1183–1195.

Ehrenberg, R. (2015). Global forest survey finds trillions of trees. *Nature*. https://doi.org/10.1038/nature.2015.18287

Elmonem, M. A. A. E., Nasr, E. S., & Geith, M. H. (2017). Benefits and challenges of cloud ERP systems: A systematic literature review. *Future Computing and Informatics Journal*, *1*(1–2), 1–9. https://doi.org/10.1016/j.fcij.2017.03.003

Farhat, M., & Owayjan M. (2017). ERP neural network inventory control. *Procedia Computer Science*, *114*, 288–295.

Ginsburgh, V., Radermecker, A., & Tommasi, D. (2019). The effect of experts' opinion on prices of art works: The case of Peter Brueghel the Younger. *Journal of Economic Behavior and Organization*, *159*, 36–50. https://doi.org/10.1016/j.jebo.2018.09.002

Guan, C., Qin, S., Ling, W., & Ding, G. (2016). Apparel recommendation system evolution: An empirical review. *International Journal of Clothing Science and Technology*, *28*(6), 854–879.

Guisea, A., Oliveirab, J., Teixeirac, S., & Silva, A. (2023). Development of tools to support the production planning in a textile company. *Procedia Computer Science*, *219*, 889–896.

Hanea, A. M., McBride, M. F., Burgman, M. A., Wintle, B. C., Fidler, F., Flander, L., Twardy, C.R., Manning, B., Mascaro, S. (2016). Investigate discuss estimate aggregate for structured expert judgement. *International Journal of Forecasting*, *33*(1), 267–279. https://doi.org/10.1016/j.ijforecast.2016.02.008

Hindocha, C. N., Antonacci, G., Barlow, J., & Harris, M. (2021). Defining frugal innovation: A critical review. *7*, 647–656. https://doi.org/10.1136/bmjinnov-2021-000830

Holm, L. S., Reutersward, M. N., & Nyotumba, G. (2019). Design Thinking for Entrepreneurship in Frugal Contexts. *The Design Journal, 22*(sup1), 295–307. https://doi.org/10.1080/14606925.2019.1595865

Huang, C., & Gong, D. (2019). How participation management influences work engagement: The mediating role of perceived fit and leader-member exchange. *International Journal of Business and Management*, *14*(12), 191–202. https://doi.org/10.5539/ijbm.v14n12p191

Hustad E., Haddara M, & Kalvenes B. (2016). ERP and organizational misfits: An ERP customization journey. *Procedia Computer Science*, *100*, 429–439. https://doi.org/10.1016/j.procs.2016.09.179

Ievtushenko, O., & Hodge, G. L. (2012). Review of cost estimation techniques and their strategic importance in the new product development process of textile products. *RJTA*, *16*(1), 103–124.

Ito, A., Ylipaa, T., Gullander, P., Bokrantz, J., Centerholt, V., & Skoogh, A. (2021). Dealing with resistance to the use of Industry 4. 0 technologies in production disturbance management. *Journal of Manufacturing Technology Management*, *32*(9), 285–303. https://doi.org/10.1108/JMTM-12-2020-0475

Kamrani, A., Smadi, H., & Salhieh, S. E. M. (2012). Two-phase methodology for customized product design and manufacturing. *Journal of Manufacturing Technology Management*, *23*(3), 370–401. https://doi.org/10.1108/17410381211217425

Kareksela, S., Aapala, K., Alanen, A., Haapalehto, T., Kotiaho, J. S., & Lehtomäki, J. (2020). Combining spatial prioritization and expert knowledge facilitates effectiveness of large-scale mire protection process in Finland. *Biological Conservation*, *241*(November 2019), 108324–108328. https://doi.org/10.1016/j.biocon.2019.108324

Kibria, M. G., Masuk, N. I., Safayet, R., Nguyen, H. Q., & Mourshed, M. (2023). Plastic waste: Challenges and opportunities to mitigate pollution and effective management. *International Journal of Environment Research*, *17*, 20. https://doi.org/10.1007/s41742-023-00507-z

Klaus, H., Rosemann, M., & Gable G. G. (2000). What is ERP? *Information Systems Frontiers*, *2*, 141–162.

Laroche, M., Bergeron, J., & Barbaro-Forleo, G. (2001). Targeting consumers who are willing to pay more for environmentally friendly products. *Journal of Consumer Marketing*, *18*(6), 503–520. https://doi.org/10.1108/EUM0000000006155

Leung, A., Xu, H., Wu, G. J., & Luthans, K. W. (2019). Industry peer networks (IPNs) cooperative and competitive interorganizational learning and network outcomes. *Management Research Review*, *42*, 122–140. https://doi.org/10.1108/MRR-02-2018-0057

Liao, Y. X., Lezoche, M., Rocha Loures, E., Panetto, H., & Boudjlida, N. (2014). A semantic annotation framework to assist the knowledge interoperability along a product life cycle. *Advanced Materials Research*, *945–949*, 424–429. https://doi.org/10.4028/www.scientific.net/amr.945-949.424

Liozu, S. M. (2013). *Pricing Capabilities and Firm Performance: A Socio-Technical Framework for the Adoption of Pricing as a Transformational Innovation* [Doctoral dissertation, Case Western Reserve University]. OhioLINK Electronic Theses and Dissertations Center. http://rave.ohiolink.edu/etdc/view?acc_num=case1364839749

Liozu, S. M., & Hinterhuber, A. (2012). Pricing orientation, pricing capabilities, and firm performance. *Management Decision*, *51*(3), 594–614. https://doi.org/10.1108/00251741311309670

Liozu, S. M., Hinterhuber, A., Boland, R., & Perelli, S. (2011). Industrial pricing orientation: The organizational transformation to value-based pricing. *Proceedings of the First International Conference on Engaged Management Scholarship* (pp. 1–27). Cleveland.

Ma, Q., Jiao, J., & Tseng, M. (2003). Towards high value-added parts and services: Mass customization and beyond. *Technovation*, *2*(3), 809–821.

Mamasioulas, A., Mourtzis, D., & Chryssolouris, G. (2020). A manufacturing innovation overview: Concepts, models and metrics. *International Journal of Computer Integrated Manufacturing*, *33*(8), 769–791. https://doi.org/10.1080/0951192X.2020.1780317

McCrevan, P. (2020). What more can my ERP do? Enterprise resource planning is becoming more integrated, efficient, and interconnected, with increased security for manufacturing operations and processes. *Foundry Management & Technology*, *148*(1), 1919–1928.

Mourtzis, D. (2018). Design of customised products and manufacturing networks: Towards frugal innovation. *International Journal of Computer Integrated Manufacturing*, *31*(12), 1161–1173. https://doi.org/10.1080/0951192X.2018.1509131

Niazi, A., & Dai, J. S. (2006). Product cost estimation: Technique classification and methodology review. *Journal of Manufacturing Science and Engineering*, *128*, 563–575.

Nikzad-Khasmakhi, N., Balafar, M. A., & Reza Feizi–Derakhshi, M. (2019). The state-of-the-art in expert recommendation systems. *Engineering Applications of Artificial Intelligence*, *82*(August 2018), 126–147. https://doi.org/10.1016/j.engappai.2019.03.020

Ninan, J., Phillips, I., Sankaran, S., & Natarajan, S. (2019). Systems thinking using SSM and TRIZ for stakeholder engagement in infrastructure megaprojects. *Systems*, *7*(48), 1–19. https://doi.org/10.3390/systems7040048

Novák, P., & Popesko, B. (2014). Cost variability and cost behaviour in manufacturing enterprises. *Economics and Sociology*, *7*(4), 89–103. https://doi.org/10.14254/2071-789X.2014/7-4/6

Nunes, A., & Kirlik, A. (2005). An empirical study of calibration in air traffic control expert judgment. *Proceedings of the Human Factors and Ergonomics Society 49th Annual Meeting*, *49*(3), 422–426.

Oikonomou, A. (2018). Assessing the impact of employee participation on team-work performance: A way to reinforce entrepreneurship. *Journal of Economics and Business*, *68*(2/3), 48–61.

Pallant, J. L., Sands, S., & Karpen, I. O. (2020). The 4Cs of mass customization in service industries: A customer lens. *Journal of Services Marketing*, *34*(4), 499–511. https://doi.org/10.1108/JSM-04-2019-0176

Paris Agreement. (2024). United Nations Treaty Collection. The Paris Agreement (TIAS 16–1104)

Paulsson, V., & Johansson, B. (2023). Cloud ERP systems architectural challenges on cloud adoption in large international organizations: A sociomaterial perspective. *Procedia Computer Science*, *219*(2023), 797–806.

Pech, M., & Vrchota, J. (2022). The product customization process in relation to Industry 4.0 and digitalization. *Processes*, *10*(539), 1–30. https://doi.org/10.3390/pr10030539

Peng, G. C. A., & Gala, C. (2014). Cloud ERP: A new dilemma to modern organisations?*Journal of Computer Information Systems*, *54*, 22–30.

Peng, W., Zan, M. A., & Yi, T. (2011). Application of expert judgment method in the aircraft wiring risk assessment. *Procedia Engineering*, 17, 440–445. https://doi.org/10.1016/j.proeng.2011.10.053

Pohludka, M, Stverkova, H. & Ślusarczyk, B. (2018). Implementation and unification of the ERP system in a global company as a strategic decision for sustainable entrepreneurship. *Sustainability*. *10*(8), 2916. https://doi.org/10.3390/su10082916

Pokojski, J., Oleksiński, K., & Pruszyński, J. (2018). Conceptual and detailed design knowledge management in customized production-industrial perspective. *Advances in Transdisciplinary Engineering*, *7*, 1064–1073. https://doi.org/10.3233/978-1-61499-898-3-1064

Quesada-Pineda, H. (2010). *The ABCs of cost allocation in the wood products industry: Applications in the furniture industry. In the wood products industry: Applications in the furniture industry* (p. 298). Virginia Cooperative Extension.

Relich, M., & Pawlewski, P. (2018). A case-based reasoning approach to cost estimation of new product development. *Neurocomputing*, *272*, 40–45. https://doi.org/10.1016/j.neucom.2017.05.092

Rodríguez, A. E., Pezzotta, G., Pinto, R., & Romero, D. (2020). A comprehensive description of the Product-Service Systems' cost estimation process: An integrative review. *International Journal of Production Economics*, *221*, 1074–1081. https://doi.org/10.1016/j.ijpe.2019.09.002

Rush, C., & Roy, R. (2001). Expert judgement in cost estimating: Modelling the reasoning process. *Concurrent Engineering*, *9*, 271–284. https://doi.org/10.1177/1063293X01009004

Sackett, P. J., Maxwell, D. J. & Lowenthal, P. A. (1997). Customizing manufacturing strategy. *Integrated Manufacturing Systems*, *8*(6), 359–364.

Sarzynski, A. (2015). Public participation, civic capacity, and climate change adaptation in cities. *Urban Climate*, *14*, 52–67. https://doi.org/10.1016/j.uclim.2015.08.002

Sharma, M., Luthra, S., Joshi, S., & Kumar, A. (2022). Analysing the impact of sustainable human resource management practices and industry 4. 0 technologies adoption on employability skills. *International Journal of Manpower*, *43*(2), 463–485. https://doi.org/10.1108/IJM-02-2021-0085

She V., & Thuraisingham B. (2007). Security for enterprise resource planning systems. *Information System Security*, 16(3), 152–163. https://doi.org/10.1080/10658980701401959

Shibin, K. T., Dubey, R., Gunasekaran, A., Luo, Z., Papadopoulos, T., & Roubaud, D. (2018). The management of operations frugal innovation for supply chain sustainability in SMEs: Multi-method research design research design. *Production Planning & Control*, *29*(11), 908–927. https://doi.org/10.1080/09537287.2018.1493139

Simmonds, D., Tadesse, A., & Murthy, U. (2018). ERP system implementation and sustainability performance rating and reputation. *Twenty-Fourth Americas Conference on Information Systems (AMCIS)*, New Orleans.

Spring, M., & Dalrymple, J. F. (2000). Product customisation and manufacturing strategy customisation. *International Journal of Operations & Production Management*, *20*(4), 441–467.

Stewart, K. (2021). An Eco-wakening Measuring global awareness, engagement and action for nature. World Wide Fund for Nature (WWF), The Economist Intelligence Unit Limited.

Syreyshchikova, N. V., Pimenov, D. Y., Mikolajczyk, T., & Moldovan, L. (2020). Automation of production activities of an industrial enterprise based on the ERP System. *Procedia Manufacturing*, *46*, 525–532. https://doi.org/10.1016/j.promfg.2020.03.075
Vestad, H., & Steinert, M. (2023). Creating your own tools: Prototyping environments for prototype testing. *Procedia CIRP*, *84*(March), 707–712. https://doi.org/10.1016/j.procir.2019.04.225
Wang, Y., Liu, C., Zhang, X., Zeng, S. (2023). Research on Sustainable Furniture Design Based on Waste Textiles Recycling. *Sustainability, 15*(4), 3601. https://doi.org/10.3390/su15043601
Wagner, P., Prinz, C., Wannöffel, M., & Kreimeier, D. (2015). Learning factory for management, organization and workers' participation. *Procedia CIRP*, *32*(Clf), 115–119. https://doi.org/10.1016/j.procir.2015.02.118
Wijekoon, R., & Sabri, M. F. (2021) Determinants that influence green product purchase intention and behavior: A literature review and guiding framework. *Sustainability*, *13*, 6219. https://doi.org/10.3390/su13116219
Wikhamn, W., Wikhamn, B. R., & Fasth, J. (2022). Employee participation and job satisfaction in SMEs: Investigating strategic exploitation and exploration as moderators. *The International Journal of Human Resource Management*, *33*(16), 3197–3223. https://doi.org/10.1080/09585192.2021.1910537
Wilkinson, A., Gollan, P. J., Marchington, M., & Lewin, D. (2010). Conceptualizing employee participation in organizations. *The Oxford Handbook of Participation in Organizations*, 1–24. https://doi.org/10.1093/oxfordhb/9780199207268.003.0001
Winterhoff, M., Wendt, T., Wright, J., Knapp, O., Zollenkop, M., & Durst, S. (2014). *Frugal Innovation - Simple, simpler, best*. Roland Berger
Yaolin, W., Liu, C., Zhang, X., & Zeng S. (2023). Research on sustainable furniture design based on waste textiles recycling. *Sustainability*, *15*(4), 3601.
Zachary, S., & Boal, G. (2012). Custom-built furniture and equipment. In *Planning and constructing book and paper conservation laboratories: A guidebook* (pp. 111–131).
Zennaro, I., Finco, S., Battini, D., & Persona, A. (2019). Big size highly customised product manufacturing systems: A literature review and future research agenda. *International Journal of Production Research*, *57*(15–16), 5362–5385. https://doi.org/10.1080/00207543.2019.1582819
Zeschky, M. B., Winterhalter, S., Gassmann, O., Zeschky, M. B., Winterhalter, S., & Gassmann, O. (2014). From cost to frugal and reverse innovation: Mapping the field and implications for global competitiveness. *Research-Technology Management*, *57*(4), 20–27. https://doi.org/10.5437/08956308X5704235
Zhang, Y., Li, Y., Sun, Z., Xiong, H., Qin, R., & Li, C. (2020). Cost-imbalanced hyper parameter learning framework for quality classification. *Journal of Cleaner Production*, *242*, 118481. https://doi.org/10.1016/j.jclepro.2019.118481
Zheng, C., An, Y., Wang, Z., Qin, X., Eynard, B., Bricogne, M., Le Duigou, J., Zhang, Y. (2023). Knowledge-based engineering approach for defining robotic manufacturing system architectures. *International Journal of Production Research*, *61*(5), 1436–1454. https://doi.org/10.1080/00207543.2022.2037025

Chapter 4

Methodological Implications Seeking To Solve Cost Estimation Issues For Customise Production Process

Birutė Mockevičienė

Mykolas Romeris University, Lithuania

Abstract

This chapter aims to give an overview of a methodological framework based on three research disciplines: customised production management and engineering, data science assumptions and conditions, and inclusive management. The framework addresses the challenge of customised production by assessing the cost of a new product during its early development phase. The methodology consists of various steps. The first step involves collecting empirical data to gain a better understanding of the phenomenon. Next, machine learning algorithms are selected for historical data mining. After that, an experiment is conducted to explore the potential of engagement management. Finally, a prototype is formalised and validated in real-life conditions.

Keywords: Prototype; methodology; synergy of the methods; customise manufacturing; data science; inclusive management

Introduction

Based on the theoretical constructs discussed in the previous chapters, this chapter presents our unique research methodology developed by integrating three research disciplines:

1. Customise manufacturing management and engineering (Pallant et al., 2020; Zennaro et al., 2019). Customisation manufacturing is defined as the

Participation Based Intelligent Manufacturing:
Customisation, Costs, and Engagement, 101–115

Published under exclusive licence by Emerald Publishing Limited
doi:10.1108/978-1-83797-362-020241004

manufacture of products with a high degree of customisation, where the customer is actively involved in the design phase. It is always a low-volume product development with a long engineering cycle, as these are extremely complex. Engineering components are broken down into modules since they have many sub-assemblies, non-standard parts and are made of complex materials. Due to the complexity of the product, the engineering processes involve a number of employees with different skills performing many different tasks, sometimes even in parallel. The product requires a large floor area and specific tools.
2. Data science (Hong et al., 2021). We take the attitude that data science is a new interdisciplinary field of research, innovation and education, which integrates traditional disciplines such as computer science, statistics and information science with many other disciplines such as the humanities and sociology. It combines fundamental computer science with applied disciplines and the arts. Data analysis, data mining and big data are becoming integral to data science.
3. Inclusive management (methods of expert decision modelling) (Jonasson & Lauring, 2017). Inclusive management practices recognise the general empowerment of people, most often employees, by encouraging participation in decision-making. This is often referred to as identity-blind inclusive management.

Such a superposition of interdisciplinary approach allows to reveal more fully the content of the process evaluation of the research phenomenon – cost at the early design stage, to explain the interrelationships between production, pricing and data access, and to create the necessary tool to address production pricing issues. The methodology is based on several experiments to collect empirical data.

Strengths of the methodology:

- Integration (IT, data analytics and social research methods are combined in a complementary way).
- Time management: the research workflow is optimised to collect/analyse data in parallel if they are not related. Or working on one data set in sequence, where the analysis of other data is related to the data analysed before.
- Comprehensibility: studies representing each field of science are concentrated in a single task, thus preserving the boundaries of the science and the ability of the relevant specialists to understand it.

Since the aim is to develop a new and Integral Early Price Evaluation System, the performance of which is to be verified by a prototype, a methodology is needed that allows the study to start by systematising fundamental knowledge, to go through practical data collected in real production processes, to check which data are the most representative, to reproduce the value of these data for machine modelling, and then to investigate to what extent the production data may be complemented by the social non-explicit knowledge of the employees, as experts. Thus, the most prominent research methods in the social and computer sciences were employed to collect and analyse the data.

The research methods used are set out in the methodology in four main steps. This four-step design facilitates time management and methodologically ensures

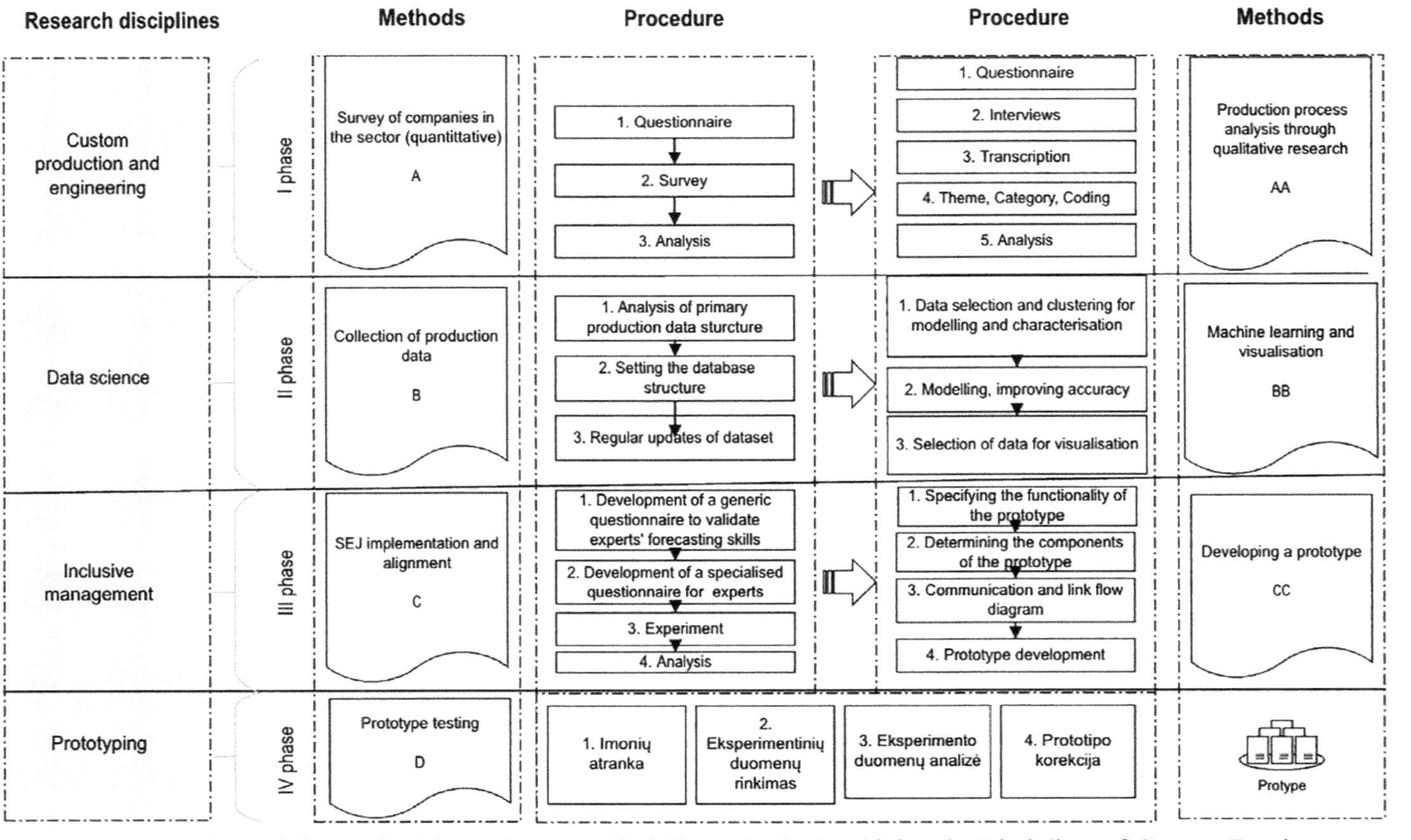

Fig. 4.1. A Flowchart of the Methodology of an Interdisciplinary Study Combining the Disciplines of Custom Furniture Manufacturing, Data Science and Inclusive Management.

the comprehensiveness of the use of the data and the transferability of the knowledge discovered. For the sake of clarity and visualisation of the methodology for the organisers of the study, the four phases are represented through the research discipline, methods and procedures (see Fig. 4.1). The methodology diagram shows the individual methods with the letters A, B, C, D, and AA, BB, CC. The double letters AA, BB, and CC indicate that these methods belong to a single phase. Specifically, the methods assigned to one phase are conditionally assigned to A and AA, while the methods assigned to the other phase are assigned to B and BB, and so on.

Outlining of the Methodology

Phase I. Analysis of Customised Production Processes

The objectives of this phase are to select the main theoretical principles of the standardisation processes of furniture production and early pricing and to match them with the requirements of customised furniture production in terms of process management.

Thus, after examining the theoretical foundations of manufacturing industry processes, the state of the art of furniture production is investigated: typical trends in the organisation of production, attitudes towards customisation, pricing principles and other organisational issues (Method A), while at the same time, the production processes of the selected companies are examined in great depth (Method AA). Here we recommend mixing the qualitative and quantitative methods used in social science. The practice of combining qualitative and quantitative methods together has long been established, as the combination of qualitative and quantitative methods could help to overcome in practice the limitations and resolve the problems associated with the application of either method (Kelle, 2008).

The survey of enterprises in the sector (Method A) was conducted. Quantitative sociological survey research is also useful for its ability to capture a very specific social context, as it is sensitive to culture-specific knowledge (Kelle, 2008). This study is useful to carry out by means of a survey – a representative survey of furniture companies – in order to find out to what extent the problem of rapid order estimation is important for regional (Lithuania, Latvia and Estonia) furniture companies, how companies in the region deal with this problem in the short and long term, and what factors determine the orientation of companies towards customised production orders.

Instrumentarium. The survey questionnaire was designed to reflect the main research questions. To facilitate the respondents, the questions have been grouped into four blocks: (a) the organisational set-up of the company under study; (b) the complexity of pricing; (c) the involvement of employees and the organisation's view of employees' competences in the area of pricing; and (d) the use of IT tools for working with production data and decision-making. The questionnaire also includes demographic questions relating to the financial and production situation of the company in the last calendar year. The questionnaire is included in Annex 1.

Respondents. The survey was carried out between January and March 2019 by the international company RAIT. A total of 146 Lithuanian, 100 Latvian and 100 Estonian companies representing furniture manufacturers participated in the survey. The representativeness of the survey was ensured by reflecting the statistical distribution of companies by size.

Principles of analysis. The 35-item questionnaire used is reliable with a Cronbach's alpha of 0.809. The internal consistency of the seven-item list is also adequate with a Cronbach's alpha of 0.818.

Analyses. The quantitative data were analysed using descriptive statistics and non-parametric statistical analysis. The Cochran's *Q* test is used to test the existence of statistical differences between the opinions of company representatives based on their agreement as dichotomous variables. The hypothesis that there will be no difference in the percentage of agreement between the experts and the level of significance is set at 0.01 and tested. In order to conduct the Cochran *Q* test, the study was designed to meet the assumptions proposed by Sheskin (2011). A correlation analysis was carried out to determine the relationship between the pairs of survey questions using Spearmen's Rank correlation coefficients with a significance level of 0.01. The relationship between the two questions is based on an interpretation of the relative strength of the significant Spearman's correlation coefficients (Corder & Foreman, 2009).

Production process analysis (Method AA). Simultaneously, production process analysis (Method AA) is recommended to carry out using a qualitative approach. In order to analyse the personalised production processes of companies in the chosen sector, the semi-structured interview method is undoubtedly the right method to use in order to gain an in-depth understanding of the phenomenon by exploring the personal perspective of the employees. During the interview, questions may be adjusted or rephrased to gain a more comprehensive understanding of the recently uncovered research findings.

Semi-structured interviews have a flexible structure, which can still be modified in the course of the interview (Adams & Cox, 2008) and help to capture in detail the critical perspectives of the respondents, or their own perceptions of possible solutions. As the method is widely used in different areas and sectors, the methodological design is easily transferable to other sectors.

The mutual trust and level of understanding between the interviewees are important for the depth of the study and directly determines the quality of the data (Kelle, 2008). Therefore, in order to have a thorough insight into existing customisation practices, and to capture the factors and driving forces, the focus was on building deep collaborative relationships. By building cooperation on mutual interest, it is easier to establish the necessary level of trust. If researchers seek empirical data through validated semi-structured interviews, companies need advice, and a third-party critical perspective in their new solutions, in which researchers can participate for no compensation.

The whole research process involved several steps: to identify the most important production processes, then to identify the problems encountered and the ways to solve them.

The interview questions addressed the following topics, covering the four main areas of the study (Annex 2):

1. demographic characteristics;
2. the production process and the pricing;

3. the prevalence and integration of support IT-based instruments in the production processes; and
4. the involvement and cooperation of employees.

Interviews took place between October 2018 and April 2019. Several follow-up interviews took place in mid-2020 in order to understand what changes have taken place in the sector during the period covered.

Respondents. The survey was carried out in Lithuania, in companies specialising in customised furniture manufacturing that are currently willing to invest in an additional price assessment instrument. Two companies participated in the study. The companies were selected geographically from the largest Lithuanian cities – Vilnius and Kaunas. The export rate of the companies is around 70% and the share of customised production is between 30% and 100% of the total production. The enterprises participating in the study can be classified as medium-sized (around 100 employees) and large enterprises (over 800 employees). The interviews were carried out with respondents representing all stages of furniture production: company owners, directors, project managers, designers, production planners, furniture assemblers, production and IT engineers, managers and craftsmen. The survey data were collected through 26 interviews.

Data analysis. All interviews were transcribed. The transcribed recordings were coded and classified. Responses were categorised and sub-categorised using NVivo qualitative data analysis software. The frequency of concepts used was identified by highlighting key issues. The categories were then coded and grouped into subcategories. The procedure was carried out twice by two different researchers to validate the original categories. In total, this amounted to 23 hours of interviews and 462 pages of text. The 905 coded concepts were grouped into four generalised categories as follows: cost estimation and estimation process, company organisational structure, employee involvement and production processes.

The analysis included 435,000 characters representing around 62,000 different concepts. Due to the complex system of the Lithuanian language, the analysis had to focus on the roots of all the related words, giving the meaning of the word. Grammatical relations were analysed in syntax word sets, which helped to point out even those syntactic relations that might not be identifiable from individual parts of sentences. In this way, the most important topics for the respondents were selected and placed in a hierarchical system according to their importance. As the respondents in the study represented a wide range of manufacturing stakeholders, this helped to highlight the different concepts used and thus to identify valid code structures.

Phase II. Historical Manufacturing Data Collection and Modelling of Manufacturing Processes Based on Machine Learning Algorithms

To predict the price and other elements of the production process, it is important to identify the key price elements that have a significant impact on the final price of the product. This requires extensive historical data on past production. Machine learning is widely used in a wide range of manufacturing applications and can be an effective and accurate tool for estimating forward prices for furniture.

Data collection and building a data structure for storage (Method B). Since furniture companies, like any other business enterprise, collect and store the data used in their business processes in enterprise resource planning (ERP) systems, data acquisition from companies should not be a complex or expensive process. However, the difficulties start when one wants to apply the collected data to a specific, completely new challenge. Often companies use different data storage tools with different database architectures and, as is most common, they apply different and quite specific costing methodologies. It is also typical, especially in small companies, that companies do not use business management systems at all or do not use all the functionality they provide and do not have any data-driven culture.

The following studies are based on historical manufacturing data provided by the two companies for the most recent production periods. It should be stressed that the companies shared only historical data. Most importantly, it is for the sake of maintaining a competitive advantage. Since the calculation of the early price requires historical data not only on production processes but also on pricing, the companies preferred to share data which they considered to be safe in respect of competitors. Sharing existing data that could reveal their pricing strategy of the moment seemed dangerous to them.

The data are periodically updated according to an individual plan agreed with the companies. The date of the first entries is Q4 2017. The companies provide the following data: unique product number, product structure, description of the materials used for each component of the product, dimensions, and production operating times applied. Company 1 uses an ERP called QAD, Company 2 uses an ERP called 1C with additional in-house programmable modules. Both companies collect information on their products at different levels of detail. It was also found that the database structures of the industrial systems used are very complex, storing a lot of redundant information, and sometimes even useless (e.g. all fields have the same meaning).

The initial dataset for the studies was built from data fields that were available to both companies (common directories/classifiers: Material types, colours, units of measurement, operations, customers, etc.). From the production data provided by the companies, a database of the study was constructed in such a way that, in a recursive way, each product can be part of another product. There is no depth restriction. This structure will then be used to create an automated template for companies to fill in the data and import the data into the system/prototype. This production data describes the furniture structures (27,260 records), the nomenclature used in the business system (14,463 records), the operational times for the production of the furniture parts (6,576 records), and the cost of materials in the form of 3,631 records, invoices (5,000 pieces) and 20,000 unique products sold.

Modelling Manufacturing Processes Based on Machine Learning Algorithms (Method BB)

The modelling task is performed in the following steps:

- Data discovery – analysis of the available data and its structure, analysis of how the order-taking, design and production processes are reflected in the data, analysis of product structures and nomenclatures, operating times, sales volumes, batch sizes and material costs.

- Data preparation. For the modelling of the production processes, an analysis of the available data was carried out, the most characteristic features and groups of features were selected, and new and derived features were developed for modelling. A linear regression method was used to select the most important features of the data, which allowed the identification of the features that correlate most strongly with the cost of products. The attributes describing a product are usually equal to the sum of the values of the corresponding attributes of the product. The key attributes were found to be, in descending order of importance, 'number of different parts' (units), 'number of parts' (units), 'surface area of parts used' (m^2), 'number of different materials' (units), 'order size' (units), 'material cost', followed by the various operations in the wood (w) and metal (m) workshops ('w50', 'w70', 'm10', 'm30', 'm20', etc.).
- Selection and training of the machine learning model. After the data analysis, further work is carried out on training, testing and model tuning (hyperparameter search) of different machine learning algorithms. The algorithms evaluated were: linear regression, decision tree, random forest, extra tree, AdaBoost, gradient boosting, K-nearest neighbours' regression and artificial neural networks. At the same time, data and data sets are selected which best visualise and represent the data content for price forecasting. The screening started with the most appropriate methods for the furniture sector.

Phase III. Inclusive Governance and Modelling of Expert Decisions

As inclusive management aims to involve the widest possible range of employees, early price assessment requires specific knowledge managed by the individual employee. One effective way to systematically capture and synthesise this knowledge is through structured expert judgement. It is therefore recommended that inclusive management should be implemented through a structured approach to expert evaluations. This requires identifying and selecting suitable experts from the company's workforce for further price evaluation. The composition of the panel of experts is an important issue for the accuracy of price forecasting. The following follows the usual structured approach to expert evaluation, which is developed through an experiment to test which expert validation questions are the most appropriate and which composition of the workforce is the most productive.

Respondents. Ten employees took part in the study: the CEO, the designer, the chief accountant, the chief financial officer, the head of the finance department, the IT manager, the designer, the senior project administrator, the project administrator and the chief product manager. Structured expert judgement involves two quantitative aspects of expert performance evaluation: calibration and information.

Questionnaires. The calibration questionnaire consisted of general questions about the depth of expertise. The information questionnaire was designed to include historical production data. To test the quality of the questionnaires, two experiments were conducted. Before the first experiment, each respondent completed a general questionnaire on production volumes, accounting and operating times (Annex 3). Subsequently, for the forecasting experiment, a questionnaire on the data for a sample of nine products were presented to the employees (Annex 4). From the available data, respondents were asked to indicate the expected final price of the product. In a second experiment, another questionnaire was designed for the calibration of the experts on the specific characteristics of the production process: total operations, and coefficients of operations. This was followed by information on 32 products. These products were selected on the basis of specific criteria familiar to the respondents, and the products also represented the boundary conditions: the most expensive products (eight products), the cheapest products (nine products), the medium-priced products (15 products), the largest products in terms of size (seven products) and the smallest products in terms of size (12 products).

Data analysis. Based on Cooke's classical model (Cooke & Goossens, 1999), the EXCALIBUR software package (Ababei, 2019) was used for structured expert evaluation, where two separate scores (calibration and information scores) are scored and weighted together to obtain an overall prediction weight for each expert (Werner et al., 2017). Experts were asked to provide their estimates with a certain bias, for example, with minimum and maximum expected values, in other words, to indicate quanta of the distribution of interest, such as the 5th, 50th or 95th. The calibration and information scores together allow to calculate the weight of each expert.

Prototype (Phase III, Method CC)

The *Integrated early price assessment system* as a prototype will be developed in several stages: identifying the required and desired functionalities of the overall system, based on which a prototype layout will be developed to visualise the functionalities and possible interfaces. The prototype helped to understand the possible usage scenarios. Subsequently, a communication flow diagram was developed to represent the functionalities and to reveal and justify the need for interfaces based on knowledge management. Based on the developed layout, the prototype is being developed. The prototyping process is based on the following steps: database design, building reports, designing screens, and designing menus.

The JAVA programming language framework 'Spring Boot' has been used to develop the prototype. The Spring Boot framework is characterised by the automatic configurations it provides using the Dependencies of the Spring open-source library. The Spring Boot application uses specific methods to store data in a database, searches the data, processes the resulting data and sends it to the system visualisation application via HTTPS requests. The open source framework React is used for system visualisation. It contains CSS and JavaScript-based design templates, forms, buttons, navigation and other interface components. The data are hosted in a MySQL database. MySQL is the most popular open-source relational database management system based on the SQL language. It is often chosen for web programming and its popularity is directly related to the popularity of JAVA, as a combination of the two is commonly used.

Phase IV. Testing

In the testing process, it is important to create as near to realistic conditions as possible, where the prototype is accessible in real-time remotely and companies use the prototype in their day-to-day operations. Testing can be realised in two ways: the first scenario where the prototype is used for public access and the second scenario where the prototype is used on a consensual basis in agreement with specific companies. However, the second scenario is easier to implement, if only because it requires less marketing and communication capacity to publicise the possibility of such an instrument, and because the testing with the selected company is more systematic and comprehensive, ensuring continuity of testing. After the testing, features, and bugs are revised.

Following this methodology, the following sections present the research undertaken and the findings.

Companies Selected as Case Studies for the Empirical Data

A quantitative research survey was conducted to evaluate the frugal nature of the furniture industry region, which includes Lithuania, Latvia, and Estonia. On the meanwhile, qualitative research was carried out with a selection of companies that could reflect the sector of the region in a more comprehensive way, covering the problems faced by SMEs while taking into account the requirements of the region. Additionally, special attention was given to the production of these companies, considering that a significant part of the production would be personalised.

Selection criteria for case representation:

- *A small or medium-sized enterprise*: SMEs have certain advantages over larger companies, such as being more flexible and open. This is because they tend to have less hierarchy, which makes research decisions easier and faster. In an upcoming study, we will investigate the readiness of customised production to use big data and test a prototype of price evaluation. Therefore, it is important that the company's processes are easily adaptable or that it is possible to intervene with the prototype without disrupting the existing systems.
- *Customised manufacturing*: It was decided to search for a small or medium-sized enterprise (SME) that prefers customised production. Customised production has become the new dominant form of production, which poses several organisational challenges. Some of the most fundamental challenges include a focus on the consumer, the urgent and emerging need for meaningful use of data, and a paradigm shift towards pricing principles.
- *Quality management system*: Preferably a company which has integrated quality management systems like Lean or ISO in their production processes. Companies that continuously work on improving their management systems are more likely to cooperate with researchers, have a keen interest in enhancing their production processes and aiming for innovation and have fewer limitations or negative connotations.

- *Selection criteria representativeness*: It is important that the criteria selected should still be representative of the different boundary conditions, including size, region, and percentage of individualised production.

Case 1 (Company A)

It is situated in the second largest city of Lithuania, which is considered to be the industrial hub of the nation. This is due to the development of free economic zones, concentration of manufacturers and location of logistics centres for supplies.

Employees. Company A had 80 employees at the time of the survey. The company is highly sensitive to regional and global market changes, which is evident from the fluctuating number of employees. In 2020, due to the COVID-19 pandemic and a temporary standstill in the order market, the company had to suspend production and consequently reduced its workforce to 50 employees. The CEO made a strategic decision to change the business model and focus less on international orders, which are difficult to win and require unbearable price reductions and short deadlines. After conducting research and testing a prototype in 2022, the company was able to bring back its workforce to 90 employees, and it showed signs of expected growth in the business sector. The company's flexibility in responding to macroeconomic triggers reflects the resilience of its management structure.

Organisational structure. The company has a simple and flat management structure, consisting only of a Chief executive officer (CEO) and a chief financial officer. Interestingly, the owner and the CEO are the same person. The core group of employees includes warehouse staff, production managers, IT support, designers, and prototype assemblers. These employees are loyal and have been with the company for a long time. In addition, the company has a culture of inclusiveness. This is achieved through informal task force groups, which are formed to solve problems, as well as the physical location of different employee groups within the company premises. For example, a representative of the company's workforce working on the assembly of the prototype works in the same shared space as the furniture designer. This helps the designer to witness and understand the assembly process, which in turn helps them improve the existing layout more easily and realistically. Due to its small size and flat organisational structure, the company does not have a separate sales department or metal or wood assembly department. The production manager controls and plans the metal and wood production, while sales are carried out by an approved person.

Type of manufacturing. The company primarily focusses on producing customised orders, which currently make up to 80% of their total production volume. Although they have plans to manufacture mass products, their main product is sales equipment that includes design, engineering, production, and supply. They actively participate in international competitions and boast a customer base that includes well-known retail chains like Adidas, Chanel, and local customers with annual sales turnovers in millions. Visual examples are provided in Figs. 4.2 and 4.3.

Approach to digitalisation and other managerial innovation. The company has a positive attitude towards digitalisation and other managerial innovations. They are always on the lookout for new automation solutions and ways to improve their

Fig. 4.2. An Example of the Products that were Ordered and Produced by the Company, Which Represents Case 1 (Photo of the Company).

management systems. They have developed a LEAN system, which although it's not given much attention and is stagnating, is still being improved and its individual elements are being applied. Additionally, the company has implemented a sophisticated and expensive ERP system that is meant to manage accounting for a larger organisation. The company encountered several obstacles in implementing the system, including IT resource constraints. However, they were able to overcome these challenges and are currently utilising the system for data collection and reporting. As a result, the company collaborated with the researchers to provide historical data arrays and pricing documentation for their products over the past decade. Additionally, they are capable of providing real-time data if required.

Approach the new pricing paradigm. This company is struggling with accurately predicting the price of new and complex products, which is causing imbalances in its planning and financial management. To address this issue, they are looking for ways to systematise their products by developing classification and categorisation principles based on both parametric evaluation and value. They have created a two-dimensional classification system (see Fig. 4.4) with scores visualisation as

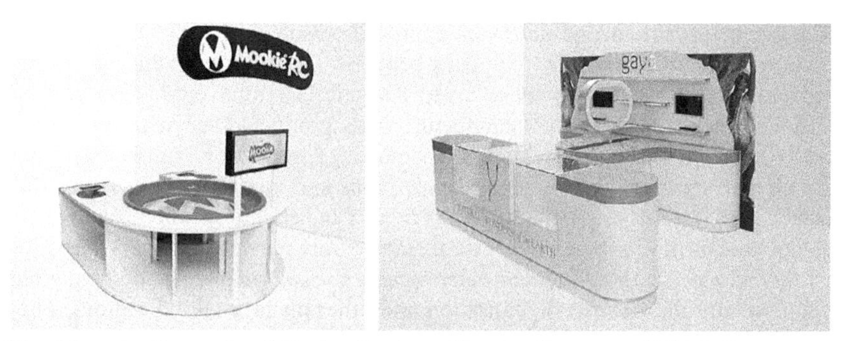

Fig. 4.3. An Example of Design Produced by the Company, Which Represents Case 1 (Photos of the Company).

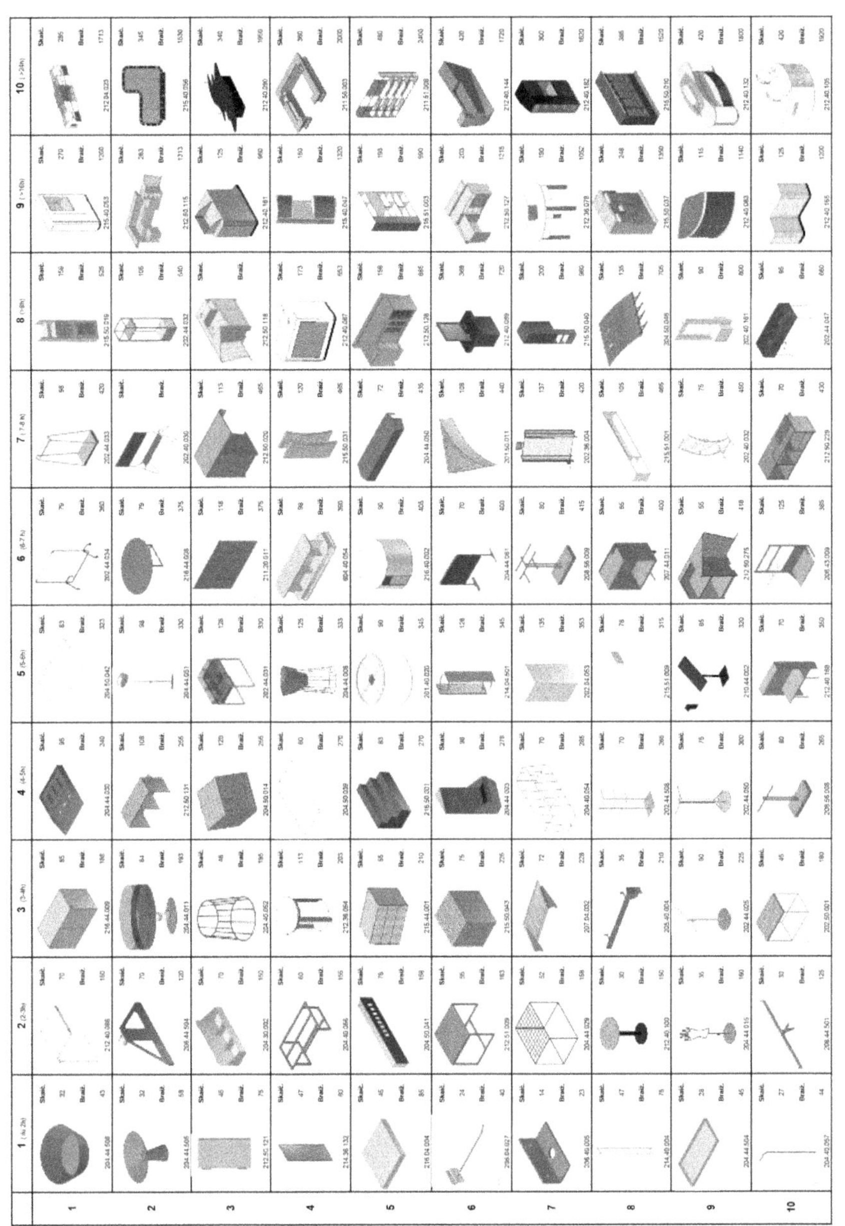

Fig. 4.4. A Two-dimensional Classification of Products (Case 1) (Provided by the Company).

a test to demonstrate their understanding of the problem, but it has not yet led to a more effective price management approach.

Case 2 (Company B)

The company, formerly a branch of the parent company, is now based in the capital of Lithuania and aims to expand geographically, while the parent company remains in the country's second-largest city.

Employees. The second company is a large organisation that has a workforce of more than 800 employees. Although there is some turnover rate at the worker level, it is not higher than the natural turnover rate. The company has been able to maintain its size and stable operation amidst the disruptions caused by COVID-19 and the mini-crises in the furniture market. These mini-crises in the furniture industry (short-term decrease in turnover) are now more frequent and are observed in a 3–4 years cycle.

Organisational structure. The company has a complex management structure that is based on a hierarchical decision-making process. It operates through different divisions, such as marketing, project management, metalworking, finance, and so on. These divisions are set up according to the functions they perform. The company has established all the necessary production and management units, which create a clear but somewhat bureaucratic management system. This system is essential to ensure effective multidisciplinary and multi-order management.

Type of manufacturing. Although the company's managers see their aim as mass production, the project's management team views it as fully customised production. However, from the perspective of customised production, if a customised product is one that uses only original and first-time parts, then the company is still operating on a 60/40 split between mass and customised production. Nonetheless, in most situations, customised orders are still brought into the framework of the mass product, and are assembled from already manufactured and designed offsets.

Approach to digitalisation and other managerial innovation. The company is open to innovation and digitalisation of managerial processes, but less so in production. To meet the emerging IT and digitalisation needs of management professionals, the company has a dedicated IT department. This availability of IT specialists creates a flexible system for addressing digitisation issues, optimising input and output, and formalising monitoring indicators. As a result, the company has implemented an ERP package that is open to internal programming improvements, affordable for input and output generation, and relatively easy and cheap to adapt to specific challenges. This approach avoids the high costs of adapting standardised and off-the-shelf ERP packages. This allows each department to generate only relevant ERP output data packages for monitoring and to forward them to the next department for the next stage of production.

Approach to new pricing paradigm. The company is committed to the accounting-based cost pricing method, which requires precise measurements of the product and materials needed for accurate pricing. Despite having a rich pool of historical data and access to it for decision-making, the company has not shown any curiosity in evaluating it. However, there is a sense of dissatisfaction regarding the pricing, which appears to be uneven and incomplete. Currently, the company is addressing the issue by hiring new employees and establishing new roles and responsibilities to act as pricing specialists and mediators between the production and ordering departments. This is in response to the implementation of a new pricing mechanism. It is indeed true that the company took great care in providing the data for the study. However, instead of simply providing a raw dataset for analysis as an ERP output, they chose to clean the data themselves. They combined two separate outputs and removed what they believed to be specific and anti-competitive information. This shows their increased caution towards the study and their desire to ensure that the data provided is as accurate and unbiased as possible.

References

Ababei, D. (2019). Excalibur. http://www.lighttwist.net/wp/excalibur

Adams, A., & Cox, A. L. (2008). Questionnaires, in-depth interviews and focus groups. In P. Cairns & A. L. Cox (Eds.), *Research methods for human computer interaction* (pp. 17–34). Cambridge University Press.

Cooke R. M., & Goossens, L. J. H. (1999). Procedures Guide for Structured Expert Judgement, EUR18820, European Commission, 1999. Available at ftp://ftp.cordis.europa.eu/pub/fp5-euratom/docs/eur18820_en.pdf

Corder, G. W., & Foreman, D. I. (2009). *Nonparametric Statistics for Non-Statisticians*. John Wiley & Sons. https://doi.org/10.1002/9781118165881

Hong, L., Moen, W., Yu, X., & Chen, J. (2021). The disciplinary research landscape of data science reflected in data science journals. *Information Discovery and Delivery*, *49*(4), 287–297. https://doi.org/10.1108/IDD-06-2020-0071

Jonasson, C., & Lauring, J. (2017). Inclusive management in international organizations: How does it affect local and expatriate. *Personnel Review*, *47*(2), 458–473. https://doi.org/10.1108/PR-12-2015-0323

Kelle, U. (2008). Combining qualitative and quantitative methods in research practice: Purposes and advantages. *Qualitative Research in Psychology*, *2006*(3), 293–311. https://doi.org/10.1177/1478088706070839

Pallant, J. L., Sands, S., & Karpen, I. O. (2020). The 4Cs of mass customization in service industries: A customer lens. *Journal of Services Marketing*, *34*(4), 499–511. https://doi.org/10.1108/JSM-04-2019-0176

Sheskin, D. J. (2011). *Handbook of Parametric and Nonparametric Statistical Procedures* (5th ed.). Chapman and Hall/CRC. https://doi.org/10.1201/9780429186196

Werner, C., Bedford, T., Cooke, R. M., Hanea, A. M., & Morales-Nápoles, O. (2017). Expert judgement for dependence in probabilistic modelling: A systematic literature review and future research directions. *European Journal of Operational Research*, *258*(3), 801–819. https://doi.org/10.1016/j.ejor.2016.10.018

Zennaro, I., Finco, S., Battini, D., & Persona, A. (2019). Big size highly customised product manufacturing systems: A literature review and future research agenda. *International Journal of Production Research*, *57*(15–16), 5362–5385. https://doi.org/10.1080/00207543.2019.1582819

Chapter 5

How The Small Country's Furniture Sector Builds Its International Competitiveness (Survey of Lithuania, Latvia and Estonia)

Julija Moskvina[a] *and Birutė Mockevičienė*[b]

[a]*Lithuanian Centre for Social Sciences, Lithuania*
[b]*Mykolas Romeris University, Lithuania*

Abstract

This chapter presents data on furniture companies in the three Baltic States. The analysis of the survey data reveal peculiarities of the production process, the organisational structure and the cost estimation practices. The survey data are valuable as it captures the main trends in furniture production in the region. The data also provide important insights into prevailing production practices, organisational structure and costing strategies. The survey data allowed capturing the importance of organisational capabilities such as optimal organisational structure, competence of business staff, performance growth and attitude to technological change in the application of IT tools to specific manufacturing operations.

Keywords: Manufacturing; furniture industry; organisational structure; IT; Baltic countries

Furniture Manufacturing Sector: Similarities and Differences Between the Baltic Countries

The wood and furniture sector in the Baltic region is recognised as a traditional industry. The vast forest resources, strategic location, manufacturing expertise and cost-efficient labour force make furniture production one of the leading

Participation Based Intelligent Manufacturing:
Customisation, Costs, and Engagement, 117–149

doi:10.1108/978-1-83797-362-020241005

exporting industries in the Baltic States. The main products of the Baltic countries' furniture industry go to North and West Europe. During the period 2010–2016, the furniture industry was shrinking in most West European countries and growing in Central and East Europe. For example, during this period the output of Lithuania's furniture industry grew 2.3 times, the fastest growth in the European Union (EU).

The growth of the sector is accompanied by the creation of new small and medium-sized enterprises in all three Baltic countries (with higher growth of micro-enterprises in Estonia). Larger companies that have been operating in the market for longer have tended to grow steadily. In the third quarter of 2018, Lithuania had 1,184 furniture manufacturing companies, a considerable increase from 872 in 2015. During the last decade, the value created in the sector doubled to exceed 3 billion euros in 2017. Though woodworking is one of the main industries of Estonia and Latvia, still at least 40% unprocessed wood and logs are exported to other countries (Latvian Business Guide, 2019; Versli Lietuva, 2019; Wood Processing and furniture production competence center, 2020). There are still a lot of opportunities for innovation and developing high-added-value products.

The pandemic has presented new challenges to the furniture industry. As COVID-19 changed our daily lives, the demand for furniture was also growing rapidly. COVID-19 has accelerated the need for office furniture and other comfortable furniture as people work and spend more time at home. In addition, many furniture stores were closed making e-commerce more important than ever. The pandemic restrictions revealed many shortcomings in production. With increasing demand, companies needed to implement new processes and solutions to increase the speed of the supply chain. This change was crucial in the manufacturing process. Most furniture companies no longer stock items. Only a small amount of material is stored, rather than semi-finished or finished products.

But how can the customer get what she/he wants as quickly as possible? When a product is purchased, the seller sends the order to the manufacturer, who orders the materials needed to produce the product. The manufacturer must figure out how to produce quickly while still making a profit. Customers expect high-quality products with fast delivery for low prices. Moreover, consumer habits of buying furniture have changed. Previous generations were buying items for decades. Today, furniture became a fashion item. Just as people change their wardrobes because clothes go out of style, furniture designs and materials also go out of style. Some consumers are ready to change it every five years or so. Since the durability of the furniture is no longer a priority, more attention is being paid to a lower price. It might seem that if people change their furniture more often, then this is a boon for furniture manufacturers. To meet the needs of fashion-conscious customers, entrepreneurs need to use innovative designs and materials. Much of the change in the furniture industry comes from the customers' wishes. The close location of the product is also an important factor as speeds up the delivery.

The furniture producers are aware of the balance between production costs and time. However, the reality is that many small manufacturers depend on traditional methods of furniture production, which are unfortunately much more expensive. Therefore, if relocation of production is not possible, automation is the only way to reduce the cost of the search. Today's manufacturers must have an automated production process and adopting the right software is the key to success here. The software and other technological tools allow manufacturers to manage each task without major errors and produce minimising costs. As a result, the stocks of finished goods and prices are falling. This transformation in furniture production has led to the emergence of so-called 'micro-factories' (smaller but highly automated and technologically advanced factories). It is an end-to-end, one-stop shop that enables seamless data transfer throughout the supply chain, saving time and money. A smoother workflow, in turn, reduces errors, and makes order fulfilment easier, leaving more time for a customer-centric approach. In conclusion, the brands, retailers and manufacturers that rely on automation and digitalisation approaches can remain flexible, allowing them to keep pace with consumer demand and market challenges.

Estonia

In 2019, 1,825 woodworking and furniture companies operated in Estonia. Industry enterprises account for about 1.6% of Estonian companies. In 2017–2019, the number of companies in the woodworking and furniture industry increased by about 10%. This industry employs more than 21,000 people (more than 5% of the employees of Estonian companies), and in recent years the number of employees has increased by about 7%. In 2019, the total sales revenue of the mentioned companies amounted to more than 2.8 billion euros, which was 4.2% of the total sales revenue of Estonian companies, and in the period of 2017–2019, the sales revenue increased by about 14%. More than half of the revenue of woodworking and furniture companies comes from exports (for comparison, the share of exports in the revenue of all Estonian companies is about 40%). Woodworking and furniture industry exports account for almost 6% of Estonia's total exports. Companies in the woodworking and furniture sector increased their exports by just over 60% in the period of 2017–2019 (TSENTER, 2021). Therefore, export is very important for Estonian woodworking and furniture companies, as well as for Estonia as a whole.

In 2020, the most important export countries for the Estonian furniture industry were Finland, Sweden, Denmark and Germany. Estonia occupied 18% of the Finnish furniture market and constituted 2% of Swedish furniture imports. In 2020, furniture exports from Estonia decreased by 8%. Compared to 2019, exports to Finland remained almost at the same level, while exports to Sweden and Denmark decreased. In Sweden, Estonia competes the most with Polish, Chinese, Lithuanian and German furniture manufacturers. Swedish furniture imports have increased by 6% over the past five years, while Estonia has reduced furniture exports to Sweden by as much as 32%. In the Swedish furniture market, cheaper furniture chains have a strong position, which also affects Estonian furniture manufacturers with cheaper prices (Estonian Business and Innovation Agency, 2021; TSENTER, 2021).

Estonian furniture manufacturers mainly export various furniture parts. In 2020, the export of furniture parts accounted for the largest share of the export of Estonian furniture manufacturers – 29%. This was followed by the export of wooden furniture for bedrooms with a share of 26% and other wooden furniture with a share of 23%. In the period of 2017–2019, the operating profit of companies in the woodworking and furniture sector fell by 12%. The value added by the woodworking and furniture sector per employee in 2019 was 29,692 euros, which is 11% lower than the average value added per employee for Estonian companies. The value added per employee decreased by 3% compared to 2018 (TSENTER, 2021).

Latvia

Forestry, wood processing and furniture manufacturing represented 5.3% of gross domestic product (GDP) in 2020, while exports amounted to EUR 2.6 billion – 19% of all exports. There is no parish in Latvia with a larger or smaller wood processing company. Often these are the most important employers in the surrounding area, thus being the main pillar of support for local economies and residents. During the period 2017–2019, the net turnover of the furniture sector was 266–294 million Euros (Latvian forest sector in facts & figures, 2022). The main export market for wooden furniture is Denmark, then Germany, Estonia and Sweden. Export markets for the main products of the forest sector (such as wood-based panels: plywood, veneer, particle board and fibreboard) in 2020 were much bigger than for furniture items. Such products are mostly exported to the UK, Lithuania and Germany.

According to Lursoft data, in 2020 more than 1,500 companies were registered in the furniture industry in Latvia (757 furniture manufacturing companies, 120 furniture wholesalers, 279 furniture retailers and 439 carpenters and joinery manufacturers in Latvia [Lursoft, 2020]). The majority of the producers deal with customised manufacturing: about 65% of the companies work on individual orders, and 45% work only on individual orders (Saksone, 2021).

Due to the small size of the domestic furniture market, companies seeking to grow are looking for their niche abroad. The largest incomes were obtained in such countries as Germany, the USA, Finland, Denmark and Sweden. It is exporting companies that are usually focussed on production modernisation. At the same time, exporting companies constantly strive to increase their productivity. The Latvian furniture manufacturers both in the national as well at international markets face risks of supplies and the lack of qualified labour. The furniture industry also suffers from a lack of capacity, inefficient production and insufficient technological development (Profesionālais izglītības centrs T-senter, 2017).

Lithuania

In Lithuania, as in other Baltic countries, the wood industry sector contributes significantly to the creation of GDP. In 2022, about 5% of country's population was occupied there, and the added value created by the sector amounted to about 4.5% of it created in Lithuania during the year. Over two-thirds of the produced furniture was sold in overseas markets comprising almost one-fifth of the total

Lithuania's export. During the period 2010–2016, furniture prices decreased by 1% in Lithuania and grew by 7.8% in the EU. It has to be noted that the level of productivity of Lithuania's furniture businesses is growing whereas the share of personnel costs in the value of output decreases. In the third quarter of 2018, Lithuania had 1,184 furniture manufacturing companies, a considerable increase from 872 in 2015. During the last decade, the value created in the sector doubled to exceed 3 billion euros in 2017. Over two-thirds of the produced furniture was sold in overseas markets comprising about 19% of the total Lithuania's export. Lithuania's furniture manufacturers sell the major portion of their products in foreign markets as the national market is too small.

The major export destinations were EU member states (77% of the exported production). The products must meet extremely high standards to face severe international competition. Some globally famous brand stores trade in Lithuanian furniture and use it in their interiors (Versli Lietuva, 2019).

The pandemic has significantly adjusted the activity of the sector. According to the data of the Department of Statistics, the export of production created by this branch of the economy in 2022 accounted for about 11% of all Lithuanian exports. The challenges of staff shortages are accompanying the furniture sector for several years as well. The continuing economic uncertainty forces manufacturers to carefully plan for possible growth; they focus more on maintaining production volumes and efficiency (e.g. reusing production waste within the production process, investing in renewable energy, implementing resource-saving initiatives, improving products and processes). The further perspective of the sector's development is also significantly influenced by economic uncertainty, with records of high prices of raw materials and energy. The changing behaviour of consumers that are choosing cheaper products is shaping the producers' further developing plans.

Before the pandemic, the rapidly developing furniture production in the Baltic countries overcame the pandemic period quite successfully. However, the challenges of transport, energy and human resources force managers to look for new technological solutions for faster and cheaper product production. Therefore, the furniture manufacturing sector is promising for implementing new performance improvement instruments, including price forecasting instruments.

Prevailing Organisational Structure in Furniture Manufacturing Companies

Understanding that the management structure of furniture manufacturing companies and its inherent features such as management, number of employees and delegation of functions are key drivers of digital technology penetration in the sector (Shrouf et al., 2014), in this section we present an analysis of the prevailing organisational structure in Lithuania, Latvia and Estonia.

The survey of Baltic furniture manufacturers was carried out between January and March 2019 by the international company RAIT. A total of 146 Lithuanian, 100 Latvian and 100 Estonian companies representing furniture manufacturers participated in the survey. The representativeness of the survey was ensured by reflecting the statistical distribution of companies by size.

As technological change is usually accompanied by a transformation of the organisational structure, the development of new systems and new management policies, we sought to identify the prevailing management structures in the Baltic States in the selected furniture sector and to identify the factors that promote or inhibit technology adoption. Below we present the main findings from a survey of 146 Lithuanian furniture manufacturers, complemented by a comparison with other Baltic countries.

The questionnaires for randomly selected companies were filled in by the following representatives: 48% of owners, 35% of directors, 6% of heads of production, 4% of managers, 2% of project managers, 2% of deputy directors, 1% of heads of a division or unit, 1% of designers and 1% of accountants. Quantitative data were analysed using descriptive statistics and nonparametric statistical analysis. The Cochran *Q* test is used to test the existence of statistical differences between the opinions of company representatives based on their agreement as dichotomous variables. The hypothesis that there will be no difference in the percentage of agreement between the experts and the level of significance is set at 0.01 and tested. In order to perform Cochran's *Q* test, the study was designed to meet the assumption proposed by Sheskin (2011). A correlation analysis was carried out to determine the relationship between the pairs of survey questions using Spearmen's Rank correlation coefficients with a significance level of 0.01. The relationship between the two questions is based on an interpretation of the relative strength of the significant Spearman's rank correlation coefficients (Werner et al., 2017).

51% of the companies surveyed employ 1–9 people, 34% employ 10–49 people, 13% employ 50–249 people and only 2% employ 250 or more people. In 70% of the furniture manufacturing companies surveyed, the most basic management structure is dominated by a manager and other employees. Lithuanian furniture makers have not grown up to have a structure of 'President, General Manager, Heads of Departments, Divisional Managers, Executives and other staff'. Sixty per cent of the companies surveyed do not have departments or divisions. These companies fall into the first level of organisational structure (see Fig. 5.1). Almost all enterprises, that is 91%, that have a Planning or Design Department (22% of all enterprises surveyed) also a Woodworking Shop (32% of all enterprises surveyed) and are therefore at one level of the organisational structure: level 2. Also, at this level, a company may or may not have an Accounting or Finance Department (19% of all companies) or a Personnel Department (14% of all companies). All enterprises with a woodworking shop have a metalworking shop (3% of all enterprises), but not necessarily, so enterprises with a woodworking shop are already at level 3. Only 10% of the surveyed enterprises have a Marketing Department and 14% have a Personnel Department, so these enterprises are at level 3. There is no reason and no need to introduce a fourth level: 60% of the surveyed enterprises are at level 1, 27% at level 2 and 14% at level 3.

The factor analysis confirmed the proposed division of the surveyed companies into three groups according to the level of organisational structure: the Woodworking Shop, the Accounting/Finance Department and the Planning/ Design Department are in one group; the Marketing Department and the

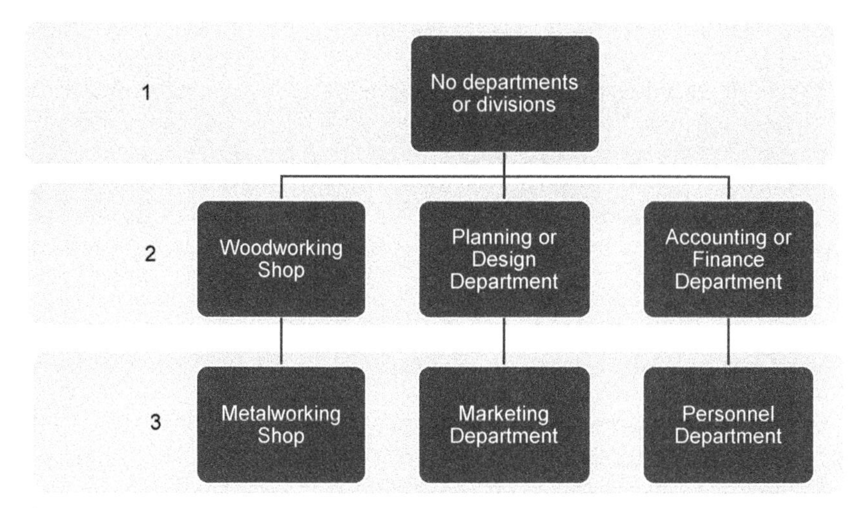

Fig. 5.1. Levels of Organisational Structure at Furniture Manufactures.

Personnel Department are in the other. As only 3% of the surveyed enterprises have a Metalworking Shop, it is not appropriate to assign a separate level to this small group (see Table 5.1).

The first tier includes companies with no departments or divisions (61%): 1–9 people work in 44% of the companies surveyed, 10–49 people in 16% and 50–249 people in 1%; the second tier includes companies with a Woodworking Shop, a Planning or Design Department, an Accounting or Finance Department (or a Personnel Department) (27%): 1–9 people work in 7% of the surveyed enterprises, 10–49 people in 14%, 50–249 people in 6%; the third level includes enterprises with a Marketing or Marketing Department, a Metalworking Shop and a

Table 5.1. Rotated Component Matrix.[a]

	Component		
	1	**2**	**3**
K6_1 NO departments/divisions	–,810		
K6_2 Woodworking Shop	,904		
K6_3 Metalworking Shop			,974
K6_4 Marketing Department		,615	
K6_5 Accounting/Finance Department	,772		
K6_6 Planning/Design Department	,822		
K6_7 Personnel Department		,888	

Note: Extraction method: principal component analysis and Rotation method: varimax with Kaiser normalisation.

[a]Rotation converged in five iterations.

Personnel Department (12%): 10–49 people in 4%, 50–249 people in 6%, 250 and more in 2%.

56% of the companies surveyed have production managers, that is, practically all of them with more than one employee (e.g. director). 51% of companies have a designer Only 10% of companies have a head of department or division.

The detailed correlation analysis showed a *strong* (Spearman coefficient values varying from 0.5) positive significant (with $\alpha = 0.01$) correlation between the questions in the group of the organisational governance structure of the company, suggesting that:

- the more people working in the company, the more developed the company's organisational governance structure is, and vice versa ($r_s = 0.551$);
- the more people work in the company, the more departments or divisions there are in the company ($r_s = 0.623$), and vice versa; and
- the more developed the organisational management structure of the company, the more departments or divisions the company has, and vice versa ($r_s = 0.498$).

Average (Spearman coefficients vary between 0.3 and 0.5):

- the longer the duration of the firm's operation, the more people are employed in the company, and vice versa ($r_s = 0.382$) and
- the longer the duration of the firm's operation, the more developed the firm's organisational management structure, and vice versa ($r_s = 0.318$).

Weak (Spearman coefficients ranging from 0.1 to 0.3):

- the longer the duration of the firm's operation, the more departments or divisions it has, and vice versa ($r_s = 0.271$).

There is a weak positive significant correlation (with $\alpha = 0.01$) between the firm's organisational governance structure and the dominant type of production in the firm (with a Spearman's correlation coefficient of 0.274), suggesting that the more developed the organisational governance structure is, the more the organisation tends to be concentrated on mixed production.

There is a moderate positive significant correlation (with $\alpha = 0.01$) between the size of the firm and the type of production that dominates in the firm ($r_s = 0.424$), suggesting that the larger the number of people employed in the firm, the more the firm is concentrated in mixed production.

Similar trends can be observed for the three Baltic countries. The furniture manufacturing sector in the Baltics is dominated by small and medium-sized enterprises. The Baltic furniture manufacturing companies are dominated by the simplest management structure (see Fig. 5.2) – A manager and other employees (68% in Lithuania and Estonia, 76% in Latvia). Lithuanian and Latvian furniture manufacturing enterprises have a similar distribution in terms of their structure. Fifty-nine per cent of the Lithuanian, 40% of the Latvian and only 5% of the Estonian companies surveyed had no departments or divisions.

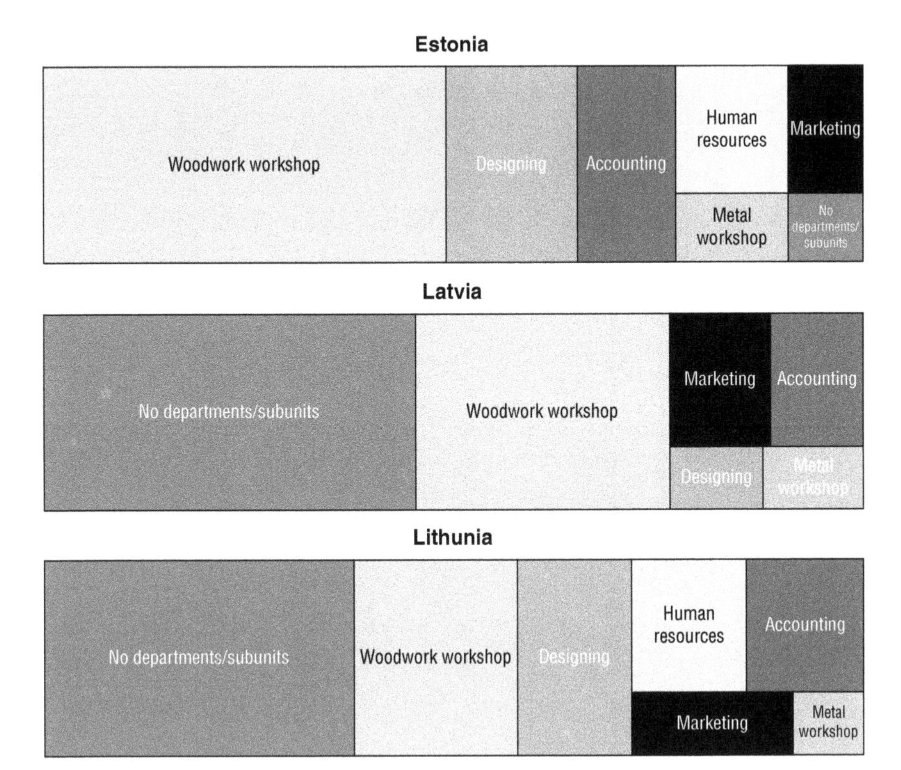

Fig. 5.2. Structure of Surveyed Companies with Divisions/Departments by Country.

Competences of the Manufacturing Team

Despite the dominance of small and medium enterprises (SMEs) in the furniture manufacturing sector, the turnover of Baltic manufacturers is quite impressive. For example, 27% of Lithuanian companies surveyed had a turnover of EUR 50,000–250,000 in 2017; 20% - EUR 250,000–500,000; 14% – up to EUR 50,000; 10% – EUR 500,000–1 million; 7% – over EUR 5 million; 6% – EUR 1–2 million; 3% – EUR 2–3 million; 1% – EUR 3–4 million; and 12% of the companies refused to give an amount or did not know. We assume that production volumes depend on the competences of the employees, their ability to choose the right technologies, and to make the most of them.

The correlation analysis shows that there is a weak positive significant correlation (with $\alpha = 0.01$) between the turnover of the company in 2017 and the dominant type of production in the company (the Spirmen correlation coefficient is 0.388), suggesting that the higher the turnover of the company in 2017, the more the company is focussed on mixed production (the correlation coefficient was calculated for only the companies that reported turnover, 88% of the companies interviewed) (Table 5.2).

Table 5.2. Turnover and Nature of Production (Crosstab).

		Nature of Production						Total
		Individual	**Small batch**	**Batch**	**Large batch**	**Mass**	**Mixed**	
Turnover	Less than 50,000 EUR	16	1	2	0	0	2	21
	50,000–250,000	35	0	0	0	1	3	39
	250,000–500,000	22	2	3	0	1	2	30
	500,000–1 million	8	1	2	1	0	2	14
	1–2 million	3	1	2	0	1	1	8
	2–3 million	2	0	2	0	0	0	4
	3–4 million	0	0	0	0	0	2	2
	5 million and over	2	1	2	1	0	5	11
	Don't know/difficult to say	12	2	1	0	0	2	17
Total		100	8	14	2	3	19	146

Sixty-eight per cent of Lithuanian companies surveyed are dominated by customised production. These enterprises are characterised by having up to 9 employees (63%); the simplest management structure of the enterprise – a manager and other employees (77%); no departments or divisions (70%); no precise operating times for production processes (55%); only 17% of these enterprises have set operating times over a long time interval, working with machine tools and other equipment; 65% of these companies calculate exact material prices, operating times, make design drawings (compared to 60% of all surveyed companies) when calculating the cost of an order, and 17% calculate average prices and add a mark-up (the same proportion of all surveyed companies). Among the indicators, 47% of the companies have seen an increase in the number of orders in the last year compared to the previous year (compared to 50% of all companies), 40% of the companies have seen an increase in the profit margin (compared to 44% of all companies), 49% of the companies have seen an increase in the revenue (compared to 51% of all companies), 30% of the companies have seen an increase in the number of employees (compared to 32% of all companies) of all companies), 53% have seen an increase in the number of sales (compared to 55% of all companies) and 58% have seen an increase in the number of equipment or technology used (compared to 55% of all companies). 35% of these enterprises had a turnover of EUR 50,000–250,000 in 2017 and 22% had a turnover of EUR 250,000–500,000. 56% of these enterprises do not produce for export; in 28% of the enterprises, export accounts for up to 10% of orders. 77% have never received orders from companies representing international brands.

Tools for Data Managing and Exchange

68% of companies surveyed fill in approved MS Word/MS Excel tables, 18% - install/implement original/specialised software (PRO 100, BAZIS, KARAVELA, AUTOCAD, IMOS and RIVILE), 12% – install/implement enterprise resource planning (ERP) system, 6% – each division/unit creates and uses tools that they find suitable.

Specialised IT process management tools are used in these processes: client management – 24% of all companies, warehouse management – 21%, client offer management – 28%, order management – 43%, production planning and management – 45%, quality management – 19%, logistics management – 15%, book-keeping – 64%, management of human resources – 15%, and analysis and reporting management – 22%.

Specialised IT tools are the most necessary in companies: to carry out calculations – 75%, to collect/process information faster – 49%, to analyse information – 44%, to collect/systematise data – 43%, to receive generalised information to adopt the solutions – 31%, to manage production stoppages – 27%, for fast information circulation between employees/divisions – 23%, to coordinate data/information between different units – 14%.

Only in 32% of companies, the specialised IT tools/software used in company completely meet respondent's needs/expectations; partially, because they have suitable IT tools/systems only in several areas – 30%; partially, because not all necessary information/data is stored properly in IT tools/systems – 6%; partially, because it requires a lot of extra work – 6%; no, because it does not deliver the result that is expected – 3%; no, because it costs too much and results are not satisfactory – 3%; it is hard to assess as they are still developing IT tools/systems – 13%; and other (don't use specialised IT tools; to calculate the order price, convenient programme for storage) – 7%.

The lack of IT solutions in the companies is felt in following processes: production planning and management (31%), order management (25%), client offer management (24%), warehouse management (15%), client management (14%), quality management (14%), logistics management (14%), analysis and reporting management (14%), management of human resources (13%) and book-keeping (8%).

Some advantages of IT solutions in the companies are pointed out: more transparent reporting process (36%), optimised production (31%), production process is managed better (30%), decrease in production errors (30%), more efficient resource management (26%), shorter operation times (25%), increased production efficiency (25%), more accurate stock management (25%), more flexible management of production (24%) and better understanding of the unused potential (19%).

Production and Experts

51% of the companies surveyed do not have precise operating times for production processes; 15% for some production processes; 34% have precise operating times for production processes.

21% of the enterprises surveyed have set operating times over a long period of time when working with machines and other equipment; 9% have experimented with machines and other equipment; 4% have been set by the director.

Companies that set precise operating times for production processes (34%) tend to set operating times over long periods of time when working with machines and other equipment (positive significant with $\alpha = 0.01$ strong correlation, $r_s = 0.883$).

Enterprises that set accurate operating times for production processes (34%) are characterised by a company structure with a manager and other employees (49%); 10–49 employees (45%); no departments or divisions (43%); set operating times over a long period of time when working with machines and other equipment (63%); calculate accurate material prices, operating times and make design drawings (65%). In the last year, compared to the previous year, 63% of companies have seen an increase in the number of orders, 49% have seen an increase in profit margins and revenues and a stable number of employees, 59% have seen an increase in sales and 51% have seen an increase in the number of equipment or technologies used. Twenty per cent of these enterprises had a turnover of EUR 250,000–500,000 in 2017, 14% EUR 50,000–250,000 and the same proportion, 14%, over EUR 5 million. Sixty-one per cent of these enterprises have a customised production profile and 14% have a mixed profile. Seventy-one per cent of these enterprises produce for export. In 29% of these enterprises, up to 10% of total orders are for export, in 21% – 75% or more. 51% never receive orders from companies representing international brands, 20% rarely and the same proportion and 20% – sometimes.

The companies that have received or receive orders from companies representing international brands on a regular basis (9%) are those employing 50–249 people (61%), 10–49 people (31%) and 1–9 people (8%). The dominant company structure is that of a manager and other employees (46%) and a general manager, heads of departments executives and other employees (31%). 46% of these companies do not have precise operating times for production processes, 23% for some production processes and 31% do (compared to 34% of all companies surveyed). 23% of the enterprises that have set precise operating times for a production process have set operating times over a long period of time when working with machines and other equipment (compared to 21% of all enterprises surveyed) and 39% calculate precise material prices, operating times, making construction drawings (compared to 60% of all enterprises surveyed) and also 39% calculate average prices and add a mark-up (compared to 17% of all enterprises surveyed). In these enterprises, mixed production (46%, compared to 13% of all enterprises surveyed) and customised production (39%, compared to 69% of all enterprises surveyed) are more prevalent.

Among the indicators that stand out in the last year compared to the previous year are the increase in the number of orders in 76% of the enterprises (whereas 50% of all enterprises), the increase in the profit margin in 62% of the enterprises (whereas 44% of all enterprises), the increase in the revenue in 69% of the enterprises (whereas 51% of all enterprises), the increase in the number of employees in 46% of the enterprises (whereas 32% of all enterprises), the increase in the number of sales in 92% of the enterprises (whereas 55% of all enterprises) and the increase in the number of equipment or technologies used in 62% of the enterprises (whereas 38% of all enterprises). 23% of these enterprises had a turnover of EUR 1–2 million in 2017 and 23% had a turnover of over EUR 5 million. 76% of these enterprises produce for export. 9% of companies receive orders from companies representing international brands on a regular basis; 12% sometimes; 14% rarely; and 65% never. 39% of the companies that never receive orders from international brands nevertheless produce for export.

The companies that have received or receive orders from international brands sometimes or rarely (26%, 12%) stand out because 37% have established operating times over a long period of time with machine tools and other equipment (compared to 21% of the total surveyed companies) and 68% calculate exact material prices, operating times and make design drawings (compared to 60% of the total surveyed companies). These enterprises are more dominated by small series production (16%, compared to 5% of all enterprises surveyed) and series production (18%, compared to 10% of all enterprises surveyed). 18% of these enterprises had a turnover of up to EUR 50,000 in 2017 and the same proportion, 18%, had a turnover above EUR 5 million. Among the indicators, the most notable in the last year compared to the previous year are the increase in profit margins (net margins) (50%, compared to 44% of all companies) and in revenues (58%, compared to 51% of all companies surveyed). 76% of these companies produce for export.

The companies that have customised production are small, they have practically no set operating times and they have not seen an increase in their indicators. On the contrary, in the companies that have set operating times, all indicators have increased.

Cost Estimation Practice

Price Evaluation Strategy

Custom furniture manufacturing companies usually calculate the exact cost of materials when pricing an order (77% of Lithuanian, 88% of Latvian and 70% of Estonian companies surveyed) (see Fig. 5.3). No differences in the methods of calculating the order price were found between the firms dominated by individual production and the other firms.

There is no difference in IT consumption between enterprises using different calculation methods.

The proportion of enterprises that have installed/deployed an original/specialised information system, implemented a business management system (ERP) or completed validated MS Word/MS Excel spreadsheets is similar between those that calculate and those that do not calculate accurate material prices, with no significant differences found ($p > 0.05$).

The analysis of IT implementation in the costing process showed that having or not having an original/specialised information system and ERP, filling in or not filling in validated MS Word/MS Excel tables does not affect the errors in costing ($p > 0.05$). There was no significant difference in the use of IT in processes for companies reporting a margin of error of up to 10%, between 10% and 20% and more than 20% ($p > 0.05$). There is no significant difference in the post-estimation cost of enterprises that have implemented/implemented an original/specialised information system, ERP or MS Word/MS Excel spreadsheets and those that do not use these tools ($p > 0.05$).

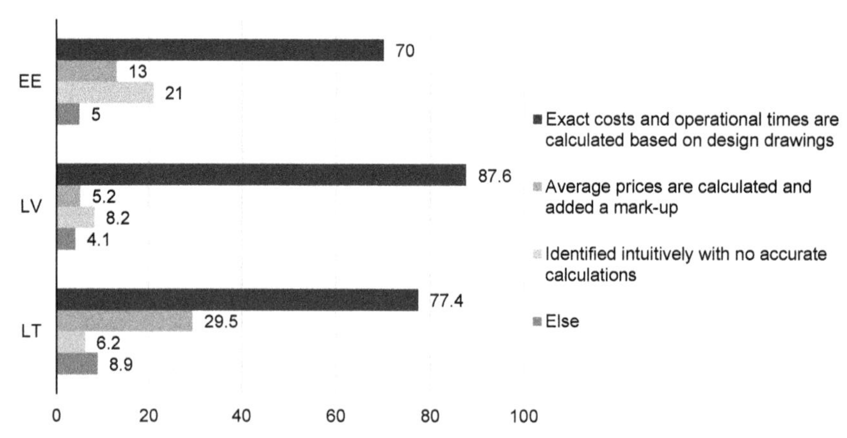

Fig. 5.3. Order Price Calculation Methods by Country.

The traditional approach of calculating exact material prices for costing may be hindered by limitations in IT efficiency. Companies lacking robust IT tools struggle to collect, store and analyse detailed production data. This hinders quality management and the ability to accurately assess pricing. By implementing efficient IT systems, customised manufacturers can gather rich production information that can be leveraged for both cost estimation and strategic pricing decisions, ultimately bridging the gap between these two crucial strategies.

The survey allowed us to identify how the organisational structure relates to the pricing principle. A nonsignificant positive weak correlation was found. It can be concluded that there is no correlation between the pricing principle and the company's organisational management structure. However, our survey revealed a positive correlation ($r = 0.287$, $p < 0.001$) between company structure complexity and the number of personnel involved in price calculations. This suggests that companies with more intricate structures, as evidenced by a higher number of pricing process participants, exhibit a statistically significant ($p < 0.001$) tendency towards lower IT integration across various business functions ($r = 0.298$). Furthermore, the analysis indicates a statistically significant negative association ($p < 0.05$) between a larger number of pricing related personnel and the utilisation of IT systems in crucial areas like order management, customer management, warehousing (all $p < 0.01$), and functions like production planning, quality management, logistics and analytics (all $p < 0.05$).

Thus, the research suggests a link between a company's organisational structure, as reflected by the number of people involved in price setting, and the need for IT tools. Companies with a more complex price determination process, often involving multiple decision-makers, tend to experience a greater gap in their IT capabilities. However, many companies often limit their investments in expensive, externally sourced IT solutions. Consequently, many companies rely on internal solutions, primarily Microsoft Word and Excel, which are utilised by both small and large firms despite their limitations in complex pricing tasks (Fig. 5.4).

Attitudes of Manufacturers Towards IT and Other Decision Support Instruments

The use of information technology not only in production processes but also in pricing can increase the value created by companies. However, the penetration and benefits of these technologies at the company level are still poorly understood. In this subsection, we will reveal the attitudes of Baltic furniture manufacturers towards IT and other decision support instruments.

IT Adoption Factors and Organisational Features

First of all, IT adoption differs by economic activity: IT adoption is greater in information intensive industries and less prevalent in low-tech sectors. Varanakis (2017) points out that the research and development factor indirectly affect investment in machinery and equipment and firm performance through manufacturing flexibility and new product innovation. It has also been observed that the

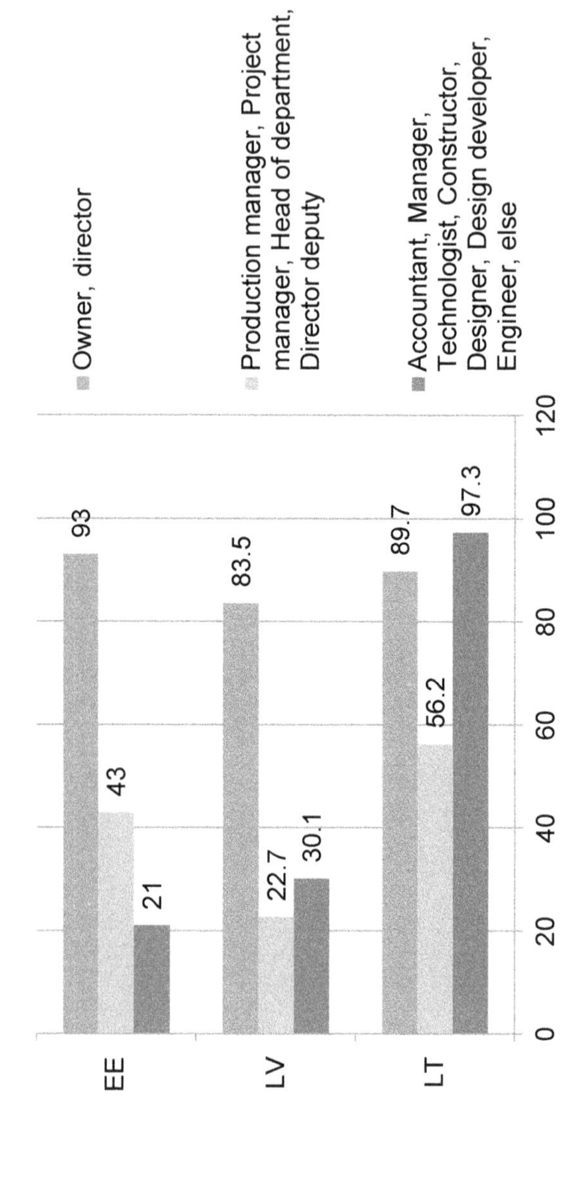

Fig. 5.4. Employees of Companies Involved in the Calculation of the Order Price (Per cent, Multiple Choice).

incidence of IT is significantly higher in the service sector than in the manufacturing industry (Drew, 2003, Fabiani et al., 2005).

Firm size is also a determining factor in terms of IT adoption (Arduin et al., 2010; Giunta & Trivieri, 2007; Ordanini, 2006; Osorio-Gallego et al., 2016). Even within the SME group itself, larger firms are more likely to adopt IT (Premkumar & Roberts, 1999). Smaller firms are often faced with a choice between investing in technology for information management activities or developing physical value creation activities (Bharati & Chaudhury, 2006). However, despite their limited capacity to invest in technology, SMEs are characterised as more flexible, responsive and with higher innovation capacity (Bliliand & Raymond, 1993; Hall & Khan, 2002; Raymond & Croteau, 2006).

The literature has explored more extensively the organisational factors that affect IT adoption. In terms of the structural characteristics of firms, it has been observed that multi-unit firms are more likely to adopt new ICTs in order to improve information transfer (Galliano & Roux, 2003). Without a proper organisational structure, the firm will not be able to experience the benefits of adopting these technologies (Song et al., 2007). As the organisational and management model can vary considerably, standardised IT solutions suitable for larger companies may not be suitable for smaller ones (Ordanini, 2006). Moreover, the application of advanced manufacturing technologies usually triggers changes in both organisational structures and organisational processes (Millen & Sohal, 1998), and such changes may be difficult to implement in the SME segment, which are family-owned businesses and are usually characterised by simple and highly centralised structures with the chief executive officers. In family businesses, managers are most often employed by family members, who tend to have a lower level of education than nonfamily managers, which may prevent the use of advanced technologies (Jorissen et al., 2005). Bruque and Moyano (2007) also include changes in the firm's hierarchy and power structure, as well as the absence of qualified personnel in SMEs, as inhibitors of IT adaptation. The intensity and speed of adoption of IT, according to them, depend on employee relations, the synchronous implementation of IT and quality systems, and the improvement of knowledge and skills of employees in family businesses. SMEs that have successfully adopted advanced manufacturing technologies were more likely to apply these technologies in the areas of flexibility, delivery and quality, but not in the implementation of cost reduction strategies (Khazanchi et al., 2007).

Research on advanced technology adoption suggests that a company's manufacturing strategy also plays a significant role. Studies (e.g. Bocquet et al., 2007; Chang et al., 2008) have identified several factors influencing IT adoption, including:

- A shift from centralised to decentralised decision-making.
- A focus on economies of scope (producing a variety of products efficiently) rather than just economies of scale (producing a high volume of a single product efficiently).
- Introduction of innovative products.

- Reliance on external suppliers for parts or processes.
- Production for targeted market segments.
- Investment in new employee skills.
- Minimisation of inventory costs.
- Focus on responsive customer service.

Applying these factors to customised production, we can expect a high level of IT adoption. Customised production inherently involves a high degree of customer focus and caters to targeted markets. Additionally, the need for precise cost calculations for made-to-order products makes information processing functions critical. Furthermore, increasing automation in furniture manufacturing necessitates 'smart manufacturing', which integrates manufacturing activities, information technology and automation. This trend highlights the growing need for manufacturers to enhance their data processing capabilities (Kusiak, 2018; Svensson & Malmquist, 2002; Thoben et al., 2017).

Technology Usage and Sector Specificity

As mentioned, the furniture production sector in the Baltic countries is highly individualised. One of the key features of customised manufacturing is the need to combine personalisation of customised products with costs as low as in mass production. Moreover, cost estimation and price-setting processes take place at the same time, therefore an ability to make early decisions on pricing is crucial for customised businesses. However, researchers of customised production (Rimpau & Reinhart, 2010; Trost & Oberlender, 2003) emphasise risks associated with early price estimation. The specificity of customised production lies in the fact that early price estimation often becomes the basis to approve the final funding scheme for the project/order. Estimates of the total costs often entail into the course of the offer specification process the risk of inaccurate data about the product and the manufacturing processes and the risk of possible post-specification fluctuations in prices and costs. Thus, the data available at the initial stage of the price determination process are usually insufficient to set an exact price. The result is often a low-level accuracy. To evaluate the costs of complex projects manufacturers are looking for the ways to embrace usually incomplete data and uncertain information on manufacturing lead time.

Another important characteristic of customised manufacturing is the uncertain production environment (Koh et al., 2005; Zennaro et al., 2019). To combat uncertainty, it is often suggested to use technological solutions such as ERP. However, empirical data indicate that increased uncertainty in the SME environment does not increase the assimilation of advanced manufacturing technologies (AMTs) (Raymond, 2005). The study of Koh and Simpson (2005) found that ERP could improve responsiveness and agility to change, but not to uncertainty. In general, advanced technologies can be used in all elements of the value chain: primary manufacturing activities (inbound logistics, operations, outbound logistics, marketing and sales or service) and support activities (accounting and finance, human resource system, computer aided design and electronic procurement)

(Chen & Li, 2010). Both main and support activities have an information-processing component, and more and more companies, especially large companies, use IT to manage this component. Bharati and Chaudhury (2006) found that the use of technology is not uniform in different stages of product production and realisation. Accounting and finance, websites, materials management, order processing, computer aided design is applied the most to IT and customer relationship management is the least.

Also, the level of technology use varies depending on the pricing strategy applied in the company, which, in turn, depends on the prevailing decision-making process, the price-setting process, the organisational structure, the diffusion of pricing capabilities and leaders' behaviour (Liozu & Hinterhuber, 2012). Liozu and Hinterhuber (2012) based on the analysis of small and medium-sized companies in the USA, found that companies supporting value-base orientation rely on market research, scientific pricing methods and expert recommendations, that is, more IT-based methods. Meanwhile, cost- and competition-based companies base their decisions on experience, previous knowledge and intuition. Thus, the use of technology in making a price decision is low in them.

Kienzler (2017) further emphasises the link between pricing strategy and IT adoption. His research suggests varying levels of effort, information subjectivity and uncertainty associated with different pricing approaches. The cost-based pricing requires the least effort from a company. Information used for decision-making is objective and has low to medium uncertainty. The competition-based approach involves moderate effort to gather information, which can be subjective and have medium to high uncertainty. While the value-based pricing relies on information that is subjective, difficult to collect and often vague. These findings suggest that a company's pricing strategy (and its associated information management capabilities) can influence the level of IT adoption.

A summary of the factors discussed earlier in this section is provided in Table 5.3.

By summarising the factors identified in the scientific literature regarding IT adoption and examination of the potential impact of customisation on IT adoption within the furniture manufacturing industry, we can assess how customisation might influence IT development in this sector. The emphasis on highly customised furniture presents a potential catalyst for IT advancement. Several factors contribute to this potential. The competitive landscape in customised furniture manufacturing incentivises companies to seek efficiency gains and product innovation. IT solutions can play a crucial role in achieving these goals. Also, the constant introduction of new and customised furniture designs necessitates robust IT infrastructure to support design, development and production processes. Such factor as the effective management of long-term partnerships requires robust IT systems as well in order to facilitate communication, data exchange and collaboration with both suppliers and customers.

However, despite the potential benefits of IT adoption the furniture manufacturing industry, on average, exhibits a lower level of technological adoption compared to other sectors. This historical trend might hinder widespread IT integration together with the traditional focus on the physical product itself that might overshadow the potential benefits of IT solutions for optimising production

Table 5.3. Summarising the Factors of the Level of IT Adoption Within Industrial Company.

	The group of factors	Individual factor	High IT adoption is expected (High)	Low IT adoption is expected (Low)	Expected level IT adoption in customised furniture manufacturing
1	**Sector specificity**	*Technological intensiveness*	High tech	Low tech	**Low**
		Industries (sectors)	IT based	Traditional	**Low**
		Market competitiveness	Targeted market	Broad market	**High**
2	**Output of business specificity**	*Product/service orientation*	Service	Product	**Low**
		The range of product and novelties	New product development	Traditional products with low variations	**High**
		Suppliers	High demand and long lasting relationship-based suppliers	Random suppliers	**High/Low**
3	**Organisational features**	*The size of company*	Large companies	Small and medium companies	**Low**
		Organisational structure	Hierarchical	Flat organisation	**Low**
		Role delegation	Dense structure of role	Few roles	**High/Low**
		Competence of human resources	High competence in IT	Absence special competence on IT	**Low**

processes and customer interaction. The prevalence of small companies in the sector can be another limiting factor. Often characterised by flat organisational structures and limited resources, these companies might struggle to implement specialised IT roles and departments, ultimately hindering IT advancement. So, customisation in furniture production presents a potential for IT acceleration due to increased competitiveness and new product development demands. However, historical low-tech industry practices and the dominance of small companies pose challenges to widespread IT adoption.

Assessment of Baltic Manufacturers' Attitudes Towards IT

The research on manufacturers' attitudes towards IT employed a two-pronged approach to identify factors influencing IT adoption in furniture manufacturing. First, a literature analysis was conducted to identify the key factors influencing IT adoption across various industries. These factors were then categorised based on the relevant sections identified in the analysis (see Table 5.1). Then building upon the established factors, the potential influence of pricing strategies on IT adoption was explored. To achieve a comprehensive view, the scope of analysis was expanded beyond the standard factors typically considered (sector specificity, business characteristics and organisational features, as illustrated in Fig. 5.5). This research additionally included financial stability as a factor, recognising its role as an indicator of the economic climate and its potential impact on business behaviour towards IT adoption. Since the study was conducted during an economic boom, this additional factor allowed for a more nuanced understanding of potential business decisions regarding IT investment.

To contextualise the key IT adoption factors within the furniture manufacturing industry, this section examined the content of pricing strategies (i.e. the approach to pricing) and identified the most prevalent information technologies utilised in the sector. This is how we formed the research strategy, which is shown in Fig. 5.5.

The survey instrument comprised 35 questions designed to capture data on five key aspects of corporate pricing and governance. These aspects included: organizational structure, employee engagement, the price estimation approach, production processes and information technology implementation. The survey implemented in three Baltic countries focusses on IT adoption and price estimation. The questions such as the structure of the organisation, the performance and the pricing strategies are taken into account.

This study defines technology/IT adoption in furniture manufacturing based on the following key variables:

- CNC machine utilisation.
- IT process management tools.
- Technologically enabled manufacturing activities.
- Advantages provided by IT.
- Compliance of IT with the needs of the company.
- Lack of IT.

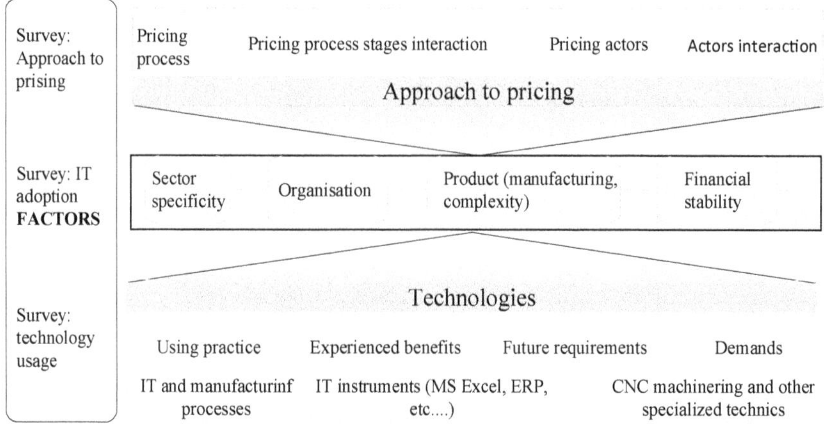

Fig. 5.5. The Approach Towards Survey's Questions Superposition of IT Adoption Factors with Pricing Strategy and Technology Usage.

This subsection analyses the surveyed Lithuanian, Latvian and Estonian companies together because, due to the extremely small number of respondents, calculations that ensure statistically significant results in some of the categories selected during the study were not possible. In addition, the Latvian sample of companies is dominated by mixed production companies, in contrast to the Lithuanian and Estonian samples, where individualised production dominates.

Equipment and Technologies Used

The survey showed that the number of equipment and used technologies grew in 38% of the surveyed Lithuanian, 32% of the Latvian and 29% of the Estonian companies. The survey results confirm the relationship between company size and technology adoption identified in the literature review. Companies with a larger workforce exhibited a significantly higher prevalence of CNC machine usage. Companies with more employees were significantly more likely to utilise CNC machines in production ($p < 0.001$). Larger companies were significantly more likely to implement original or specialised information systems ($p < 0.001$). The use of business management systems (ERP) also showed a significant positive correlation with company size ($p < 0.001$).

Interestingly, the survey found no statistically significant difference ($p > 0.05$) in the use of basic productivity tools like Excel and Word tables across companies of different sizes. This suggests that basic IT skills and tools are likely implemented regardless of company size. Furthermore, the survey revealed that only a small portion of the companies outsource IT services. Notably, the decision to outsource IT services did not exhibit a statistically significant correlation with company size ($p > 0.05$). This finding suggests that even smaller businesses may consider outsourcing their IT needs, potentially due to factors beyond simply the number of employees. However, companies purchase IT services much less often than production-related services (Fig. 5.6).

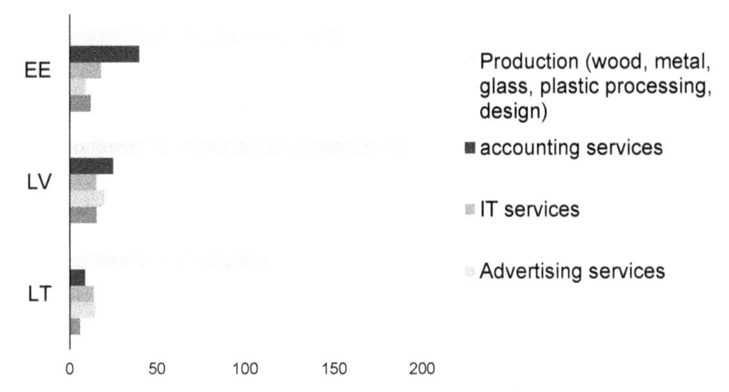

Fig. 5.6. Purchase of Services (Per cent, Multiple Choice).

The growth of equipment or used technologies in companies dominated by individual production was similar compared to others, no statistically significant differences were found ($p > 0.05$).

Purpose and Need for IT Usage

The survey shows a clear link between company size and how much employees rely on technology. In larger companies, with more employees, staff reported using IT tools more frequently across a variety of tasks. This included customer management, warehouse operations, managing customer offers, order fulfilment and various aspects of production planning and management. Additionally, functions like quality control, logistics, accounting, human resources and data analysis/reporting all showed increased reliance on IT tools in larger companies. The statistical significance of these findings was very high ($p < 0.001$).

This trend went beyond just using more technology. Employees in larger organisations also expressed a greater need for IT tools to help with specific challenges. Faster information flow between departments ($p < 0.01$) was a critical need, along with access to combined data for better decision-making ($p = 0.01$). Additionally, managing disruptions in production ($p < 0.001$), efficiently collecting and organising data ($p < 0.001$), analysing information ($p < 0.001$) and performing calculations ($p < 0.001$) were all areas where larger companies felt that IT tools were particularly valuable. The need for faster information processing in general was also more pronounced in larger companies ($p < 0.01$).

Interestingly, despite the increased use of IT tools, some gaps were identified in larger companies. Respondents from companies with bigger number of employees reported a greater need for IT support specifically in warehouse management ($p < 0.01$) and logistics ($p < 0.05$), suggesting that these areas might benefit from further technological investment. Similarly, a significant need for IT tools was reported for analytics and report management processes in larger companies ($p < 0.05$).

Companies with higher revenue (turnover) exhibited a stronger internal push to develop and utilise IT tools ($p < 0.001$). This connection was particularly strong for areas like faster information flow, informed decision-making, production disruption management and data management in general. Additionally, companies with higher revenue were more likely to use a wider variety of specialised IT tools (like digitised and robotic machines, original/specialised information system, business management system (ERP)) across different functions ($p < 0.001$). The specialised IT process management tools were particularly prevalent in areas like warehouse management, customer offer management, order management, production planning and management, quality management, logistics, accounting, human resources and data analysis/reporting. The increased complexity of manufacturing activities for companies with higher revenue explains their greater reliance on IT tools to manage both core and supporting functions.

The survey also revealed interesting differences in how companies that manufacture customised products use specialised IT tools compared to other companies. While customised manufacturers used these tools less frequently for warehouse management ($p = 0.001$), they reported using them more often for managing customer offers and orders ($p < 0.05$). There were no significant differences in how these companies used IT tools for other functions. These findings suggest that customised manufacturers prioritise using IT to manage customer interactions and orders, while relying less on IT for warehouse management compared to companies with a more standardised production flow.

Companies with higher revenue (turnover) demonstrated a stronger internal drive to develop and utilise IT tools. This connection was particularly significant for areas like ensuring faster information flow between departments ($p < 0.01$) and obtaining aggregated data to support informed decision-making ($p < 0.01$). Additionally, managing production disruptions ($p < 0.01$), efficient data collection and systematisation ($p < 0.001$) and information analysis ($p < 0.001$) were all identified as areas where companies with higher revenue felt that specialised IT tools were most crucial.

Despite the wide use of IT in companies with higher turnover, the need for these technologies is not fully satisfied. These companies more often note that they feel a lack of information technology in the warehouse management process ($p < 0.001$), in the production planning and management process ($p < 0.01$), in the logistics process ($p < 0.01$) and in the human resources management process ($p < 0.05$).

No significant differences were found ($p > 0.05$) in the use of IT tools between companies with a turnover of up to 1 million and those with higher turnover. This includes practices such as using pre-approved MS Word/Excel templates or allowing individual departments/employees to create and utilise their own preferred tools. Nor was it determined differences according to the need for specialised IT tools to perform calculations ($p > 0.05$) and to collect/process information faster ($p > 0.05$). No differences were observed in companies of different turnover according to whether they feel a lack of information technology in the customer management process ($p > 0 .05$), in the management process of offers to customers ($p > 0.05$), in the management process of offers to customers ($p > 0.05$), in the quality management process ($p > 0.05$), in the accounting process ($p > 0.05$), or in the analytics/report management process ($p > 0.05$).

Performance

The survey results revealed that approximately one-third of furniture manufacturers in the Baltic countries invested in new equipment and technologies during the analysed period. The positive impact of such investments varied across the region. In Estonia and Lithuania, companies that increased their equipment or technology usage experienced a greater rise in both income and profit margins. However, in Latvia, investments during the analysed period did not yet translate into improved financial indicators. Further analysis revealed no statistically significant difference ($p > 0.05$) in the performance indicators (profitability and income growth) between companies dominated by individual production and those with other production structures. This suggests that production structure may not be a major factor influencing financial performance.

While the survey participants reported increases in revenue, profits and the number of orders, a noteworthy trend emerged. For the majority of companies, investments in equipment remained stagnant despite the reported financial growth. This could indicate that companies are not prioritising short-term investments in technology, even as their financial capacity increases. A similar pattern was observed with the number of employees, although the magnitude of this trend was less pronounced compared to technology investments (Fig. 5.7).

The survey explored the IT tools used by furniture manufacturers in the Baltic countries for process management (Fig. 5.8). The most common solutions were

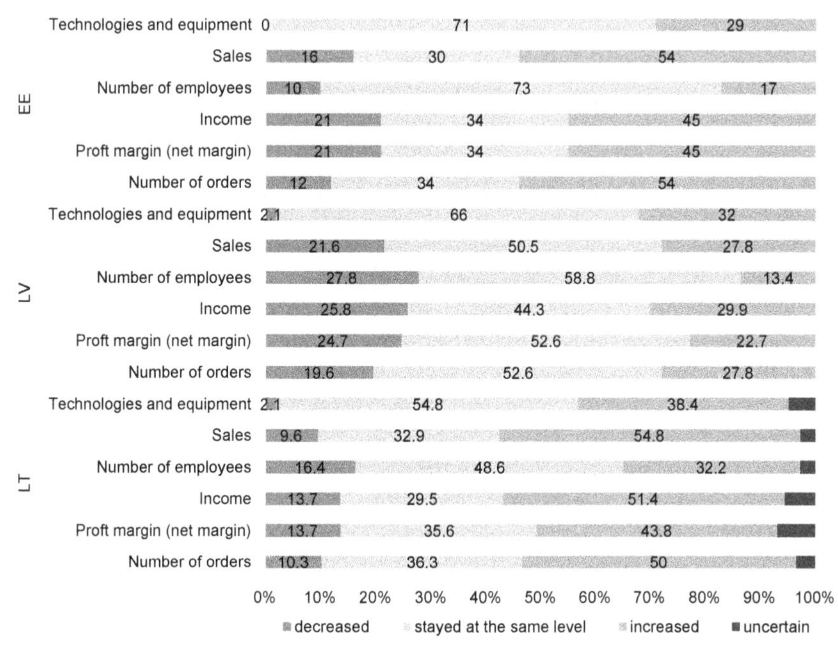

Fig. 5.7. Changes in the Performance Indicators of Some Companies Over the Past Year, Compared to The Previous Year, Per cent.

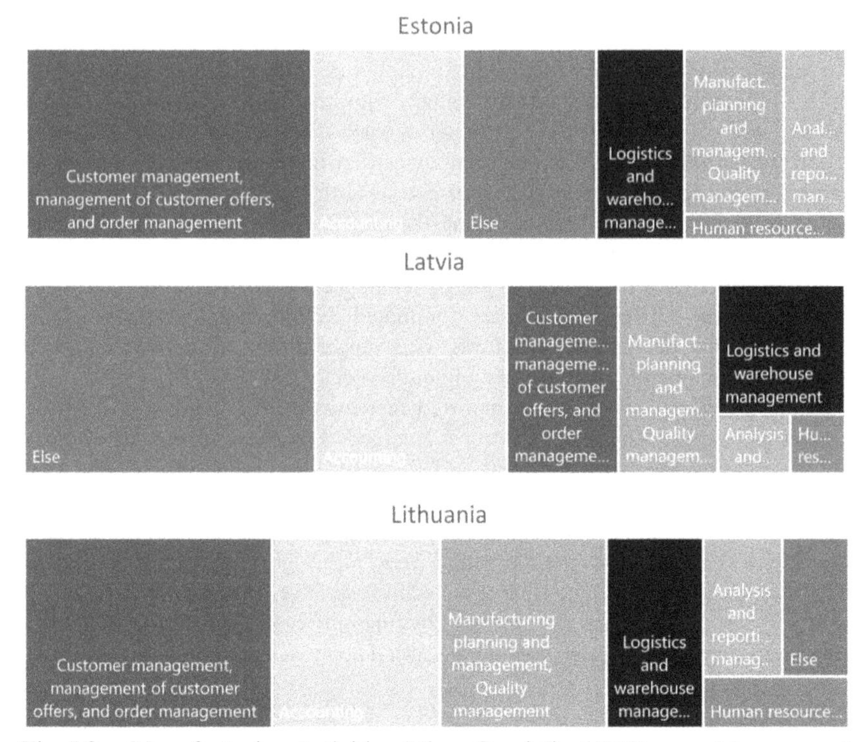

Fig. 5.8. Manufacturing Activities, Where Specialised IT Process Management Tools Are Applied (Per cent, Multiple Choice).

basic tools like MS Word and MS Excel (68% in Lithuania, 54% in Latvia, 49% in Estonia). A smaller, but still significant portion of companies utilised original, specialised information systems (18–21% across the region). Business Management Systems (ERP) were the least common solution, with adoption rates of 12% in Lithuania, 2% in Latvia and 1% in Estonia. Interestingly, companies that implemented ERP systems reported statistically significant improvements in financial indicators over the past year ($p < 0.05$) for number of orders, profit margin and income.

The survey also investigated the impact of specialised IT process management tools on specific functions. These tools were used in customer management by 24% of Lithuanian companies, 7% of Latvian companies and 22% of Estonian companies. Companies utilising these tools in customer management experienced significant increases in orders ($p < 0.05$), profit margin ($p < 0.05$), income ($p < 0.05$) and even sales ($p < 0.01$) over the past year. A similar pattern emerged for companies using specialised IT tools in managing customer offers (27% in Lithuania, 2% in Latvia and 25% in Estonia). These companies saw statistically significant growth in orders ($p < 0.05$), income ($p < 0.05$) and sales ($p < 0.05$) while profit margins remained unchanged ($p > 0.05$). Similar positive correlations were observed between using specialised IT tools and improved performance in order management (42% in Lithuania, 10% in Latvia and 32% in Estonia) and logistics management (15% in Lithuania, 3% in Latvia and 14% in Estonia). Companies utilising these tools reported significant increases in orders ($p < 0.05$ or $p < 0.01$), profit margin ($p < 0.05$), income ($p < 0.05$) and sales ($p < 0.01$ or $p < 0.05$).

Conversely, no significant differences in growth rates were found between companies using specialised IT tools in production planning/management, quality management or human resources management compared to those who did not. The exception was an increase in sales for companies employing specialised tools in production planning/management.

These findings suggest a strong correlation between the use of specialised IT process management tools and improved financial performance in Baltic furniture manufacturing companies, particularly for customer-facing functions and core operational processes. However, the impact on production planning, quality management and human resources management appears less pronounced.

Our findings suggest a nuanced relationship between IT adoption and financial performance in furniture manufacturing companies. The findings suggest a differential impact. While companies using IT tools for customer and order management activities experienced operational success and improved financial indicators, no significant association was found between IT adoption in production planning, quality management and human resource management and financial performance. This implies that the benefits of IT tools are more pronounced in areas directly interacting with customers and fulfilling orders.

Furthermore, the research revealed a production type-based differentiation. Make-to-order furniture companies benefitted most from investments in technology that manage customer interactions, quotes, orders and logistics. In contrast, customised furniture companies primarily gained an advantage from utilising IT tools for managing customer offers and orders. These findings suggest that the optimal functionalities of IT tools may vary depending on the specific production model employed by a furniture company.

The evaluation of IT adoption factors in the furniture industry revealed several trends that align with expectations. Our research highlighted the need for customisation-driven adjustments to IT implementation strategies, confirming our pre-assumptions about the influence of organisational structure and industry sector. Additionally, the analysis of prevalent pricing strategies has broadened our understanding of the motivations behind IT adoption within the furniture manufacturing industry.

Sector Specificity and IT Adoption

The traditional Baltic furniture manufacturing sector invests minimally in production-related IT. Even for day-to-day production functions, IT is used minimally, there are no specialised instruments for price estimation, and MS Excel remains the most popular instrument. Large companies that successfully identify a target market and subsequently invest in ERP systems exhibit the potential for increased competitive advantage. This strategic approach is often followed by observations of growth in order fulfilment and profitability. However, large companies engage in customised production only up to 10–20%. More than 70% of the companies engaged in customised furniture production specialise in product production, but implement IT instruments more to strengthen the client relationship management function.

Organisational features and IT adoption

The more complex the company's structure, the more widely (more processes) it uses specialised IT management tools ($r = 0.260$, $p < 0.001$). Organisational development (existing structure, grouping into departments according to functions, distribution of roles and delegation of functions and specialisation) is closely related to the organisation's need to invest in IT instruments, which is directly related to the development of pricing strategy and the freedom to use accumulated knowledge (Fig. 5.9).

Technology and Company Size

The study showed that larger companies are more willing to invest in IT. It is natural that a larger company (after hiring more employees) starts to raise questions of efficiency, therefore a larger part of companies starts to invest in production automation. And what is most interesting, together they invest in the development of the original information system, thus aiming for the most accurate possible benefit from the newly created product, responding precisely to its individualised needs. In addition, the size of the company also determines the financial stability of the company by accumulating the necessary investments that can be directed towards IT development. Meanwhile, all other companies (both small and large) are still successfully meeting their needs using the Excel/Word spreadsheet tool, showing that their market situation is quite tight, and focussing

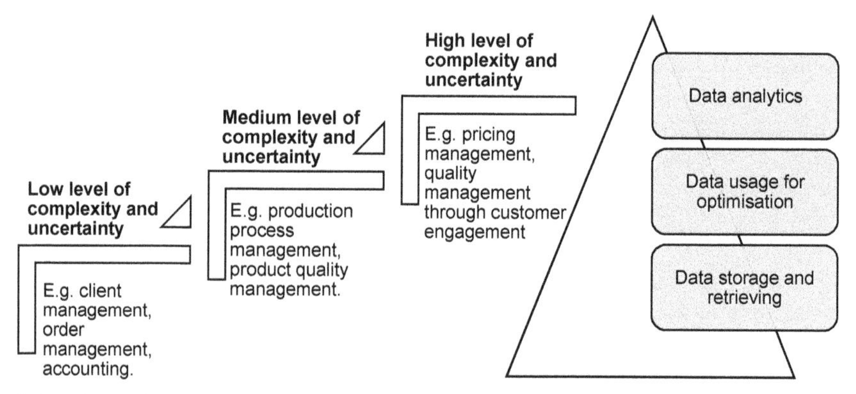

Fig. 5.9. The Model of Factors of IT Adoption in Furniture Manufacturing, Linking Data Management Levels.

on a customised product is taking away all available resources, leaving no room for IT development.

The more employees a manufacturing company has, the more IT is needed to manage information. And here comes the need for customer, order and offer, warehouse management, production planning and quality management and logistics management solutions. The progressive thing is that in such companies there is a need not only to share information between departments or the need to implement more complex accounting functions, but analytics and reporting functions are already mentioned.

In conclusion, it can be said that furniture companies with a tendency towards individualisation absorb technological innovations at a rather low level, and this is primarily shown by the prevalence of MS Excel in companies of various sizes in the furniture sector.

Technology and Financial Stability

It should not be surprising that higher turnover creates greater opportunities for the company to plan investments more widely. The study shows that companies, especially larger ones with growing turnover and orders, are more likely to use both automated machines in their production processes and invest in customer order management systems. A rather new finding is that the same successful customised companies still avoid standardised IT tools and invest more in unique systems, or in the worst case, improve their own systems or individual modules based on MS Excel. However, IT penetration is quite low. Only about 20% surveyed companies improve the specialised system they have created. And standardised ERP systems have units. It is true that those who decided to implement ERP demonstrate better growth parameters (growth was more pronounced, orders and profits increased). If you analyse it more carefully, then the number of orders ($p < 0.05$), revenue ($p < 0.05$) and sales ($p < 0.05$) increased more often for those who had offer management modules. Therefore, customer orientation becomes one of the factors of successful growth. Unfortunately, with means of production, we no longer observe such a connection between growth and IT.

In Conclusion

Thus, the survey helped reveal the success factors of the furniture industry in the Baltic States. The furniture industry has been experiencing stable growth in the last year in the countries under review: the number of new companies is increasing, the old ones are growing and strengthening, profitability and order flows are increasing. At the same time, the competitiveness of the sector lies in the flexible ability of companies to adapt to market changes and to respond sensitively to the needs of customers, while maintaining fairly high product quality standards, and most importantly sustainable order management, to achieve the highest order time, quality and price standards.

The structure of the organisational relationships of the companies explains the orientation of the furniture sector towards individualised production. 51%

of surveyed companies employ 1–9 people, 34% – 10–49, 13% – 50–249 and only 2% – 250 and more. Seventy per cent of the surveyed furniture manufacturing companies are dominated by the simplest management structure – a manager and other employees. Lithuanian furniture makers have not grown to such a level that there is a structure of 'President, general manager, department heads, department heads, executors and other employees'. Even 60% of surveyed companies do not have divisions or departments. Almost all companies viz. 91% that have a Design or Design Department (22% of all surveyed companies) have a Woodworking workshop (32% of all surveyed companies) and therefore are at the same level of the organisational structure. All companies with a Woodworking Workshop have a Metalworking Workshop (3% of all companies), but not vice versa, so companies with a Woodworking Workshop are already at the third level. Only 10% of the surveyed companies have a Marketing Department and 14% - a Personnel Department.

Industry 4.0 trends inevitably permeate the development trends of furniture companies. First of all, in process automation. Companies invest in automation systems, and for a while they are content with that, ignoring other knowledge management systems. But this is explained by limited resources, as the sector is dominated by small companies. Therefore, the sector is dominated by small companies with a flat management structure. In such a company, there is no clear distribution of roles, employees are involved in all stages of the company's production, their position is questioned and agreed upon. Personalisation helps companies better respond to customer preferences and maintain their appeal. At the same time, the study revealed that mixed production, which requires appropriate knowledge management tools, is steadily taking its share in the growing portfolio of companies. This is where there is a greater need for data exchange. It is true that data exchange is very limited for now. Practically limited to popular unified products. But when faced with difficulties in the integration of unified IT ERP products, investments in the development of IT systems begin, and here again the need for expert opinion and sophisticated data management systems (classification of products and classification of working hours by complexity) appears.

It is true that the spheres of management and production collide rather controversially. This is especially felt when evaluating sources of expert judgement. If management systems recognise as experts only other managers with greater responsibility, then production engineers recognise other engineers. The overlapping of managerial and engineering experience is possible, but in rather limited cases, where the employee, who also works as an engineer, has greater managerial responsibility.

The constructors/designers are the most isolated in the system. Although they are consulted both in the pre-design phase and in the production phase, they are the least involved of all phases, especially in the price evaluation phase.

The larger the companies, the greater the need for an expert knowledge management strategy. Companies start to trust the power of data analysis more, they start looking for a way to know the necessary data, what to do with it. And this is where the task of cost management comes to the fore, which links the issues of optimising production after the order is placed with the issues of getting to know the customer before the order is signed. This is precisely the place where the most uncertainties arise (order management and customer relations, design phase and production), especially if it will be individualised production. However, even in the case of mass production, the lifetime of mass production is decreasing, so there are uncertainties when starting a new mass production process.

References

Arduin, D., Nascia, L., & Zanfei, A. (2010). *Complementary approaches to the diffusion of ICT: Empirical evidence on Italian firms* [Working paper, 2010-02 Series in Economics, Mathematics and Statistics]. University of Urbino Carlo Bo, Department of Economics, Society & Politics. http://www.econ.uniurb.it/RePEc/urb/wpaper/WP_10_02.pdf

Bharati, P., & Chaudhury, A. (2006). Current status of technology adoption: Micro, small and medium manufacturing firms in Boston. *Communications of the ACM*, *49*(10), 88–93.

Bliliand, S., & Raymond, L. (1993). Information technology: Threats and opportunities for small and medium-sized enterprises. *International Journal of Information Management*, *13*(6), 439–448.

Bocquet, R., Brossard, O., & Sabatier, M. (2007). Complementarities in organizational design and the diffusion of information technologies: An empirical analysis. *Research Policy*, *36*(3), 367–386.

Bruque, S., & Moyano, J. (2007). Organisational determinants of information technology adoption and implementation in SMEs: The case of family and cooperative firms. *Technovation*, *27*(5), 241–253.

Chang, M. K., Cheung, W., Cheng, C. H., & Yeung, J. H. (2008). Understanding ERP system adoption from the user's perspective. *International Journal of Production Economics*, *113*(2), 928–942.

Chen, J. S., & Li, E. Y. (2010). The effect of information technology adoption and design customisation on the success of new product development. *International Journal of Electronic Business*, *8*(6), 550–578.

Drew, S. (2003). Strategic uses of e-commerce by SMEs in the east of England. *European Management Journal*, *21*(1), 79–88.

Estonian Business and Innovation Agency. (2021). *Eesti ekspordimaht Skandinaaviasse ulatus kevadkuudel 1,3 miljardi euroni*. EASi ja KredExi ühendasutus. https://eas.ee/eesti-ekspordimaht-skandinaaviasse-ulatus-kevadkuudel-13-miljardi-euroni/

Fabiani, S., Schivardi, F., & Trento, S. (2005). ICT adoption in Italian manufacturing: Firm-level evidence. *Industrial and Corporate Change, 14*(2), 225–249.

Galliano, D., & Roux, P. (2003). Spatial externalities, organisation of the firm and ICT adoption: The specificities of French agri-food firms. *International Journal of Biotechnology*, *5*(3), 269–296.

Giunta, A., & Trivieri, F. (2007). Understanding the determinants of information technology adoption: Evidence from Italian manufacturing firms. *Applied Economics*, *39*(10), 1325–1334.

Hall, B. H., & Khan, B. (2002). Adoption of new technology. Working Paper 9730. NBER working paper series adoption of new technology. National Bureau of Economic Research, Cambridge.

Jorissen, A., Laveren, E., Martens, R., & Reheul, A. M. (2005). Real versus sample-based differences in comparative family business research. *Family Business Review*, *18*(3), 229–246.

Khazanchi, S., Lewis, M. W., & Boyer, K. K. (2007). Innovation-supportive culture: The impact of organizational values on process innovation. *Journal of Operations Management*, *25*(4), 871–884.
Kienzler, M. (2017). Does managerial personality influence pricing practices under uncertainty? *Journal of Product & Brand Management*, *26*(7), 771–784.
Koh, S. C. L., Gunasekaran, A., & Saad, S. M. (2005). A business model for uncertainty management. *Benchmarking: An International Journal*, *12*(4), 383–400.
Koh, S.L., & Simpson, M. (2005). Change and uncertainty in SME manufacturing environments using ERP. *Journal of manufacturing technology management*, *16*(6), 629–653.
Kusiak, A. (2018). Smart manufacturing. *International Journal of Production Research*, *56*(1–2), 508–551.
Latvian Business Guide. (2019). http://www.liaa.gov.lv/files/liaa/attachments/liaa_lvbusinessguide_a4_05062019.pdf
Latvian Forest Sector in Facts & Figures. (2022). *Zaļās mājas*. https://site-335431.mozfiles.com/files/335431/Figures_and_facts_2022.pdf
Liozu, S. M., & Hinterhuber, A. (2012). Industrial product pricing: A value-based approach. *Journal of Business Strategy*, *33*(4), 28–39.
Lursoft. (2020). *Enterprise register of Latvia* [Database]. Latvia. https://www.lursoft.lv/
Millen, R., & Sohal, A. (1998). Planning process for advanced manufacturing technology by large American manufacturers. *Technovation*, *18*(12), 741–750.
Ordanini, A. (2006). *Information technology and small businesses: Antecedents and consequences of technology adoption* (p. 174). Edward Elgar Publishing.
Osorio-Gallego, C. A., Londoño-Metaute, J. H., & López-Zapata, E. (2016). Analysis of factors that influence the ICT adoption by SMEs in Colombia. *Intangible Capital*, *12*(2), 666–732.
Premkumar, G., & Roberts, M. (1999). Adoption of new information technologies in rural small businesses. *Omega*, *27*(4), 467–484.
Profesionālais izglītības centrs T-senter. (2017, August 18). Kokapstrādes un mēbeļu industrijas produktu inovācijas un eksporta spēju paaugstināšana Veru apgabalā un Vidzemes plānošanas reģionā. *Tirgus ziņojums*. http://jauna.vidzeme.lv/upload/WF/I_LV.pdf
Raymond, L. (2005). Operations management and advanced manufacturing technologies in SMEs. *Journal of Manufacturing Technology Management*, *16*(7/8), 936.
Raymond, L., & Croteau, A. M. (2006). Enabling the strategic development of SMEs through advanced manufacturing systems. *Industrial Management & Data Systems*, *106*(7), 1012.
Rimpau, C., & Reinhart, G. (2010). Knowledge-based risk evaluation during the offer calculation of customised products. *Production Engineering*, *4*(5), 515–524.
Saksone, E. (2021). *Latvijas mēbeļu ražotāju un tirgotāju konkurētspēja vietējā tirgū un pasaulē*. In *Summary of the conference presentation*. Rīgas Tehniskā Universitāte.
Sheskin, D. J. (2011). *Handbook of parametric and nonparametric statistical procedures* (5th ed.). Chapman & Hall/CRC.
Shrouf, F., Ordieres, J., & Miragliotta, G. (2014, December 9–12). *Smart factories in Industry 4.0: A review of the concept and of energy management approached in production based on the Internet of Things paradigm* (pp. 697–701). In Proceedings of the 2014 IEEE International Conference on Industrial Engineering and Engineering Management, Bandar Sunway, Malaysia.
Song, J. B., Dai, D. S., & Song, Y. Q. (2007). *The relationship between change of organizational structure and implementation of advanced manufacturing technology: An empirical study* (pp. 103–119). In International Conference on Management Science and Engineering, 05-07 October 2006, Lille.
Svensson, D., & Malmqvist, J. (2002). Strategies for product structure management at manufacturing firms. *Journal of Computing and Information Science in Engineering*, *2*(1), 50–58.

Thoben, K. D., Wiesner, S., & Wuest, T. (2017). Industrie 4.0 and smart manufacturing-a review of research issues and application examples. *International Journal of Automation Technology*, *11*(1), 4–16.

Trost, S. M., & Oberlender, G. D. (2003). Predicting accuracy of early cost estimates using factor analysis and multivariate regression. *Journal of Construction Engineering and Management*, *129*(2), 198–204.

TSENTER. (2021). *Puidu- ja Mööblisektori Osatähtsus Eestis.* EWERSi Puidutöötlemise ja mööblitootmise kompetentsikeskus. https://tsenter.ee/puidu-ja-mooblisektori-osatahtsus-eestis/

Versli Lietuva. (2019). *Business sector.* Furniture *industry*. Retrieved June 29, 2020, from https://www.enterpriselithuania.com/en/business-sectors/furniture-industry/

Varanakis, S. (2017). The critical link between products related factors and machinery investments. *European Journal of Applied Business and Management*, *3*(3), 101–119.

Werner, C., Bedford, T., Cooke, R. M., Hanea, A. M., & Morales-Nápoles, O. (2017). Expert Judgement for dependence in probabilistic modelling: A systematic literature review and future research directions. *European Journal of Operational Research*, *258*(3), 801–819.

Wood processing and furniture production competence center. (2020). https://tsenter.ee/eesti-moeoebli-populaarsemad-ekspordimaad/

Zennaro, I., Finco, S., Battini, D., & Persona, A. (2019). Big size highly customised product manufacturing systems: A literature review and future research agenda. *International Journal of Production Research*, *57*(15–16), 5362–5385.

Chapter 6

Frugal Innovation as Intersection between Complexity of Early Cost Estimation, Machine Learning and Expert-Based Decision System

Julija Moskvina[a], *Anca Hanea*[b], *Tomas Vedlūga*[c] *and Birutė Mockevičienė*[c]

[a]*Lithuanian Centre for Social Sciences, Lithuania*
[b]*The University of Melbourne, Australia*
[c]*Mykolas Romeris University, Lithuania*

Abstract

This chapter discusses the empirical data analysis that will form the basis of the early pricing framework. It focusses on the complexity of furniture production and describes the historical production data collected from companies, along with the potential applications of machine learning for knowledge management purposes. The chapter then presents the results of machine learning for early cost estimation as part of a lean innovation that is affordable and accessible for small and medium-sized enterprises (SMEs). Finally, the chapter describes an experiment on the structured expert evaluation methodology, which shows that a well-formed panel of experts can increase the predictive power of machine learning solutions, particularly at extreme points.

Keywords: Early cost estimation; data usability strategies; pricing process; structured expert judgment; the Classical Model

Participation Based Intelligent Manufacturing:
Customisation, Costs, and Engagement, 151–238

doi:10.1108/978-1-83797-362-020241006

Complexity Management in Customised Furniture Manufacturing

The increase in complexity is driven not only by the overlap of traditional production processes, but also by new requirements, shorter product lifetimes and the customisation (uniqueness) of production orders.

The problem of optimising production costs is related to the complexity of customised production in furniture companies, which complicates pricing and other related issues such as planning and management. This chapter aims to explore how custom furniture companies identify the complexity of their operations and what strategies they adopt to deal with the complexity of pricing of new products. Research has become significantly important for furniture companies. It helps them understand the trends and priorities for sustainable development and resource utilisation in a region. It also assists in identifying production challenges that arise due to technological advancements and changing complexity perceptions. The importance of the regional dimension is growing significantly. Some regions are more inclined to invest in complex projects, particularly manufacturing. Additionally, regions like Lithuania, abundant timber resources and a low-cost yet skilled workforce, can offer competitive pricing in the market. Therefore, it is crucial for companies to adapt quickly and effectively to the ever-changing demands of customised production, in order to accurately set prices and remain competitive.

Complexity in Manufacturing

Complexity is a term frequently used to describe problems that are hard to define, measure, and quantify. This dynamic and intricate phenomenon is often characterised by adjectives such as unpredictable, unexpected, and subject to personal interpretation. It does not conform to conventional standards and is difficult to predict due to its constantly changing nature (Elmaraghy et al., 2012). There are four factors that contribute to the issue of complexity: diversity, heterogeneity, dynamism, and opacity. Diversity refers to the analysis of the number of possible elements, while heterogeneity focusses on the differences between the elements. Dynamism refers to the observation of the environmental conditions, and opacity involves the analysis of the perception of the differences between the different situations, identification of the elements, and their dependence (Colangelo et al., 2018).

The simplest way to define the dimensions of complexity is to distinguish between the effects of external environmental and internal factors on the specific complexity factors in manufacturing (Elmaraghy et al., 2012). These two key dimensions allow the dividing line between the factors that determine complexity (Vogel & Lasch, 2016) and facilitate the prioritisation of management measures. Intrinsic complexity arises from organisational structure, intra-organisational processes and product complexity, while extrinsic complexity is primarily related to market fluctuations, customer behaviour and legal regulation. Similar principles are recommended for the analysis of product complexity, which is often found in customised production. It is recommended to distinguish between two levels when evaluating and controlling product complexity: external factors such as functionality determine the first, and internal factors determine the second (Marti, 2007). This approach to complexity makes it easier to quantify and manage complexity.

Complexity management (strategies to cope with complexity). Comprehending the sources and factors that contribute to complexity can enable the management of complexity. Although complexity management is commonly viewed as an effort to minimise complexity, it is not always feasible to completely eliminate complexity in actual production scenarios. Furthermore, attempts to evade complexity can occasionally have detrimental effects on the quality of a product or service. Nevertheless, 'current approaches to complexity management focus on product complexity, in particular product modularisation and variation management' (Kluth et al., 2014).

There are different ways to deal with complexity. According to Jäger et al. (2016), the four most common strategies are: avoiding complexity, reducing complexity, generating complexity, working with complexity, and pricing complexity.

- *Avoiding complexity* involves defining performance in a standardised, predefined manner, such as standardised approaches, breaking down processes into smaller components, and creating standardised management units.
- *Reducing complexity* can be achieved by reviewing the process to identify non-value-added elements and gaps in decision support tools. The most popular approach to complexity reduction is to remove unnecessary elements from the process. However, this strategy may not always be feasible and may not necessarily lead to process simplification. Another way to reduce complexity is to group process steps into manageable best achieved through a hierarchical organisational structure that enables excellent control of complexity, especially if the structural units are independent of each other and their functional roles can be kept entirely separate.
- *Generating complexity* is a term to describe situations where new innovations are introduced, new products are launched, or new markets are explored. Whenever an innovation is introduced, it typically adds a new element of complexity, which in turn needs to be managed and controlled.
- *The pricing complexity* strategy can be a useful approach to determine the optimal price for a product based on the customer's perceived value. By understanding the customer's willingness to pay, businesses can set prices that effectively balance profit margins and customer satisfaction.

To effectively manage complexity in a corporate environment, it is essential to consider the level of corporate maturity. This can help determine the company's ability to handle complexity and make use of advanced management tools. A higher level of maturity determines the company's readiness to adopt more sophisticated management tools (Kluth et al., 2014).

There are various IT solutions that can aid in the implementation of a complexity management strategy. These solutions include Product Data Management (PDM), Product Lifecycle Management (PLM), Enterprise Resource Planning

(ERP), and Customer Relationship Management (CRM). They are also known as knowledge management. These solutions complement each other and add complexity that can be challenging for companies to manage.

IT solutions are usually developed through modules, each of which is designed to accomplish a specific objective. Since these modules are created by specialised IT experts, modularisation can improve the quality of IT solutions and reduce complexity. However, it also increases the cost of the management system compared to integrated architecture. Therefore, it is important to choose the optimal degree of modularisation that does not significantly increase the cost of the system or the cost of adapting the product (Jäger et al., 2016).

In the field of product design, the most effective approach to simplify complex systems is to create modular control systems, which can then be integrated into a larger integrated architecture. Although this may appear to be an ideal method for managing complexity in customised manufacturing, in reality, modularisation introduces additional challenges that are closely linked to product cost, design quality management, the manufacturing process, supply chains, and potential risks throughout the product's lifespan (Umeda et al., 2008).

Cost Estimation as Complexity Management Tool

Managing complexity is a significant challenge in customised production, especially as the number of customised orders continues to rise. To tackle this problem, companies need to control complexity effectively, especially at the early design stage, as inaccuracies in cost estimation can have a negativity on production later. It is, therefore, crucial for companies to make reliable cost estimates at the early design stage while selecting from different design options.

According to scientific sources, the most commonly used costing method involves breaking down a product into its individual components and calculating the cost of each part based on the time and money required to produce it (Helbig et al., 2014). There are three categories of data, with varying levels of uncertainty, required to calculate the price: factual data, estimated values and application parameters (Helbig et al., 2014).

Facts are data that are well-defined and have no hidden uncertainties. They are accumulated at the end of production and can be obtained from fixed databases. Estimates, on the other hand, are based on data that cannot be directly collected from databases, such as working hours. However, additional observation data can be used to estimate these values. It is important to note that data based on experiments may introduce a bias that can lead to uncertainties. When constructing possible forecasts, application parameters often include several scenarios.

Custom production has always relied on traditional cost estimation methods, which mostly involve using analytical tools to deal with complexity.

Controlling complexity through manual data analysis has proven to be unsuccessful. Such studies are time-consuming, lack precision and completeness, and heavily rely on the experience of the staff. However, depending on the employee's experience can put the company at risk when the employee leaves. While regression analysis techniques are commonly used for pricing in most manufacturing sectors, a growing number of industries are now turning to artificial intelligence-based approaches for better accuracy and completeness. These include manufacturing, software engineering, process engineering, the construction industry, and research. The most popular analytical tools currently being used are analogue cost estimation (ACE), bottom-up estimation methods, and computational technologies combined with artificial intelligence (AI).

Specificity of Complexity Management in Customised Furniture Manufacturing

Some Methodological Highlights Coding and Clustering of Qualitative Empirical Data

To explore the complexity reduction strategies adopted by firms in the studied furniture sector, we conducted a qualitative content-based interview as described in the Methodology section. In this section, we will focus on discussing only the dimensions of qualitative research that are directly related to the research question (complexity management).

After reviewing the data, it was found that certain statements were more significant to the respondents than others. The most frequent statements were related to the customer's role in furniture production (68 responses), information systems (59 responses), and the need for information systems to resolve installation and production issues (129 responses). These issues are particularly related to the production of nonstandard furniture.

The most significant criteria relevant to furniture production were communication between customers and managers (46 responses), production time and quantity of materials (71 responses), materials used (78 responses), cooperation (41 responses), and innovation (49 responses). Respondents also highlighted pricing and price-related problems (99 responses). Among the most commonly mentioned topics, the ones that stand out are price (377 mentions), data (238 mentions), manufacturing/product (282 mentions), and material (187 mentions). Further examination of character units exceeding eight in length revealed that mentions of cost, percentage, and calculation procedures were the most frequent. Notably, respondents tended to use verbs such as need, can, must, and do in conjunction with time specifications (e.g. need more time and must do now), often referring to themselves in the first-person plural. In the context of cost estimates, the most commonly discussed factors were manufacturing time and volume (count = 71), material (count = 78), guessing (count = 32), calculation (count = 99), and client-producer communication (count = 46). These are the main aspects associated with cost estimation. The analysis of data indicates that the primary issue is the estimation of the price of customised furniture. This category also includes troubleshooting in the manufacturing of customised furniture, such as pricing (count = 8), offer formulation (count = 1), and pricing participants (count = 15). The major groups in the category of organisational structure are cooperation (count = 41) and management (count = 17). Within the category of inclusion, most discussions arise over communication (count = 22), expectations (count = 28), implementation of innovations (count = 49), planning (count = 30), and meetings (count = 30). The manufacturing category is the weakest and the least prominent, with a single distinct group of product complexity (count = 12). The low number of responses in this category suggests that respondents are hesitant to reveal the advantages, deficiencies, and solutions related to manufacturing. The analysis of the relationships between the respondents' statements reveals that some concepts are strongly linked while others are weakly linked. The concepts of working environment, collaboration, the similarity of pieces of furniture, management, team building, pricing participants, and independence are the most prominent and have the highest number of links with other concepts. Additionally, there are strong nodes for the concepts of counting, meetings, director influence, customer-manager communication, and procedural compliance.

The concepts of making an offer, dependence on other professionals, and five-minute meetings had the least influence and the fewest links. The concepts of change agents and markups also showed the lowest correlations. Two interpretations are possible here. The first one is that the actors involved in proposal generation are likely to have only one relationship with the communication between the client manufacturer, and other processes are not affected by this criterion. Change agents also have a minimal impact, except for a few processes such as innovation, meetings, production adaptation, communication, team building, etc.

The second interpretation is that criteria such as calculation, management, collaboration, and working environment have an impact (both negative and positive) on all the other criteria.

Formulating Strategies for Managing the Complexity

It has been noted that the topics discussed most frequently are not random, but rather point towards areas that are becoming increasingly complex and need attention. These conversations suggest that the identified issues, like materials or design, have been discussed in detail and agreed upon at all levels, and that they have been taken into account.

When it comes to managing time and volume of materials, things can get complicated. Factors such as need, cost, lead time, and quality also come into play, making the situation more uncertain. With the level of uncertainty, people are willing to discuss these issues, showing that they understand how to address them in order to promote further development. Despite the high level of uncertainty, the respondents show a willingness to discuss these issues, which indicates a clear understanding that uncertainty is a barrier to further development.

- *Complexity Management Strategy No. 1: ERP Implementation:*

  Companies that work with customised manufacturing often make the mistake of adopting a complexity reduction strategy without realising the extent of complexity and the possibilities to control it with existing tools. Companies often adopt existing ERP systems that focus on planning and resource monitoring. However, the implementation of such systems can reveal areas where they simplify complexity and where they add to it a question that possesses significant expertise in enterprise resource management, specifically in managing material resources. Consequently, their production data is seamlessly integrated into their ERP systems. Colangelo et al. (2018) have put forward that ERP is an effective tool for identifying and predicting material resources. However, there are certain areas including planning, forecasting, design and technology compatibility with customer feedback, which are managed by independent departments and fall outside the scope of the ERP system. These departments do not share the data collected by them. Although modularisation is a growing trend in IT system design, it can actually lead to an increase in the number of functions in the ERP system, it is not possible to identify the number of functions in the ERP system. Modularisation is a popular trend in IT system design that aims to address complexity issues in manufacturing. However, it has been found that this approach reduces the efficiency of ERP systems. The study confirms that the traditional linear ERP design is only suitable for capturing deterministic production factors and is no longer capable of addressing the complex challenges of customised manufacturing, including the challenge of price prediction.

- *Complexity Management Strategy No. 2: Marketing and Customer Relationship:*
  Customer relations is a structured process that mainly involves communication. The customer usually expresses their preferences, needs, and budget, which leaves little room for negotiation, especially when dealing with famous brands. The organisation provides feedback to the client, usually focussing on the essential factors, while omitting the details that may affect production complexity. From the organisational viewpoint, the customer relationship process concludes with a planning phase, which is then followed by staff's more targeted participation and reduced interaction with customers. This clustering of planning and encounters suggests that complexity is experienced and understood, but not fully identified.

- *Complexity Management Strategy No. 3: Search for Similar Patterns in Product Design (Analogous Recognition)*
  Experts are currently searching for ways to fix prices faster due to the increasing uncertainty of new orders. Although it may appear that using similar models could help classify previous experience and reduce uncertainty, it may not always be possible to identify similar patterns, especially in customised production. Due to the unique design and production process for each order, identifying, classifying and creating similar models can be a complex task. The search for similar patterns is closely tied to identifying tangible elements, which can be endless in a custom production process. This is why experts refer to the process of examining the accumulated experience of production as 'feeling', emphasising that the complexity is not fully recognised.

- *Complexity Management Strategy No. 4. Organisational Structures:*
  Due to the failure or shortcomings of ERP and similar models in managing complexity, companies are now turning to soft management tools such as flexible organisational structures. Such structures promote employee

involvement and ensure the compatibility of production processes. They are often supported by a physical infrastructure in which different professionals work closely together, including designers, supply managers, product managers and planning managers. This organisational structure helps control complexity and enables further specialisation of customised production. One selected company observed that this structure has a positive effect. Another company operates based on a clear system of function allocation, hierarchy and defined roles which reduces customised production and operates via classified analogue models.

Early Cost Estimation: Managerial Perspective

Early cost estimation plays a significant role in both traditional manufacturing and customised manufacturing. However, early cost estimation is particularly important for producers of customised products due to the unique challenges and complexities involved in meeting individual customer requirements while maintaining profitability for the producer. The time limits challenge producers to assess the multiple factors related to assessing feasibility, providing accurate pricing, optimising designs, controlling costs, managing the supply chain, and ensuring customer satisfaction. So, the complex task of early cost estimation requires special tools, skills, and approaches. Those aspects are reviewed in the chapter in general and more specific ways, reflecting the experience of Lithuanian furniture producers.

Managerial Challenges Related to Uncertainty

The production of unique products involves a high degree of uncertainty, which requires the manufacturer to make immediate decisions regarding the timing and pricing of production. In order to meet customer requirements, manufacturers tend to reduce production costs and production times. The product configuration process determines the production process itself and has an impact on price and lead time as well as on product quality. The production time and cost of a project/order are often influenced by factors such as changes in the production process, delays, which lead to a high degree of process uncertainty (Zennaro et al., 2019).

Dean et al. (2009), based on a literature analysis, identified the main stages of making a product according to a customer's drawing in mass-customisation manufacturing. First, the aim is to identify customer requirements and translate these requirements into design specifications. Subsequently, in the product design phase, families of representative products are modelled and individual products meeting the customer requirements are identified. Finally, a production planning and control system is applied to the new product. All the elements described above were recognisable in the activities of the companies we have analysed in this chapter, with the exception of the second phase, where a single version of the product to be manufactured is usually presented to the customer.

Authors on customised production (Rimpau & Reinhart, 2010; Trost & Oberlender, 2003) emphasise the risks associated with early pricing. Early pricing often becomes the basis for determining the final financing of a project/order.

The calculation of total cost may be subject to risks during the offer specification process due to incomplete information about the product and production processes, as well as possible price and cost fluctuations after specification. Pugh (1992) used a top-down approach to estimate costs for complex projects with incomplete information and uncertain lead times.

The information available in the initial phase is insufficient to determine the price more accurately: the accuracy of determination is often low. When it comes to pricing and costing, there is the problem of uncertainty inherent in individual production and so far, difficult to overcome. This uncertainty relates to the need to provide the customer with a price before production begins (Kingsman & de Souza, 1997; Zach & Olsen, 2011; Zennaro et al., 2019). As Kingsman et al. (1996) stated, the costing and pricing phases are intertwined and typically occur simultaneously. When it comes to custom-made furniture, the initial price quoted to the consumer often serves as the final selling price. This is because the consumer is usually involved only in the planning phase, as there is no room for negotiation with the consumer. Trost and Oberlender (2003) identified five factors, based on multivariate regression analysis, to predict the accuracy of the early price: basic process planning, team experience and cost information, time available for the preparation of the estimate, space requirements, offer/bidding, and the working environment. Authors (Song et al., 2002; Wang et al., 1998) writing on time and pricing also point out that the individual production environment is heavily influenced by uncertainty, so that offer and bid preparation becomes a strategic phase on which firms' performance depends.

According to contemporary literature, there are three main pricing strategies that can help overcome pricing-related issues: cost-based pricing, competition-based pricing, and value-based pricing. The value-based pricing strategy, which aims to create value for the customer, is believed to be the most profitable in any industry. However, it has not been fully operationalised thus far due to its lack of rigour and definitiveness (Liozu & Hinterhuber, 2013). Liozu and Hinterhuber (2012) reveal that businesses differ in their decision-making process, the factors influencing price-setting, the organisational structure, the diffusion of pricing capabilities and leaders' behaviour depending on the chosen pricing orientation. Based on their analysis of small and medium-sized enterprises (SMEs) in the USA, the researchers discovered that companies that adopt value-based pricing strategies tend to depend on market research, scientific pricing methods, and expert advice to set their prices. On the other hand, firms that use cost- and competition-based pricing strategies rely on their experience, knowledge, and intuition to make pricing decisions. Liozu et al. (2012) noted that speaking about the value-based approach one should bear in mind the fact that estimations of pricing capabilities are impossible without an understanding of manager knowledge about the customer value. Their research has disclosed that manager knowledge about and understanding of value-based pricing may considerably differ. Paraphrasing the findings of Liozu et al. (2012), before developing a company's pricing orientation and strategies, it is crucial to assess the company's actual pricing capabilities and the managers' understanding of the pricing process. All pricing decisions require a certain amount of data to support them. According to

Kienzler's (2017) research, companies that follow a cost-based approach require less effort, subjective data, and have low to medium uncertainty. Firms that follow a competition-based approach face a medium level of uncertainty, subjectivity, and data collection attempts. On the other hand, a value-based approach typically involves uncertain, subjective, and difficult-to-collect information. This chapter considers information management as a key factor in a firm's pricing capability.

Role of Data in Early Price Estimation

Smart manufacturing enables the proactive management of manufacturing companies by leveraging data. Data play a crucial role in providing companies with insights onto what actions to take and when to take them. This data-driven approach to business management facilitates real-time decision-making (Santos et al., 2018; Wang et al., 2018). To meet customer needs better, quicker and cheaper, manufacturers have little choice but to turn to information technologies. The recent decades have been earmarked by the transition from 'the information age' to 'the age of big data'. The difference between the two manifests in technological achievements, namely the shift from data storage in computer systems and data processing by information systems to the rise of the Internet of Things technologies, cloud computing, big data analytics, and artificial intelligence (Balakrishnan et al., 1999).

The need to combine personalised customisation with low costs, similar to mass production, is a fundamental aspect of made-to-order manufacturing. Additionally, the processes of cost estimation and price-setting occur simultaneously, highlighting the importance of making early pricing decisions for made-to-order businesses. Therefore, it can be assumed that a thorough analysis of data forms the foundation for estimating product costs in made-to-order production. Researchers working on time and pricing tend to work on indicators that are particularly relevant in individual production: due time, lead time, and job processing time. The likelihood of a successful order depends to a large extent on these variables. Kingsman et al. (1996), in a cross-sectoral study, constructed a Strike Rate Matrix to estimate the probability of a successful order. The most important factors here are price, lead times, and a decision support system to coordinate sales and production. Zorzini et al. (2008) and Hegedus and Hopp (2001) have proposed models to determine the optimal due date in make-to-order production. Their model included elements such as service level, delay costs, inventory costs, uncertainty during the production process. Elfving et al. (2005) analysed the causes of long lead times and their relation to bidding as it can significantly increase resource consumption and lead to waste in the order fulfilment process. Grabenstetter and Usher (2015) confirmed that timing is a key factor in an engineering-to-order environment.

To achieve successful operation, companies are adopting a comprehensive approach to production that combines advanced IT solutions, automation, and increased production speed and reliability. This integrated approach, known as smart manufacturing, involves the integration of manufacturing activities with information technology and automation, thereby requiring manufacturers to enhance their data processing capacity (Kusiak, 2018; Svensson & Malmquist, 2002; Thoben et al., 2017).

With the advent of big data, the advancement of information technologies has greatly enhanced the producer's capacity to collect, store, and analyse data. As a result, manufacturing enterprises of all sizes, including small and medium-sized ones, can now harness the value embedded within this data (Brynjolfsson et al., 2023; Chou et al., 2010; Zolas et al., 2021). Recognising that effective data management can significantly improve competitiveness by providing deeper insights into processes, competitors, suppliers, customers, and capabilities, we decided to examine the data collection, storage, and processing solutions within selected Lithuanian customised production enterprises.

Researchers (Rimpau & Reinhart, 2010; Trost & Oberlender, 2003) warn about the potential risks associated with early price estimation in customised production. Early costing is usually the foundation for approving the final financing scheme for a project or order. However, estimating total costs at an early stage can lead to the risk of inaccurate data about the product and the production processes, as well as the risk of possible post-specification fluctuations of prices and costs. Data available in the initial phase is usually insufficient to set an exact price. The result is often a low-level accuracy. To evaluate the costs of complex projects manufacturers are looking for the ways to embrace usually incomplete data and uncertain information on manufacturing lead time (Pugh, 1992). On the basis of multivariate regression analysis, Trost and Oberlender (2003) distinguish five factors to predict accuracy of early cost estimates: the basic process planning, experience and data about the costs, time necessary to produce estimates, local requirements, bidding and working environment. This suggests that in order to benefit from the value of data, manufacturers should be able to organise automated data collection and manage and analyse a large amount of data.

Scientific literature on information management support systems in mass customisation production emphasises the necessity of centralised data storage facilitating the process, planning and control of manufacturing (Zennaro et al., 2019). Boothroyd and Fairfield (1991) noted that major IT instruments in large scale customised product manufacturing are used to make descriptions of product specifications and develop product simulation models. On the other hand, IT tools were unusual in planning and control in the analysed sectors as their use takes significant investment and recurring standardised data while engineering-to-order and made-to-order manufacturing is earmarked (Azevedo & Sousa, 2000) by numerous uncertainties and diversity in terms of time and used components.

In the context of pricing and cost estimation, specific IT solutions can further enhance smart manufacturing capabilities. Variety of IT solutions and methods are available for the manufacturers that can contribute to pricing and cost estimation optimisation. Advanced data analytics techniques such as machine learning and predictive modelling can help analyse historical pricing and cost data, identify patterns, and generate accurate pricing estimates. The bigger manufacturers use real-time data integration solutions that can be applied within production systems, supply chain management systems, or market data. The most suitable

IT solutions usually are based on the specific needs of manufacturers, industry requirements, and available resources. However, the sufficient knowledge about the latest advancements in technology and research can help manufacturers make informed decisions regarding their pricing and cost estimation strategies.

In view of the ideas listed above, the qualitative analysis of the price estimation process and practice at different managerial level was performed. Before presenting the results obtained from the interviews, we present methodology that stands beside: sampling methods, instrument for data collection, and analysis method.

Methodology of Qualitative Analysis

This research is based on qualitative data described in the methodology section. Further specifics are provided for clarity.

Two furniture manufacturers were selected for a study on pricing strategies. They were chosen based on their territorial affiliation (region, city, from Vilnius and Kaunas respectively), specialisation of the business (setting up shops), and specific sales strategies used by the companies (direct sales, sales through intermediaries). The companies produce similar products (shop counters), but use different sales channels, which means they use different business models and are likely to apply different pricing principles. One of the companies has an in-house IT department responsible for IT services (computing, data transfer), while the other relies on outsourced IT services. In order to collect detailed data on the IT tools used in the pricing processes, representatives of the outsourced IT services company were also interviewed. The IT company provides relatively expensive services, which is likely to be reflected in the manufacturer's pricing system. The scope of the survey is limited, but the interviewees represent all levels of management and production in order to clearly define the organisation's management, production and pricing procedures.

Data Analyses Instrument

We developed a tool for collecting qualitative data based on organisational theory and literature on pricing strategies. The tool was constructed using the price bidding and production-to-order process, which reflects the manufacturing business processes and supports the marketing, bidding, and production phases (Chou et al., 2010). We considered the pricing process as an integral part of production, despite occurring before the actual production process and only manifesting after delivering the product to the customer. Therefore, we collected data on cost/pricing orientation to identify who is involved in pricing assessment, when, and how.

The diagram of the instrument is shown in Fig. 6.1.

In order to create comparability and superposition effects, the researchers used identical questions to learn about each individual production process. They also designed specific questions to identify the primary categories of pricing research, which include cost/price estimation and decision-making factors. The flowchart used to develop the research questions is presented in Fig. 6.2, and the questions themselves are listed in Table 6.1.

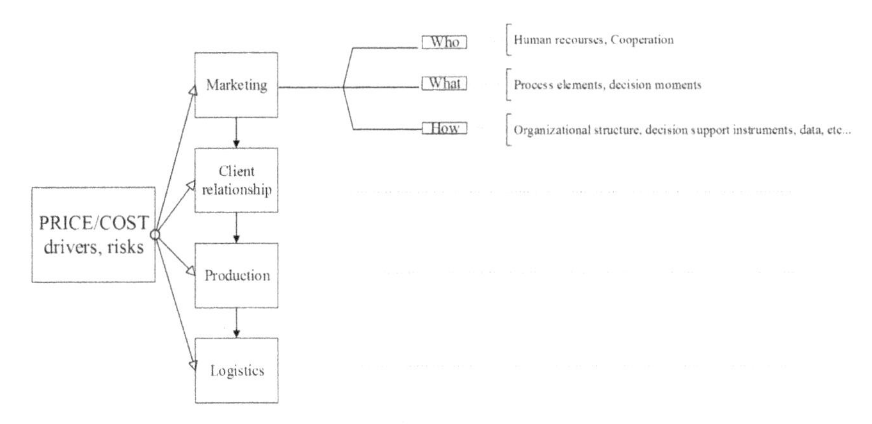

Fig. 6.1. The Diagram of the Qualitative Data Categorisation.

Fig. 6.2. Word Frequency Cloud.

Table 6.1. Interview Questions Made in Compliance with the Research Instrument Scheme.

DEMOGRAPHY	1. What is the size of the company? 2. What does the company produce? 3. What product/service are you proud of? 4. Who is your client (customer profile)? 5. Who are your competitors? 6. What proportion does your made-to-order product comprise in the total output?			
CATEGORIES PRICE SETTING	**PRICE/COST**	**ESTIMATION**		**DECISION MAKING FACTORS**
		Cost Evaluation	**Cost Tooling**	
Who	What specialists are involved into cost data collection?	Who performs the initial cost evaluation?	Who initiates IT innovations applicable in cost calculations?	What could/should be employee contribution to cost estimations?
Human resources Leadership	What specialists contribute most to price setting?	Who is in charge of cost/ price estimation?	Who is the most likely to use available IT tools?	Which employees make estimations most often taken into consideration?
What	What pricing associated marketing/ manufacturing/logistics stages take place in the company?	What does the cost/price depend on?	What IT support tools do you use?	Which stage of manufacturing or a non-manufacturing sphere is the most important for the success of the project/order? How does the accuracy of pricing affect the success of manufacturing?

Process elements Decision moments	When does it become important to have the price set?			
	When is the preliminary price/cost set?			What factors predetermine inaccuracies in price estimation?
How	In what order are the stages arranged?	How important is the accuracy of the estimates?	What instruments do you use in each stage of pricing?	What factors are the most important in integrating IT and arrangement of stages?
Process flow Organisational structure	How successful are you in setting preliminary prices/costs?	What problems arise from product complexity and price estimations?	What data do you collect?	How do IT innovations and data storage help in price estimation?

Data was gathered through semi-structured interviews with 18 respondents between October 2018 and April 2019. To ensure completeness, some respondents were re-interviewed as needed. In some cases, multiple respondents were present during the same interview. The respondents represented all aspects of the bespoke production and management chains, including the Bespoke Furniture Production Manager, the Finance Director, the Project Unit Managers, the Project Managers, the Production Manager, the IT Support Manager, the ERP Engineer, the Design Engineers, the Designers, the Production Planner, the Production Preparation Specialist, the Craftsmen, the Furniture Assembler, and representatives of IT product developers (product manager, manager). The idea was to have representation of all possible organisational levels of furniture manufacturing businesses. Previous research (Navickienė & Mikulskienė, 2019) found that furniture manufacturers often evolve by developing their organisational structure and establishing a three-level hierarchy (the first level is management and leadership, the second level represents specialised management functions, such as managing individual departments of control and production, and the third level comprises specialists). Jobs of and level representation by the respondents are given in Table 6.2.

The interviews were conducted with an independent expert who have experience in both mass and customised production to validate the data and ensure the uniqueness of the pricing procedures of the chosen companies. The goal was to determine if these procedures can be used to describe the overall situation at a company level. To ensure consistency and accuracy of the investigation, the same person conducted the interview. Out of the 29 audio-recorded interviews, 26 were chosen for analysis. The interviews had a total duration of 22 hours and 40 minutes and were transcribed, resulting in 462 pages of transcription.

Table 6.2. Classification of the Interviewed Representatives of Furniture Manufacturing Businesses. Organisational Levels Are Identified as Described by Navickienė and Mikulskienė (2019).

Organisational level	Company A	Company B
I	CEO Finance director	Head of Project Department
II	Project manager	Project manager Pricing manager Production manager Head of IT support
III	ERP engineer Designer-constructor Production planner Production preparation specialist Forewoman Furniture assembler	Designer-constructor

Analysis Methods

Qualitative content analysis of the interviews was conducted using a number of steps, including cluster analysis, comparison of themes in both cases and summarisation. The text of responses was grouped into categories and subcategories by means of a qualitative data analysis software package NVivo. Then, the frequency of the used concepts was measured to identify major problems in the field (Fig. 6.2).

In order to ensure the quality of the coding, a double coding approach was used where two different researchers coded the dataset simultaneously. The results of the coding were the each other to ensure accuracy. To validate the selected categories, the authors read the transcribed texts and notes. They tagged the frequency and forms of expression of the identified categories chronologically. Then, the categories were coded and grouped into subcategories. This entire process was repeated twice to validate the categories. While the coding and grouping were carried out by researchers using inductive logic, any differences between the primary and secondary data between the two groups of categories in the original and in the second analysis were handled by the researcher on the basis of abductive logic. During the initial stage, the analysis identified the most frequently reiterating words and word clusters. The identified dependencies between words (cluster analysis) allowed making assumptions about the key elements that constitute pricing process in the sample companies:

(a) identification of the price/cost of an order;
(b) availability of the data on the operating time to employees responsible for production (managers and designers-constructors);
(c) identification of the restrictions on materials, their processing/production and time.

The interview texts were coded according to the following categories (nodes): who undertakes the price calculation, how the price is calculated and what is calculated.

Inductive analysis was used to compile a list of codes in open coding, without a predetermined research subject. As the code list extended, the need to compile additional categories, namely, the intuition and client liaison, arose. Having compared the obtained results with those of previous research during the next stage of analysis, we noticed that the aspect of time had remained undisclosed and therefore, the list of coded was extended to include the time category.

The research results are discussed further presenting characteristic respondent quotations, which may best reflect the identified dependence and the pricing interpretation. Research categories are represented by quotations by representative of the sample companies and validated by the independent expert position.

Findings from the Selected Enterprises

The Estimation Preliminary and Final Price: Stages and Participants of the Procedure

Medium level (II) business representatives point to the specific nature of made-to-order manufacturing in terms of time and price. Thus, markups applicable in mass production naturally lose their importance here. Like all sectors of customised manufacturing, manufacturing of make-to-order furniture is specific, that is, it produces unique products (e.g. 'it was a really complex item, hemisphere, curved to fit a curved window, including sliding curved drawers' BII production manager), and the number of such products can vary from one to several units (e.g. 'we produce a great deal various products, but never identical' BII production manager).

As each order is unique, manufacturers have no choice but to set prices for each item individually. The process is rather stressed as it predetermines the prospects of further cooperation with the client. It also requires prompt reaction, a certain amount of precedent data, specialist expertise and the sensation of market. The job often takes endeavours of several people from all organisational levels. The ability of the designers-constructors (III level) to match customer's vision with the lower production costs must be stressed as an important part of the price estimation process.

Besides the expenses involved, the manufacturer needs to consider the potential benefits of the transaction, despite the reduced sales value. These benefits may come in the form of successful collaborations in the future or increased volumes of future orders. This is especially important when dealing with a potential new customer. AI CEO: 'It sometimes goes that the client and the order is attractive and we make our best to give an attractive price …'.

Regarding pricing, the independent expert emphasised the importance of considering business interests and long-term plans. Before making a decision on a specific order, he evaluates

> whether the company interested in the order or not, whether the product is viable or not, whether the company needs the order or not, how much investment the order takes, whether the company is likely to benefit in terms of know how.

During the interview, the expert suggested a simple algorithm of how to make a decision on an order: 'You simply evaluate: if you need it or not; if you have the necessary capacity or not; if you can make it or not; if you can take the risk or not'.

According to the analysis, the price calculation process of the analysed companies can be divided into two stages. The first stage involves estimating the costs of inputs and determining the production lead time. The second stage involves determining the markup factor based on the agreement with the customer, the customer's brand, and the number and complexity of the units to be produced (usually expressed in terms of labour time).

The interviews reveal that the cost estimation phase and the price setting phase happen simultaneously and are practically indistinguishable. These findings are in line with other findings in the literature. In the first stage, the selected firms provide information about material costs, production and processing time. Then, in the second stage, the markup is determined based on the agreed upon factors (Kingsman et al., 1996).

- AII project manager: 'They [AII designers-constructors] estimate supply costs and operating times. Then [AI CEO] feeds the numbers to the system and adds a certain markup. <…> The markup is added depending on the operating time'.

- BII project manager: 'Having received the cost evaluation, I make an analysis of the numbers and of how much work the production takes <...> analyse<...> what the final sales price is likely to be if I add a certain markup coefficient. <...> It depends on the brand'.

Manufacturers determine final prices based on customer demand and the unique nature of the product, rather than comparing to market prices:

- BII pricing manager: '... speaking of the key pricing factors in customised manufacturing, when your client says "I want my own design", the importance of the market factor decreases dramatically'.

In later stages, the manufacturer has to evaluate the cost of the project implementation once again in view of the company's capacities:

- BII project manager: 'It is not enough just to negotiate the order. You still have to turn it into a product'.

Analysis of the frequency of the words *price, prime cost* and *costs* in quotations by business representatives reveals that only a single person in a company access to the entire range of data on pricing (Table 6.5). However, it is not the same person who is responsible for setting the final price. It has already been mentioned that the pricing process involves a great variety of specialists and managers, such as designers-constructors, ERP engineers, project managers, directors and pricing managers. Regarding Company A and Company B, there are differing levels of responsibility when it comes to determining pricing. In Company A, the production planner, who is at organisational level III, is responsible for considering the main factors that should be taken into account when calculating the price. However, in Company B, it is the pricing manager at level II who has this responsibility. Upon closer examination, it becomes clear that in both companies, the final decision on pricing is made by a single individual who is at the highest organisational level (level I).

- BII project manager: 'having the material data, the [BI] knows certain price variations and uses their competence to set the final numbers. They have a better understanding of what they can charge ...'.
- AIII ERP engineer: 'The prices are only approved by [AI CEO] '... and the calculations are sometimes reviewed by [AIII production planner]'.

When it comes to creating a customise furniture, production price is often the same as the final selling price. While it can be challenging to determine a price due to the many unknowns involved, the pricing is typically based on the estimated cost of creating the product. The process usually involves two key employees: the designer and the cost estimator, who often work together in teams. The markup on the final price depends on factors such as the complexity of furniture, the product, the number of products ordered, and the specific needs of the customer. Additionally, the company's leadership usually needs to be assessed to ensure that the customer can take into account the cost of the customer's needs.

Pricing Procedure

Respondents generally listed components of the net price when asked about how the price is formed. Components of the net cost also constitute the core of corporate resource, that is, indexes registered in the company primarily reflect resources necessary to produce the product. According to the study, there are three key elements that need to be controlled at the beginning of the pricing process: materials, labour time, and production readiness (planning). The most detailed information gathered during the interviews was related to the estimation of costs for the materials used, including the types of materials and additives, groups of materials, and material characteristics. The estimation of materials is particularly cost estimation since it occurs at the initial stage of price/order estimation and predetermines the subsequent cost and cost estimation. AIII designer-constructor: '… if any questions or uncertainties [how much and what materials are needed], the process stops'.

An independent expert has acknowledged that material costs, including materials and rates, are a crucial factor in calculating the cost of a product. Other factors that contribute to costing include pre-production and logistics. Manufacturers take into account the complexity of a project, as higher complexity requires longer production times and may result in higher salaries. Longer transport distances can also increase waiting times and lead to additional time costs. The independent expert acknowledged that for the managers to make a decision on pricing, it is essential to know 'what is it takes to produce the article as it may entail certain design solutions, limited access, specific tools …'.

The above mentioned factors are directly related to the issues of time costs. Measuring production operation time is a strategic factor that directly affects company competitiveness:

- BII pricing manager: '… a minute spent on the same operation may have a different cost in different companies'.

Time is considered to be a key price determinant in customised manufacturing (Zennaro et al., 2019). The assumption has been acknowledged by the interviewed managers as well (BII pricing manager: 'time multiplied by a coefficient equals a price'). The word *time* is one of the most frequently used in the interviewee quotations. It has been identified in most of the larger clusters of words. The category of time is appreciated by all representatives independent on their organisational level. However, conversion of time costs into monetary costs only takes place on the operational level (operating times are used to calculate wages).

The interviews revealed that among all the possible time measures, only production time is collected and stored, despite the significance of the time category. It is only in company B that some production procedure time is captured and stored automatically. However, at the same time it is emphasised that the remaining time values are not recorded:

- BII pricing manager: 'we refer to certain operating times that are collected by manufacturing control software. < … > no other time values are available'.

The storing of operating times is not systematic, particularly where the used tools (loom) are not automated. It is common for companies to speculate about the time value of operations. This was highlighted by the Head of Finance (AI) and pricing manager of BII. These uncertainties are more often observed by senior management and those who are responsible for making the final pricing decisions. Also, a conversation with AIII ERP engineer has revealed that durations of certain operations are often set in advance ('preparation can take no longer than... this amount of production must be completed no later than...') or, as we said, are predicted:

- BII pricing manager: 'There are no set operating times in our company. [BII production manager] uses common sense to evaluate how long the process takes and sets workloads accordingly'.

It is important to provide realistic estimates of the time required for a task to avoid complications in later stages of production. Unrealistic estimates could problems and delays:

- AIII forewoman: 'For example, they would give you a week to complete the works and at the same time, take up three weeks to complete the drawings. Three weeks are spent on drawings and you are given a tight one-week deadline to produce'.

According to Company A's director, an inaccurate estimation of production operations can greatly affect the forecasted price of a product.

- AI finance director: 'Very few businesses make estimations of labour costs. When they compete in project tenders, most companies simply take the cost of materials and multiply the number by a certain coefficient, let's say 3, to arrive at the product price. Mistakes occur when, for instance, <...> product takes a very limited amount of materials and a great deal of works, such as polishing, bending, milling... and the coefficient 3 is all taken up by labour costs, which means the coefficient must be ... 5, 7, 8'.

The difference between material costs and operating times is that operating time values entail more uncertainties and doubts. Also, there are fewer details the respondents are willing to talk about. The aforementioned makes a good reason to look for actual causes of uncertainties in operating times.

It has to be noted that some of the price components (including net costs) have not been mentioned by the interviewees. These were talked over by IT product developers and by the independent expert. Speaking about pricing elements for instance, apart from the cost of materials and raw cost, the independent expert points out organisational aspects ('depends on how good at organising you are'). Good planning, in his opinion, depends on the strategy chosen by company's management.

- The independent expert: 'If you run a manufacturing unit, the question is whether you want to run it at its full capacity. Or do you want to make one of the shifts redundant? Obviously, you look for a solution of how to arrange manufacturing, how to handle the flows, ... use efficiently, adjust shift performance'. '... how much of the so-called free capacity is available and check if you can take up the order' 'That's the point where the risk assessment starts: you have to decide if you can or cannot accomplish', 'agree with the production chain if they have free capacity or not'.

Operating times are one of the elements of company's production planning. However, seems that the planning process is not systemised in the companies. In company A, the practical collection of data on operating times that may be used in planning is problematic as it is a labour consuming process.

- AIII production planner: 'I've been working on planning every day for almost a week already. And it takes hours to get into the on-going situation and finish calculations. Every day, the same situation. Actually, I need a system <...> of operating planning to allow seeing of what is done and what still has to be done to get the control of the process. Every morning on arrival to work, I spent about an hour to get into the situation and learn what assignments are still pending'.

Independent survey participants also mention various additional costs to be included into the net cost:

- An external IT manager: 'labour cost, additional costs, various occasional costs that are later dispersed upon the order randomly', discount exclusions'.

It has to be noted that additional costs in customised furniture manufacturing excludes the safety aspect:

- BII IT manager: 'Furniture production has a significant advantage: you don't have to account for safety requirements. For instance, in aircraft or motor vehicle manufacturing it is a serious problem as safety requirements incur significant costs'.

The independent expert added to the list some risks associated with various technological losses ('certain technological losses have to be included into the cost;' tools and machinery have to be updated from time to time'). However, issues of manufacturing modernisation were rarely mentioned by the top (I) organisational level representatives.

In summary, the interview provided a deep understanding of the interviewee's extensive knowledge of the materials used in manufacturing. The material component is a well-defined, described, monitored and controlled part of the product cost. However, the time index, which is essential to all interviewees, remains largely unmonitored. The operating time values, which are a sensitive issue in both companies, are set with no experimentation.

Instrumental Pricing Capabilities

After conducting interviews with actors, we were able to identify the various tools and techniques they use for data storage and processing. The list of software includes Excel, GVS, QAD, Odoo, 1 C, Rivilė, Microsoft Dynamics AX, SAP, and AIVA. Whiteware listed are accounting applications, others are standard enterprise resource planning tools. The IT applications make it easier to gather and calculate tender prices. Also, the instruments are used for monitoring (AI CEO: 'you can compare the planned and actual numbers'). However, expediency and efficiency of the instruments are questioned by most of the interviewees. Firstly, data analysis-based price calculation and introduction and use of specialised instruments do not seem to be an obvious necessity in the sample companies. Also, the companies have no uniform approach towards the goals of application of IT solutions and are not convinced that such solutions may facilitate corporate processes and in particular the pricing process.

- AI finance director: 'I have no idea if the data is necessary or not'.
- BI head of project department: 'I think processing makes a significant workload and… the possible benefit has not been estimated yet'.

In reality, it is challenging to evaluate the benefits of IT solutions, data storage, and data analysis, when primary prices are set, and products are sold without any references to the collected data.

- BII pricing manager: 'the data for the product is never available when the process of pricing takes place. That is, the numbers are acquires post factum, when the manufacturing is already in progress, and then it is already too late to change the price'. '… Sometimes a product is sold without even knowing the actual cost, but the price is set and fixed'.

Although the use of complex and advanced IT solutions in the pricing process is not necessary for profitable performance, companies may still choose to invest in data analysis solutions to improve their business indexes. However, if the available software does not perform adequately, companies may be discouraged from investing in pre-set applications with the expectation of possible benefits.

- BII pricing manager: 'we make our own solutions to fix the price. Are they effective? Yes, they are. They help us to survive. I mean, we are still in the market and the company is profitable. Can it be better? Maybe. Who knows? At the moment, it probably can'.

Also, business representatives believe that data storage takes significant efforts with no payback guarantee. The most prevalent obstacles that we encounter time and labour consumption, technical limitations, and lack of knowledge and skills.

- BI head of project department: 'we are afraid [data collection] might be a long and complex process and fail to grant expected immediate results'.

- AI finance director: 'according to our estimations, six to eight designers-constructors spend about 30% of their working time on calculations, the results of which have never been used'.
- BII pricing manager: 'data storages takes time and space', 'the storage of data on, let's say, just time, endeavour and labour resources. Yes, that would be a tremendous work', '... the data is abundant and you have to process about a million entries in thousands of sections just to get an idea of which of them are useful. Processing takes powerful servers and days, even weeks'.

The fact that data collection takes more effort may be explained by the non-standard nature of manufacturing.

- BII pricing manager: 'where numerous non-standard products are made, we have never had the goal to collect all the data'.

The data collection is non-systematic. The use of various data collection and analysis tools, along with different ERP applications, often leads to data incompatibility, making automation and improvements in automation of processes, including costing/price calculations, not feasible.

- AI CEO: 'We have a kind of two incompatible things. The price is calculated in Excel. The data then have to be transferred to another system, let's say QAD. Then, the data may be compared automatically. But this is not our case'.
- BII pricing manager: 'I can tell which order the product was included into as the Excel spreadsheet doesn't expose it. ... the home base should be checked'.

Individual departments and/or individual employees in the company take control of specific data portions and share the data on demand. If the need/necessity to share information does not arise, the calculations are made independently. Also, in certain cases, no calculations are made at all.

- BII pricing manager: '*Do you have the actual costs for each product*? No, no for each, but for most of them, yes. *Do you produce any product without calculations? Can a product pass your control without being checked*? Yes, it is possible'.

Summarising, after conducting interviews on the use of IT solutions in pricing, it was found that the companies in the sample use the collected data moderately. This is due to a lack of understanding regarding the potential value of IT solutions, and the inability of IT products to adapt to the specific needs of these companies. One company invested in an overly complex ERP, while another lacked sufficient IT resources to meet programming needs. As a result, IT solutions are primarily used to calculate the value of materials used and labour time rather than to make informed decisions. Instead, reliance is placed on intuitive decisions by individual officials rather than IT solutions. Even if a company does not use advanced IT solutions in their cost estimation process, they can still achieve profitable results. Sometimes, positive business indicators might make a company less motivated to invest in data analytics solutions. On the other hand, software malfunctions may be a disincentive to invest in predefined applications in anticipation of the potential benefits to be gained. An external IT manager (ERP supplier to furniture manufacturers) gives one more reason for not applying more sophisticated cost-estimation solutions - comparatively expensive services of IT companies: '… the costs may be simply absent here and as the payback term is long, the software is simply irrelevant. Excel spreadsheets are cheaper'.

Contingency Control Measures – Authority, Competences, Jobs

The major aspect of pricing, revealed in the course of the interview analysis, is the absence of connection between the cost estimation and pricing. During the interviews, a thorough evaluation of the pricing process and content was conducted, but the ultimate price was not disclosed. The interviewees typically provided vague explanations of the pricing methodology, lacking any specific information. Commonly used phrases included 'intuition', 'perception', 'conjecture', 'deduction', 'approximation', 'speculation', 'theoretical' or 'educated guess'. As head of BII project department stated, when he was asked if they face the need to have an IT solution for pricing: 'it would be good to improve accuracy of cost estimations, but the sales price is dictated by the market anyway and you have to *feel it*'.

We assume that respondents may use the word 'guess' or its synonyms in the survey as a way to avoid disclosing sensitive information. BI head of project department: '… this part, to my mind, always has to be confidential. I mean the pricing'.

One reason for relying on intuitive decisions is time constraints, such as the need to make quick decisions based on early prices:

– AI finance director: 'In fact, instead of wasting time on calculations, we prefer to hook the client'; 'the idea is, <…> you charge an approximate price leaving space for negotiations. You have always an opportunity to promise your customer to adjust the price…'

On the other hand, the concept of guessing may conceal complex pricing mechanisms that are not strictly defined due to little knowledge and are not clearly structured or shaped into a usable instrument:

– AI finance director: 'I have a written description of "guessing methodology" I am using at the moment. <…> an order acceptance methodology. It is about what you take into account. For instance, you are given a task… Then you visit a shop <…> You go and check what segment the product comes from and what interior the product is sold in and then you understand who is their potential client'.

The interview revealed that top managers are increasingly aware of the benefits of using knowledge and skills instead of intuition when pricing. AI finance director, who is greatly in favour of guessing, expressed the need to deepen his knowledge about pricing subject to client characteristics.

During the interview, it was revealed that only highly experienced managers, including those at the I and II organisational level who manage a significant amount of information, may be able to make educated guesses. However, even these experienced managers tend to verify their estimates with colleagues. It's important to note that the verification of prices set during the primary stage doesn't just occur on a peer review level, but also includes the submission of actual calculations to ensure the reliability of intuition-based algorithms. Intuition based pricing partly associates with the specific nature of customised manufacturing, namely with time constraints and data uncertainties. Partly, intuition was referred to by the interviewees to conceal their reluctance to reveal confidential information. Also, an assumption that managers are likely to shift to more precise decision-making styles when they have knowledge of scientific pricing techniques can be made.

Pricing Processes in Made-to-Order Manufacturing Businesses

Great attention to materials, endeavours to use ERP solutions to monitor and analyse used materials and operating time values testifies the prevailing cost-based pricing orientation of B2B customised manufacturing businesses. Upon comparing the analysis results with Liozu and Hinterhuber's process for product price point definition (Liozu & Hinterhuber, 2012), it was discovered that the analysed companies perform cost analysis utilising costing tools and attempt to leverage estimating software and advanced cost models/calculations. Additionally, middle organisational level (II) representatives attribute the company's success to the evaluation of costs. However, there are noticeable discrepancies between the sample customised businesses and the small- and medium-sized industrial firms analysed by Liozu and Hinterhuber (2012) with regard to both the pricing process and the elements of pricing. Fig. 6.3 provides a graphical representation of the pricing process utilised by the sample companies.

Company A employs a more complex pricing procedure that involves cost estimation, field and capacity testing, and a preliminary price that determines expected financial return, customer relationship, and future benefits. Representatives from all levels of the organisation are involved in this process, but the final decision is made by top-level managers who rely on their intuition and experience. The target markup estimation and customer acceptance test are used to determine the price point, taking into account team-based decision-making and cost analysis results.

On the other hand, Company B does not calculate a preliminary price. Instead, the price point is based on cost evaluation, target margin markup, and customer acceptance test. The final price is determined by customer feedback and team review, with occasional voluntary validation by level II managers.

Unlike companies analysed by Liozu and Hinterhuber (2012), customised manufacturers pay considerable attention to preliminary pricing. Despite early price estimations, the development of price points continues in tandem with the manufacturing process. Additionally, the pricing process has highlighted the elements identified by Liozu and Hinterhuber in their competition-based pricing

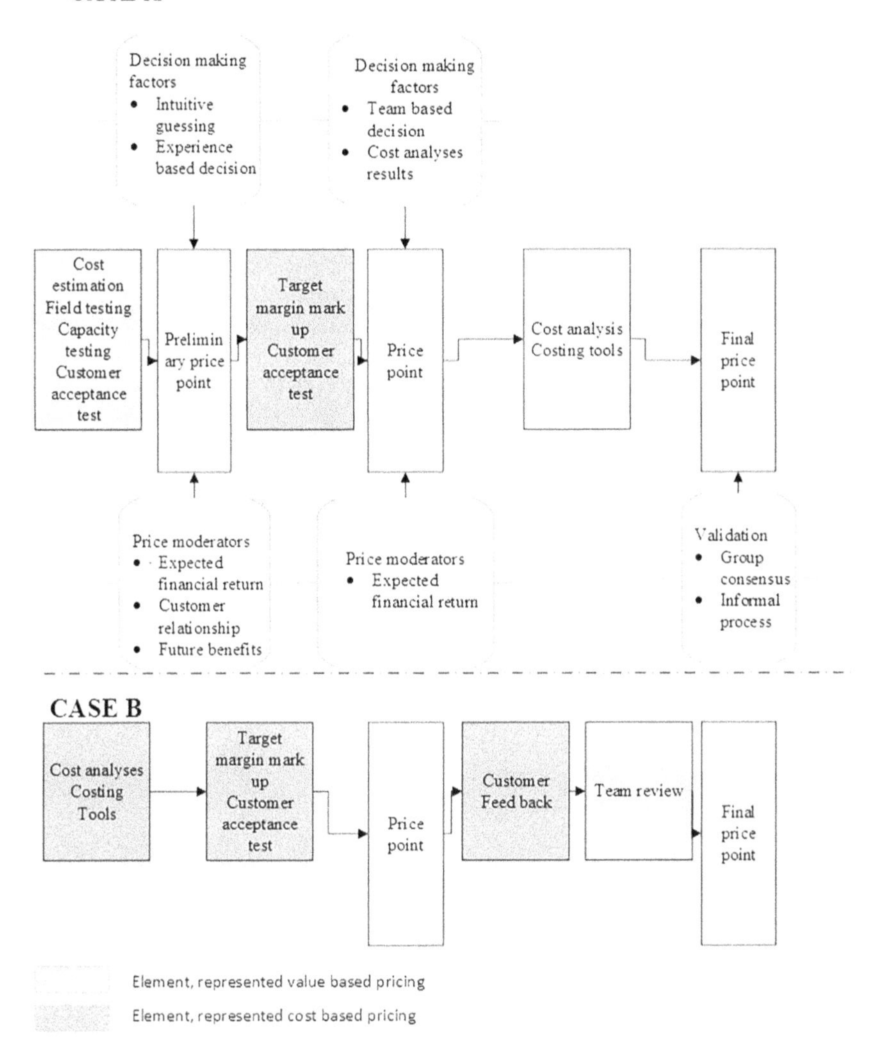

Fig. 6.3. Pricing Processes in Made-to-Order Manufacturing Businesses.

model, including intuition, group consensus, and experience-based decision making, as well as those in the value-based model, such as field testing, expected financial return, and customer relationship management). The qualitative interview analysis allowed identification of important features of the pricing process which, in our mind, predetermine the present mixed pricing orientation of customised furniture manufacturers. The features include the necessity to decide on the product price before its manufacturing is launched, the configuration of the price components, data management capacities, the level of manager knowledge about pricing strategies and the distribution of information necessary for decision making among individual actors and organisational levels.

Speaking about B2B (Fig. 6.4) customised manufacturers, value-based pricing orientation (despite its commonly acknowledged advantages) is difficult to introduce in a business similar to analysed in this chapter. In the first place, it is preconditioned by the common presence of intermediaries between the manufacturer and the end user. For intermediaries, who bare no burden of production process control, the value-based approach is much easier to adopt. Second, as the present survey reveals, middle management of manufacturing businesses focus primarily on the cost estimation associating it with competitive pricing. In their understanding, cost estimation, as a means of setting the most attractive price, is the best way to generate profit. The perception of the client as a profit generator can be seen only among the highest-level managers. However, the survey can say very little about the level of customer orientation in the sample companies as data about the client is either strictly protected or not completely structured and realised. Companies often fail to implement effective pricing strategies due to their underdeveloped data management capacities, which also explains the little value companies derive from their IT tools (as the third reason). The interviews conducted shed light on the intricate nature of implementing IT solutions for price optimisation. In addition to the factors already discussed, which include manager scepticism regarding the necessity of such tools, the intricate nature of

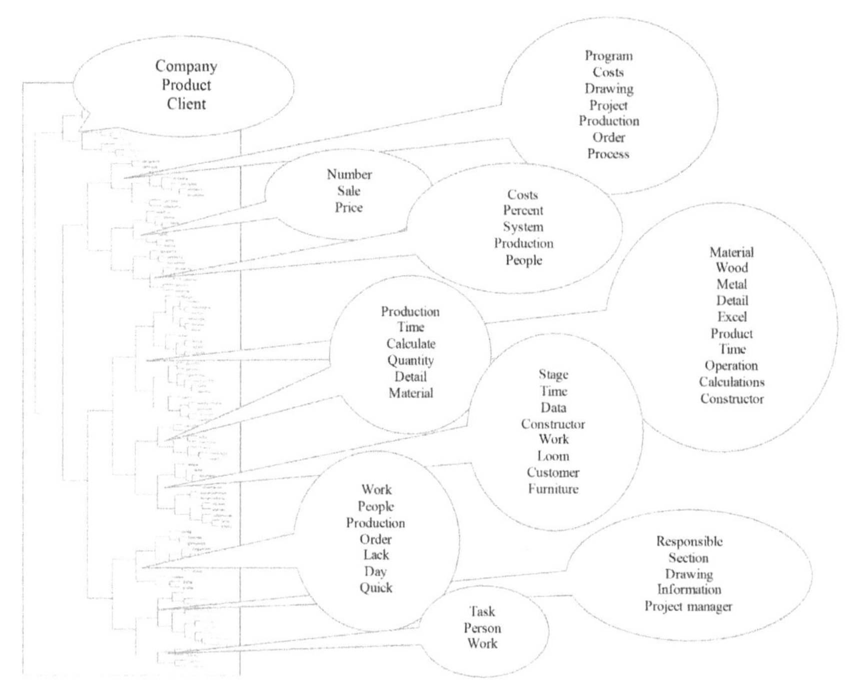

Fig. 6.4. Hierarchical Category Tree and Category Group Clusters.

customisation and implementation, and the inadequacy of staff knowledge and skills, the human element must also be considered. The study revealed the vital players who control the essential mass of information required for final pricing, specifically the AIII and BII level representatives. It is plausible that their hesitance to relinquish control over pricing decisions could have an adverse effect on the integration of IT solutions. The survey also disclosed that despite the importance of the time variable to all actors involved in the process of manufacturing and pricing, cost estimates only include operating time values. Unfortunately, the values provided are frequently inconsistent and unreliable, failing to accurately reflect the time required for operations as predetermined by managers. It is our belief that these operating time values have become a critical tool for regulating costs, allowing companies to decrease expenses related to wages. This may explain why businesses are hesitant to invest in broader applications of IT solutions for pricing, as reducing operating times presents a straightforward and effective means of cost reduction. Better transparency of the process reduces business opportunities to use their key competitive advantage, that is, cheap labour force. It is assumed that a company shifts from cost-based to value-based orientation only when there are no alternative cost-optimisation solutions. However, further research is required to verify this hypothesis. In our assessment, the pricing process within corporations may be influenced by various factors at different levels of the organisational hierarchy. For instance, the number and responsibilities of managers who interact directly with clients could impact internal competition and manager reluctance to adopt accurate cost estimation methods, preferring instead to rely on intuition-driven pricing approaches in order to prioritise product sales over improving the company's pricing capabilities. These assumptions are based on our analysis of the primary survey's pricing stages, which identified client characteristics as the key pricing element, while operational time and materials used were found to be less significant than initially believed. This is partly supported by our content analysis of interviews, which revealed that the final price is often 'invented' by the top manager, who typically possesses the most in-depth knowledge about the client. However, further studies are needed to establish a more detailed relationship between organisational structure and the pricing process. The use of databases, conventional algorithms in data analysis, and data management by information systems indicates that make-to-order manufacturers realise the advantages of the information technologies. Despite the fact that cost-estimation in make-to-order manufactories is a strategic element of the production the evolvements in information technologies in this process face major obstacles. The problems were identified in all stages: The storage of data is fragmental and messy, arbitrary decisions often used in place of data analysis, data transfer is complicated due to different calculation units, data management is labour intensive process.

The nature of customised production raises the specific requirement for the technologies that could deal with unstructured, scattered, repetitive, and isolated data. Still, cost-estimation is mainly based on the information about materials used in production. Together with persistence of the numerous obstacles in the application of data collection and management technologies, the little enthusiasm of the managers about technological enhancement and unwillingness to invest in smart manufacturing suggest that there are not enough incentives for the transition to more advanced data management techniques.

Strategies for Using Furniture Manufacturing Data to Promote Industry 4.0

Custom manufacturing faces constant challenges in optimising production to meet consumer demands (Dhaniawaty et al., 2020). This is first related to the management of the customised order, then to the management of the production process and the resulting challenges such as early cost estimation, process automation and quality assurance (Mikulskienė & Vedlūga, 2019).

The latest Industry 4.0 trends, which include the digitisation of processes and the use of robotics, are making their way into the furniture manufacturing sector. This means that customised furniture production is becoming automated and processes are being digitised. There are also higher standards in the monitoring and control of next-generation industrial equipment. Smart solutions in furniture manufacturing are being developed that use artificial intelligence to improve production. This includes centralising data or using cloud computing for data storage (Lu, 2017). To remain competitive in the global market and ensure profitability, furniture manufacturers must monitor production status and order history, conduct real-time cost analysis, evaluate factory performance, offer product customisation, and manage extensive catalogues and options (Mikulskienė et al., 2019). Efficient management of deadlines is crucial in custom furniture production, from design to delivery (Mikulskienė et al., 2019).

Data management issues can become less of a priority in low-tech sectors such as the furniture industry. However, it is important to note that the relevance of data management is not diminished. In fact, competition in such sectors is driving companies to seek lower-cost solutions, as the options available in the market are often too expensive. In business, data and its management are valuable assets for gaining a competitive advantage. Efficient data management enables faster and more effective decision-making in planning activities, streamlining processes, or generating profits through the exchange or sale of data (Santoro et al., 2019). In order to make use of data, it is essential to have a system that can collect and manage data effectively. This system should contribute to business performance by generating profits, reducing costs, and more. The performance of data employment systems starts with the process of data collection and integration. To use data effectively, it is critical to ensure that the data collected has value, and to integrate all data sources, both internal and external.

The second step is to host, append, and store the collected data. All data must be managed and stored according to the organisation's data management procedures, ensuring easy access to the data and secure long-term storage. Nowadays, the most promising data management methods are the data cloud, data lake, or data warehouse (Ren et al., 2019).

Sharing and publishing data is essential to ensure efficient and organised internal processes. It provides employees with a comprehensive and systematic access to the required information they need to perform their duties and make informed decisions (Harris et al., 2021). The effect on the business is evident in the expanding possibilities of data usage, as it increases efficiency and accuracy in dealing with customers (Harris et al., 2021). The most critical part of data management is utilising data and analytics to transform data and information into knowledge and insights that can be applied to make strategic, operational, and day-to-day decisions (Kang et al., 2016). During the process, the decisions made in one phase will affect the subsequent phases, creating a circular link between them. This means that new data will emerge and changes will be necessary in all phases. To ensure that the available data are used effectively and provide insights for the business, it is important not to skip the analysis step. This step is crucial in creating value and gaining a broader perspective on the business (Kang et al., 2016).

Small and medium-sized businesses often face greater data governance challenges than larger enterprises (Europos ekonomikos ir socialinių reikalų komitetas, 2020). Smaller companies are often viewed as less reliable for ensuring proper data protection and sharing information. This can hinder market transparency for entities that exchange information. It has been noted that smaller companies may lack the resources to manage large amounts of data, which has led them to form self-help organisations for data management and exchange (Europos ekonomikos ir socialinių reikalų komitetas, 2020). Small businesses often face challenges related to data integration, where they need to combine both internal and external data sources. They may also need to analyse this data to gain useful insights for making strategic, operational and daily decisions. Moreover, for customised production, the issue of yield management can become even more complex.

The objective of this section is to evaluate the available information on the manufacturing of personalised furniture and develop effective strategies for managing the data needed for this purpose. This study focusses on presenting opportunities for businesses to observe and analyse production data, which will enable them to respond immediately to customised orders and accurately evaluate the production processes, expenses, and timeframes. Additionally, we suggest various data management strategies that companies can implement, not only those that significantly alter pricing approaches but also those that utilise data for other objectives.

Considering Industry 4.0, it is necessary to rethink the role of early price assessment strategy in relation to other data management objectives. This integral approach provides opportunities to bring frugal innovation closer to the customer.

The chapter's analysis is focussed on Lithuania, a Member State of the European Union (EU) that presents an interesting case for multiple reasons. Firstly, the Baltic furniture manufacturing sector is currently experiencing significant challenges that are common to all EU nations, as a result of the growing global competition in the production of customised furniture (Elmaraghy et al., 2012). The industry is required to comply with certain standards, which can increase production costs, production time, and even lead to quality errors. It's worth noting that the manufacturing sector is the largest sector in the Lithuanian economy, contributing to 20.4% of Lithuania's gross domestic product. Industrial goods make up more than 80% of Lithuania's exports of goods and services. There are approximately 900 companies in Lithuania that manufacture furniture, providing employment for over 27,500 people (Versli Lietuva, 2019). This information is relevant for countries that have a similar economic structure. The furniture manufacturing industry is mainly composed of small and medium-sized businesses with up to 50 employees, which make up around 94.5% of the industry. This is a labour-intensive industry with a wage share of 59.8% in the value added (at manufacturing prices) generated by the furniture manufacturing sector (Versli Lietuva, 2019). Lithuania has a highly developed IT sector, which is one of the fastest growing industries in the country. The country has a well-established IT infrastructure and a skilled workforce to support it. The manufacturing industry, particularly furniture manufacturing, plays a significant role in Lithuania's economy. By studying Lithuania, we can explore how the management of production data through business intelligence applications can help respond to individual orders, assess production procedures, costs, and deadlines, and contribute to overall economic performance. In Central and Eastern European (CEE) countries like Lithuania, economic performance is crucial and depends on efficient and timely production processes.

The information presented in this section is based on the methodology described in the Methodology section, and specifically on two types of research data: (1) Production data from the past six years provided by two furniture companies, and (2) qualitative data obtained from semistructured interviews. The research process involved several steps, including: (a) Identifying the most important production processes, the problems encountered, and their solutions; and (b) assessing the views of experts on the complexity of the projects, the pricing process, and data management.

The State of the Art of Data Analytics in Furniture Manufacturing

The furniture manufacturing industry is set to become one of the most data-intensive businesses. Custom furniture production is particularly challenging due to non-standardised data. Traditionally, the furniture industry believed that standardisation of processes, systems, operations, management, and quality led to greater efficiency and utility in furniture production. However, a new approach to standardisation needs to be developed. Standardisation can help operational management become independent of other stakeholders. It is characterised by compatibility, interoperability, safety, repeatability, and quality (Singh, 2018).

Management decisions to implement standardised systems are based on the desire to coordinate the solution to a problem in close proximity, where all stakeholders involved can perceive mutual benefits. However, the production of customised furniture is characterised by non-standardised solutions that generate new data from the past and, therefore, generate large volumes of data. These problems are partly addressed by business analytics applications that allow companies to be in closer contact with their supply chain and customers than ever before. Each connection provides access to new analytical techniques that can reveal hidden trends, unknown linkages, market trends, customer preferences, and other useful information. Therefore, modern custom furniture manufacturing is no longer possible without data analysis.

Business analytics (BA) involves using data, statistical and quantitative analysis, predictive models, and evidence-based management to facilitate informed decision-making and actions (Chen et al., 2012). All furniture production generates vast amounts of data that can be utilised for efficient reporting and decision-making systems. In the past, decision-makers relied on manual audits and statistical reports for various purposes (Reddy & Sujith, 2018), but today's business realities require the use of software and software tools to analyse the data in real time, and the historical data helps to decide on the right way to use the data for decision-making. Analysing historical data on the production of custom furniture allows for better forecasting, quicker response to unexpected circumstances, full control for superior quality and optimised use of resources.

Analysing the data related to the business processes involved in furniture manufacturing can help you to plan and forecast these processes more effectively, both within and outside of the production line. For instance, you can use real-time data to identify the materials required to produce a specific piece of furniture, which can be easily tracked to determine when and how much to produce (Chen et al., 2012).

To thrive in a highly competitive market, business solutions for custom furniture manufacturers are now prioritising Business Analytics (BA). It's important to note that BA is distinct from Knowledge Management (KM) approaches. Knowledge Discovery from Databases involves the process of recognising valid, new, and potentially valuable data, which can be understood through data mining and statistical analysis (Fayyad et al., 1996). The BA methodology has undergone significant changes from traditional data mining and statistical analysis to include techniques for handling unstructured data and processing it in real time. By doing so, BA is capable of providing timely and accurate information, which contributes to organisational agility and flexibility (Gunasekaran et al., 2018). In addition, the widespread use of data provides transparency, helps to discover market needs (Azvine et al., 2006), reveals process or service variability, improves productivity (Brynjolfsson et al., 2011) and helps to choose more sustainable production practices (Yadavalli et al., 2019).

Competing between furniture companies relies heavily on the time taken to develop and sell new products. A shorter lead time is advantageous because it allows companies to be more flexible in responding to changing customer trends. This, in turn, increases customer satisfaction and loyalty, leading to higher sales. Shorter lead times also result in lower product development costs, faster profitability, and better operational and business outcomes. Chesbrough (2003) proposed a different approach to developing manufacturing known as open innovation. This approach involves using both internal and external ideas, as well as both internal and external routes to the market, in order to improve technology. Additionally, customer involvement is crucial in this approach, as manufacturers can benefit from knowledge of customer needs and preferences to make better decisions. To achieve this, data *analysis might be* used to gain customer insights and to adapt, adjust, and develop new service offerings (Brinch et al., 2018).

The participation of suppliers is an important external factor that can help reduce development costs, standardise components, and maintain consistency in the project. When suppliers are willing to be involved in the production process, it is seen as a positive sign of their flexibility and responsiveness to constantly changing customer needs. On the other hand, if a supplier is not willing to participate, it may be viewed as a lack of flexibility and responsiveness. According to recent studies, Business Analytics (BA) can help alleviate the challenges posed by big data volume, variety, speed, value, and accuracy. However, it is important to note that these applications require stakeholders to cooperate and willingly share all relevant data. Perfect production matching requires accurate knowledge of customer preferences in terms of which furniture features are perceived as most valuable and important and which are not. Hence, the most important approach to furniture development is one in which customers are actively involved in the development of new furniture in collaboration with manufacturers in a voluntary, creative, social, and sometimes competitive environment (Piller et al., 2012). The approach aims to provide the manufacturer with increased information about the customer's needs, applications, and technological solutions.

Pricing. In a competitive market, the evaluation of prices is becoming increasingly important compared to other phases of production, and the knowledge generated during this phase is becoming critical for the whole organisation. The relevance of price is among the first of the five stages of customised production - marketing, price quotation, construction, delivery, and assembling. The most important factor that determines the calculation of price is the retention and preservation of knowledge, which is critical in custom manufacturing. It is crucial to identify, store, and analyse critical data for pricing because the more accurately this is done, the less it will be necessary to rely on experience (intuition) or to look for product similarity (analogical reasoning). The assessment of the price can be based on the analysis of detailed critical data using either parametric or analytical methods, depending on the level of detail of the critical data (Niazi et al., 2006).

Pricing custom furniture is a challenging task as it involves considering various factors such as the company's operational costs, the availability of supply, brand awareness, and future demand forecasts. The objective is to set a price that maximises both sales and profitability.

Pricing custom-made furniture can be a challenging task, given that every product is unique and customer expectations are continually evolving. Furniture manufacturers use various cost modelling and forecasting techniques to estimate the final product's price. These techniques help them to systematically record all the costs associated with activities such as materials, labour, production, administration, sales, research, and development. By leveraging data modelling, furniture makers can accurately estimate costs and price their products competitively, ensuring long-term profitability (Elmaghraby & Keskinocak, 2003). Dynamic pricing strategies are becoming increasingly popular due to the widespread availability of data, new technologies that make it easier to calculate and modify prices, easy access to demand data, and the growing popularity of analytics and business intelligence applications (Elmaghraby & Keskinocak, 2003).

New technologies enable manufacturers to merge sales data with demographic information and customer preferences to optimise prices and analyse production data. While many customers support dynamic pricing that adjusts to fluctuating market conditions, pricing based on individual customer demand, also known as individual pricing, has a negative impact (Kedmey, 2014). This is strongly influenced by inputs, quantities and resources. It is therefore natural that the custom furniture manufacturing sector invests in optimising raw material resources in order to identify opportunities for reducing and accurately calculating raw material volumes. It also tries to calculate the resources needed for a specific task (cutting, machining, grinding, etc.) and the amount of energy consumed by the machines responsible for carrying out such a task.

There is a general consensus that quality is an essential component in the production of customised furniture and is integral to the digitisation of production processes. To improve quality, it is necessary to focus on enhancing the quality of custom furniture and processes by minimising the use of materials, equipment, and labour (Omar et al., 2019). According to Omar et al. (2019), 42% of the budget associated with quality improvement is invested in predictive tasks, such as assessing the quality of products, their physical properties, and process parameters. In contrast, other quality investments are relatively low, such as classification (25%) and optimisation (23%), which is understood as modelling cause-and-effect relationships. Traditional business intelligence related to descriptive analysis (10%), such as actions affecting profits, operating conditions, waste use, potential causes of failures, and process variation and causes of poor quality, also receives less investment (Omar et al., 2019).

Big data and business analytics. Data collection is crucial when it comes to using BA for custom furniture manufacturing. However, obtaining quality data can be challenging since some categories have very small datasets, while others have very large ones. Furthermore, the data is usually inconsistent, incomplete, and unstructured, making it difficult to determine which variables should be considered as inputs and outputs. Furniture companies that have already integrated data collection systems have a competitive advantage as they can easily combine quality data with production data and select the right dataset. To utilise the data effectively, human capital needs to be adequately trained for data analysis and interpretation. To overcome these challenges, robust algorithms and accessible and user-friendly software tools can be used. Integrated data analysis systems with manufacturing systems functions can help to reduce the cost of manufacturing a customised piece of furniture, improve the transfer of knowledge, improve the quality of the data, allow easy integration of other complementary data management applications, and provide data administration functions. This facilitates order processing and makes it easier to collect the data needed for analysis (Reuter et al., 2017).

Selecting the right business intelligence application for a furniture manufacturing company can be a challenging task. Each application needs to be compatible with the furniture manufacturing platform and the data volume and formalisation should be in harmony. This process requires a lot of effort and cost. The range of business intelligence applications for manufacturing is vast, with some applications designed to create integrated systems that are more capable of meeting the requirements of customised production. For instance, some systems create the prerequisites for a future piece of furniture, such as 3D printers. Other platforms allow customers to pre-select from pre-configured models and implement them as services. Additionally, some applications enable customers to connect to data sources and create orders and calculate prices themselves (Colangelo et al., 2018).

Many customised furniture manufacturing companies use business analytics applications that cover all production processes. These applications predict the need and duration of order materials based on previous product configurations. The production is classified into certain operational groups based on product and production characteristics. The data configurator is the data source for these predictions.

Business analytical apps have several functions that are crucial to managing a business:

1. One of these functions is *collecting and searching data* in real-time from the plant, equipment, and workers. This process helps monitor the production process effectively.
2. Another important function of business analytical apps is *processes control*. It uses data obtained from smart equipment to supervise furniture production, automatically adjusting and improving processes or providing advice to workers. This helps ensure that the production process is running smoothly and efficiently.
3. The business analytical apps also facilitate *resource allocation*, which is an essential tool for monitoring the resources used by the company. This includes keeping track of skilled labour, materials, machinery, and capital goods. By keeping tabs on these resources, businesses can effectively manage their expenses and improve their overall productivity.
4. Operation and *detail planning* is a function for the organisation and planning of operations, taking into account their importance, their dependence on production equipment and the sequence in which they are to be carried out.
5. The final function, performance analysis, is a key tool in manufacturing systems, used to compare results with the company's past performance and future business objectives in order to support decision-making, efficiency improvement and positive progress (Menezes et al., 2018).

Methodological Considerations in the Search for a Data Usability Strategy

To identify effective data use strategies that can help furniture companies leverage the technological advancements of Industry 4.0, we utilised a qualitative research approach and historical data collection methodology as described in the Methodology section.

Qualitative Research

To effectively evaluate production data, it's essential to understand personalised production and how the data can be utilised. This study utilised unstructured interviews to identify important production processes, problems faced and their solutions, as well as to carry out an expert evaluation of respondents' views on data management, project complexity, and pricing process. The interviews were conducted in 2018 and involved respondents from all levels of furniture production, including business owners, directors, project managers, designers, production planners, furniture assemblers, production and IT engineers, managers, and craftsmen, most of whom had a university degree. The study collected data from 26 interviews, which were transcribed, coded, and classified using qualitative data analysis software NVivo. The collected responses were categorised and sub-categorised to identify the frequency of concepts used and highlight key issues. The categories were then coded and grouped into subcategories, and this process was carried out twice.

The Historical Production Data Analysis

Real data was analysed using a BA. The data was analysed in different dimensions according to data type, time, demographics, order characteristics, etc. The period analysed is from 2014 to 2021 orders made. The historical data analysis was conducted using data provided by the companies, including the unique product number, product structure, materials used for each component, dimensions, and production operating times. Both companies use ERP tools, one with a standard commercial package and the other with a software package with additional programmable modules, and collect different levels of detail on furniture products. The raw data was obtained in excel format from the ERP. The analysis covered 24,853 furniture production records, including furniture description, price, order number, customer, resource, invoicing, and production data (component dimensions, procedure types, and procedural times, etc.). The data were organised according to order characteristics such as material quantity, part quantity, different part quantity, material type, material name, assembly units, length, width, thickness, slotting, covering, fittings, price, supplier, work, work time, equipment adjustment quantity and time, packing materials, product quantity, packing price, and other similar criteria.

The information is categorised based on its data type into four categories: numbers, symbols, texts, and images.

- Numbers are represented solely by numeric values.
- Symbols contain a combination of numbers, letters and other symbols, and represent an object or phenomenon.
- Texts consist of various signs, symbols, words, phrases, sentences, paragraphs, tables, footnotes, or any other combination of signs, and are designed to convey meaning and are presumed to be understandable by readers who are proficient in a specific natural or artificial language.
- Finally, images consist of any visible field and represent a visual representation of information.

For the initial analysis, data on 1,016 products were used, and the number of structure records describing them was 27,260. These data describe the furniture structures (27,260 records), the nomenclature used in the business system (14,463 records), the operating times for the production of the furniture parts (6,576 records), and the cost of materials – 3,631 records.

Factual Data Representing Furniture Manufacturing Companies

Approach to Pricing. Discovering the most effective pricing strategy is crucial for the success of any business. That's why we conducted an in-depth investigation into the process of pricing through interviews to better understand how data is generated, its meaning, and its role in furniture production. By analysing the transcribed texts, we've uncovered the principles of a well-defined operation. For instance, the cost of producing customised furniture can be predicted with a high degree of accuracy. Our research provides valuable insights that will enable businesses to optimise their pricing strategies and increase their profitability. The process takes into account a variety of factors including the size of the furniture (11.6%), lead time (25.2%), similarity of design (8.8%), materials used (26.1%), marketing (9.0%) and the creativity involved in the process (19.3%). This is the key information for the manufacturer when setting the price. It is acknowledged that estimating prices is often done based on intuition rather than rational analysis. The accuracy of this approach is heavily reliant on the proper management of data and the methodology used to select relevant data points. Many organisations determine prices based on previous orders and the experience of employees, with the assumption that the most important knowledge is stored in their heads. Data management systems are viewed as simply a tool for storing information or performing basic functions. While respondents admit that estimating prices based on hunches is a common practice, it is not always the most effective method. It is widely accepted that every detail is crucial in determining the price of a product. It is not possible to find analogous products, and it is therefore essential to prepare a design, and a prototype and only then achieve the ideal price accuracy at the pre-order stage.

Approach to Data Demand. According to research we conducted with respondents, data management is constantly referred to in terms of materials, processing techniques and operating times. a key concern for most respondents. Specifically, data management is understood in terms of three factors: materials, processing techniques, and operating times. These factors play a major role in determining the price of a product, and companies collect, store, and analyse this information in order to optimise their production processes. However, automating and standardising these processes presents new challenges, such as managing large amounts of data and applying it to customised production. While standardisation can help coordinate problem-solving and maintain quality standards, it can also create problems when a standard is unsuitable for a specific application. To address these challenges, companies are focussing on standardising material handling techniques, such as material preparation, assembly, painting, packaging, kitting, and storage of finished products. They also use appropriate units of measurement for pricing, such as meters, square meters, and units.

Insights for Data Analysis

An analysis of the available data and its structure was carried out which revealed how the processes of order taking, design and production of furniture are reflected in the data, analysing product structures and nomenclatures, operating times, sales accounts, batch sizes and material costs. The question of how to group individual orders so that standardisation actions can be applied is an ongoing issue for companies themselves. During our study, we observed an organisation attempting to group furniture orders into analogue groups based on the complexity of the order. While it was a nice attempt at representativeness, unfortunately, it was not effective. The company failed to find a connecting link for analogue grouping and ended up with a random clustering. However, the company was not able to explain the grouping principle, so the next step was to select the most relevant features of the data from the data available to the research team. The attempt indicates a shift towards more powerful data analysis mechanisms to take advantage of the available data.

Nature (Content and Type) of Data. A linear regression method was used, which allowed the identification of the attributes that correlate most strongly with product cost. The attributes describing a product are usually equal to the sum of the values of the corresponding attributes of the product. The key attributes were found to be, in descending order of importance, 'number of different parts' (units), 'number of parts' (units), 'surface area of parts used' (m^2), 'number of different materials' (units), 'order size' (units), 'material cost', followed by the various operations in the wood (w) and metal (m) workshops ('w50', 'w70', 'm10', 'm30', 'm20', etc.) (Fig. 6.5).

Analysing company data, we have discovered that certain data tend to have linear dependencies on other data. This type of data is typically used for regular day-to-day operations. Examples of such data include furniture order number, the type of furniture, times of work operations, order times, and time.

- *Original data.* Effective data management is contingent upon various factors such as the nature of the data, the complexity of the furniture components, the production volume, the intricacy of the new design, and the specificity of the production process. For instance, managing data for furniture with multiple components or complex parts may require additional efforts as compared to simpler designs. Similarly, managing data for large production volumes may require a more robust system as compared to smaller batches. Additionally, managing data for the production process that involves a lot of manual work may differ in complexity as compared to one that is predominantly machine-based. The database structures used by industrial systems have been found to be complex, which results in storing a lot of redundant and sometimes even useless information. The data itself includes dependent and independent variables, with the latter mainly used to define routine tasks. In furniture manufacturing, the most popular datasets consist of the following attributes of independent variables: operating times, numerical properties of materials (such as metal processing time, wood processing time, material quantity in meters, material quantity in square meters, material quantity in units, and material quantity in kilograms), number of parts, number of different parts, number of different materials, and order size. It has been observed that some original data are not always recorded in the form in which they are generated. For example, operation times do not always correspond to the

FIRM
Rėmas Side Frame BA-13-00
price: 41.11 order quantity: 2 amount: 82,22

vn	m	m2	m10	m20	m30	m40	w50	w70	w80	w90	kaina	Id
0	9.3618	0	0.0167	0.0167	0.0167	0.0167	0	0	0	0.0167	41.11	36

Fig. 6.5. Data Sample (Modified).

actual operation times but are predetermined during the experiment or decided at the design stage as controls in systems. Sometimes, datasets do not include design time, production time, and product lead time. Manufacturers engaging in such practices will undoubtedly make data management more complex.

- *Derived data*. Dependent data types are variables that are derived from the characteristics of an order, such as the type of furniture, type of materials and certain characteristics of the materials. These variables include the price of the product and the delivery time, which is the time between placing the order and receiving it.

Customised Manufacturing Data Specificity (Fig. 6.6)

According to our study, individual orders are much more detailed than mass production orders. They specify the dimensions, structure, and colour of each furniture component. As a result, manufacturers who specialise in individual production face greater challenges in managing their inventory. Here are a few examples. On average, the lead time for mass production is twice as short as for individual orders. This is because individual orders usually have more requirements that the supplier needs to fulfil. The more requirements a product has, the more complex the production process becomes, and the greater the number of operations needed. A study revealed that the number of operations for custom orders increases by 25% compared to those for mass production. It was also found that custom-made

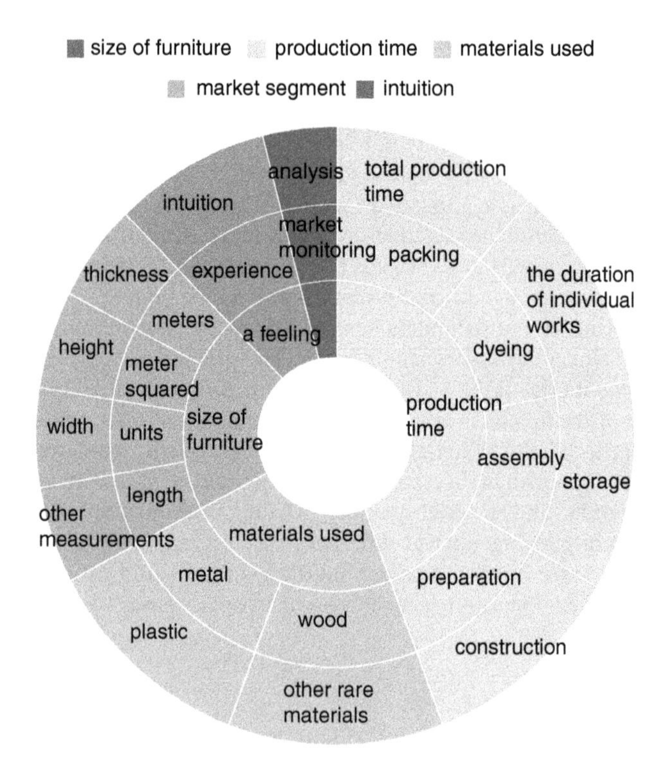

Fig. 6.6. Customised Manufacturing Data Specificity.

furniture is, on average, one third more expensive than mass-produced furniture. However, the value of custom-made furniture is also higher than that of conventional mass-produced furniture. The higher prices of custom-made furniture are due to several factors, such as stricter dimensional requirements, a greater variety of operations, longer delivery times, and the specificity of materials.

Our study has revealed that there is a certain *seasonality* to sales. The highest number of orders for individual furniture is observed in October and amounts to EUR 250,000, while the lowest is in March and amounts to EUR 75,000. When we look at the order volumes since 2012, we can see a clear downward trend in orders. For example, in 2019, they accounted for only around 7% of the total orders in 2013. This consistent decline in orders reflects the overall shrinkage of the market, and in recent years, it has been compounded by the consequences of COVID-19 pandemic.

All orders can be categorised based on the size (*dimensions*) of the furniture into three groups: small orders (such as supports, chairs, etc.), medium orders (such as tables, cabinets, etc.), and large orders (such as furniture sets, etc.). Although this classification is temporary, it is one of the determining factors in the final price. The size of a furniture piece directly affects the production time, which is also included in the final price. On average, small orders were produced within five working days, medium orders within 7–15 days and large orders within 20–30 days.

An important factor that affects the price of furniture is the *materials* used to make it. According to a survey, 92% of furniture is made of wood and metal, while only 8% is made of plastics or other materials like stone. Plastic is one of the cheaper materials used, but it is mainly used for furniture parts, which are not the main component of the furniture. The price of wood used in furniture depends on the type of wood used such as oak, beech, ash, linden, aspen, and alder. While some furniture products are produced at exceptional values, such as orders over €1 million, most orders are priced below €10,000, with an average price of €1,122 per piece of furniture. However, the majority of products are priced below € 300 (see Fig. 6.7). Businesses are the main customers, with around 200 different customers served since 2014.

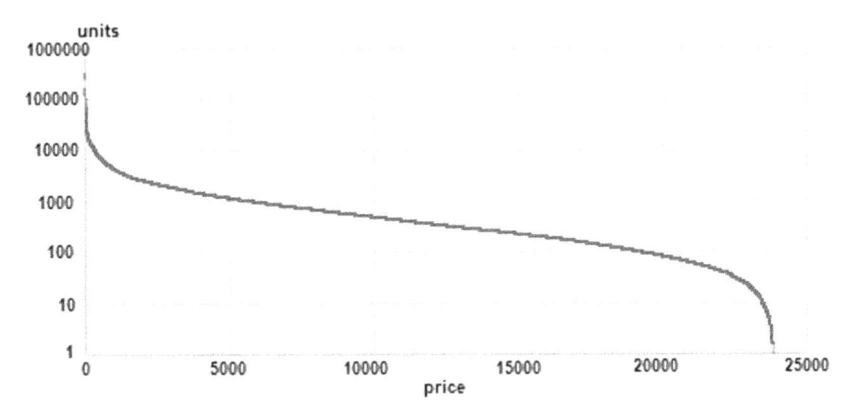

Fig. 6.7. Final Prices of all Furniture Products Manufactured by Company A During the Analysis Period.

Price Range and Predictability. If we categorise the available data on furniture products into at least six groups based on factors such as price, features, and size of customised order, we can observe that the majority of the production belongs to the first two groups. These groups consist of simple products with one or a few parts, basic structure, and low-cost materials. They are easy and quick to produce, don't pose any production difficulties or quality issues, but are also the least profitable. On the other hand, the most profitable orders come from groups five and six and represent only 5% of the total orders received in 2019 and 4% in 2020.

Range and Predictability. If we categorise the available data on furniture products into at least six groups based (Table 6.3) on factors such as price, features, and size of customer order, we can observe that the majority of the production belongs to the first two groups. These groups consist of simple products with one or a few parts, basic structure, and low-cost materials. They are easy and quick to produce, don't pose any production difficulties or quality issues, but are also the least profitable. On the other hand, the most profitable orders come from groups five and six and represent only 5% of the total orders received in 2019 and 4% in 2020.

According to the research, estimating the cost of producing expensive furniture is challenging due to various factors such as design features, complexity, and special structure of furniture elements. The cost depends on several aspects such as construction and materials of the product, technical description, timber materials, gluing areas and waste materials, cost of sanding materials, surfaces and areas to be finished, materials used for upholstered furniture, technological process, productivity of tooling, and other factors.

The quality classes of finishing complexes are used to assess the price of furniture. To determine the price, complexity groups are established based on the areas of finishing. Complexity group I includes assembled cabinet furniture products

Table 6.3. Hypothetical Grouping of Furniture Products by Price, Features, and Customer Order Size.

No	Features	Price range
1	completely simple, one-piece, simple structure, low-cost materials	1 iki 10
2	completely simple, one or more parts, simple structure, cheap materials	10 iki 100
3	simple, single or multi-part, varied structure, cheap materials	100 iki 1000
4	medium complexity, many parts, complex structure, expensive materials	1000 iki 10000
5	complex, many parts, complex structure, expensive materials	10000 iki 100000
6	extremely complex, many parts (components), extremely complex structure, expensive materials	100000 iki 1000000

such as wardrobes, secretaries, sideboards, chests of drawers, cabinets, desks, dressing tables, etc., which are mainly made up of panel elements. Complexity Group II consists of panel and point elements in various configurations. Complexity group III includes assembled products and assemblies made up of parts less than 100 mm wide, including chairs, armchairs, sofas, benches, sideboards, cupboards, door frames, hangers, cornices, etc. It has been found that 77% of the furniture produced falls into complexity groups I and II.

Data About Customers. The production data of a customised furniture manufacturing company contains valuable management information that can help improve customer relations. Fifty-five per cent of orders come from regular customers who have been using the company's services for many years. These customers come from a broad range of medium-sized Lithuanian companies. The company also has a number of large and foreign businesses as customers, including well-known international brands such as Sportland, Viva Shoes, Vodafone, and Maxima.

Possible Strategies for Using Data Based on the Level of Complexity of the Tasks Customised Manufacturing

The Course of Action Related to the Avoidance of Data Management

The realisation that certain commercial and production indicators of a company can be optimised has made mass production of furniture possible. The primary objective of mass production of furniture is to increase sales volume and profitability by optimising production costs. This sector typically employs cheaper production materials and seeks other means to reduce costs, including automation of data management according to industry standards. By observing the most characteristic indicators, a common practice of making rational decisions to optimise mass production has emerged. However, this has resulted in the addition of performance management indicators that are specific to mass production. As a consequence, production loses a certain attention to detail that is out of control.

In the case of producing custom furniture, traditional performance management principles are no longer enough to ensure profitability. Instead, success is now dependent on different factors. The difference between custom furniture and mass-produced furniture is that it is made to order for a specific customer, meeting their unique design requirements. The customer's needs and preferences are prioritised, and they are free to select the materials and techniques used in production. The main advantage of this approach is the high quality of the furniture produced, which exceeds the expectations of even the most demanding customers. The downside is that this type of furniture is usually more expensive than mass-produced furniture. To ensure long-term success, it is important to maintain and improve the quality of the products, regardless of production volumes. This may require sacrificing some profit due to reduced mass, but will lead to higher profits in the future. It will also require a new approach to data analysis principles.

Transitioning from mass to customised production requires a significant shift in data analysis and interpretation.

It is evident that transitioning towards individualised production is a complex and challenging process. This can be observed from the compensatory mechanisms that are established within companies that undertake such a transition. These mechanisms aim to overcome the limitations of a performance model that is focussed on mass production. These measures may include strategies to reduce complexity, such as:

- avoiding custom orders, improving standardised operations that are no longer suitable for customised orders;
- standardising operating times without considering the complexity of the product;
- increasing management hierarchy to act as an additional inspection mechanism for the mitigation of unperceived risks;
- developing optimisation models; and
- modularising individual activities such as warehouse residue optimisation.

Even with managerial complexity reduction, the long-term development of customised furniture production cannot be guaranteed. However, adapting data management strategies can enhance the resilience of customised production.

This discourse aims to explicate the proposed data management strategies along with the performance management practices. It is imperative to comprehend the importance of effective data management in modern organisations, which can be achieved by adopting suitable performance management practices. By delving into the intricacies of these practices, we hope to provide insights that can assist in optimising data management processes, thereby facilitating the attainment of organisational goals. In the following section, we will explore various strategies for utilising data in customised manufacturing. These strategies will range from traditional business approaches to innovative and transformative methods for leveraging data.

- *Data Utilisation Goals*

In proposing the data lifecycle, we aimed to create a smooth and natural transition to customised manufacturing without disrupting established data practices. Our goal was to make data more manageable and fit into existing ERP structures. The case studies of manufacturing companies provide tangible evidence that the management and use of data is becoming increasingly important in promoting customised manufacturing, in particular by overcoming dysfunctional compensatory mechanisms. In addition, companies are proactively seeking out ways to utilise data to optimise production efficiency, including monitoring production processes, adjusting costs, and improving production times.

Hence, the initial step is to expand and enhance the implementation of data analytics to *inform the state of production* (Fig. 6.8). This is accomplished through performance management practices, the same ones that aid in streamlining production towards mass production. The data collected for performance management can provide valuable insights into all the processes of a plant, and even suggest areas for improvement. This information can be used to optimise existing

Fig. 6.8. Three Data Utilisation Goals for Customised Manufacturing.

values and parameters, including quality standards and financial performance. By analysing production data, we can also improve our material consumption and achieve other objectives. This type of data analytics is based on linear programming and can effectively capture specific reporting points.

Another important data utilisation function that can be distinguished as distinct for customised production is a distinct type of information such as *alerting* (Fig. 6.8). Customised production benefits from alerting as it addresses complex issues and capacity questions of new orders/products. In other words, it's important to check whether the product specifications meet the company's quality standards and technological capabilities before accepting an order. A proper data management system can help safeguard the company and its team from the potential setbacks and difficulties of taking on a project that exceeds their capabilities, without compromising the long-term sustainability of the company.

One of the most unique functions of data usage, particularly in the context of Industry 4.0, is *forecasting/estimation/prognosis* (Fig. 6.8). Content-based forecasting involves predicting certain production parameters from the limited data available. This is the most valuable function of data analytics, as it can provide protection to manufacturers when dealing with highly complex customised products, for which there is no other way to estimate the parameters with complete certainty.

In the following section, we will provide a detailed explanation of how to implement these functions in the design of data utilisation strategies.

- *Data Utilisation Strategies*

The First Strategy: Performance Management

Performance management is classical approach to any activity when performance is based on measuring the efforts and outputs. So that kind of decisions going to be made regarding process and output improvement. Performance management is about evaluating processes and ultimately improving the performance of your employees. A likely consequence of performance management is the improvement of efficiency, processes, and products, resulting in cost reduction.

The production of furniture involves a high number and variety of individual operations, as each piece is made to individual orders and requires non-standard manufacturing processes. When an order is placed, several steps are necessary to start the production process, such as the customer's visual design, the dimensional diagram of the site where the furniture will be placed, and the assembly of materials. Precise joining of individual elements, matching of contours, and matching of joining points are also crucial steps in the process. Most of these procedures are automated and computerised.

This type of data use corresponds to incremental innovation.

The Second Strategy: Go/No-Go Strategy of Data Exploitation

Go is the ability to deliver the required quality in the least amount of time and with available resources no – too much uncertainty and too high risk to profitability:

- Go – the ability to deliver the required quality in the least amount of time and with available resources.
- No – too much uncertainty and too high risk to profitability.

A data management strategy called 'go/no-go' can be used to address the problem of incomplete data regarding orders. The main goal of this strategy is to determine whether an order meets the company's standards and can be produced within the required timeframe, or whether it is an order that does not meet the company's capabilities and should be rejected. This decision-making process enables companies to identify projects with a high or low probability of success based on the project's life cycle. The strategy involves making decisions in incremental steps and conducting a preliminary analysis of existing data to determine whether a project has the potential to be successful. The go/no-go strategy is effective in any situation where formal verification is required in the production of individual pieces of furniture, making it ideal for managing product or human resources data. It is particularly useful for navigating corporate constraints such as rules, regulations, policies, or terms of acceptance. However, it is important to recognise that the strategy is only effective when the furniture manufacturing company has a clear understanding of the types of orders and data management practices that match them.

A go/no-go strategy for data management should be based on relevant data that has been selected during analysis for measuring the company's success, even if it is expert opinion or socially sensitive information. In other words, the decision to not proceed should be respected, even if it is explained only by a lack of interest or passion for the project. Apart from the obvious benefits in terms of workload, cost, and frustration of not having a customised piece of furniture, there are many other reasons why a go/no-go strategy is acceptable in data management. This strategy may encourage activity-based solutions, such as targeted customer search and limiting orders that require new expertise and qualifications, which are unlikely to be implemented at an early stage of the order. The more information you have about the individual piece of furniture, the customer, and the decision-makers, the better you can adapt the strategy. This is commonly referred to as pre-positioning or fixation planning.

The Third Strategy: Early Prices Estimation Strategy (Breakthrough Innovation)

The early prices estimation as breakthrough innovation trying to solve prognostic and estimation function in the sake of use machine learning and other estimation techniques to minimise uncertainties and based price estimation not just on material prices only.

The competitive position of a furniture company is determined by the sum of the factor costs involved in the production of each individual piece. This process starts during the planning phase and ends with the realisation of production. It is important to minimise these costs through various means. We are proposing a data management strategy called early product pricing, which can be considered an innovative breakthrough. This approach involves a paradigm shift that requires changing the existing approach to costing. It is a third data management strategy that we are proposing, and we believe it has great potential.

The costing steps involved in this process typically include planning, information gathering, material collection, manufacturing operations, assembly, reclamation, disposal, labour billing, and other billing. The cost of an individual piece of furniture is traditionally made up of both direct and indirect production costs. Direct production costs refer to the expenses that are directly related to the production output. These costs usually include the cost of raw materials, consumables, and direct labour. The cost of basic raw materials encompasses the segmented products or materials that form the basis of the output.

Indirect production costs, on the other hand, are the expenses associated with production that cannot be allocated to individual products or product groups. These costs typically include ancillary raw materials, depreciation, and other production-related expenses. Auxiliary raw materials are raw materials used in the production process that are either incidental or not essential to the final product.

The classification of production costs as either direct or indirect will depend on the production data management process used by each enterprise. In the case of furniture production, for example, since multiple items are produced simultaneously, the costs cannot be determined separately for each product. As such, cost allocation is necessary during and after the production run to properly allocate costs.

Proper development and management of data on customised furniture can help to partially control the production process. However, due to the complexity of producing custom furniture, significant errors can occur, which are not acceptable in a competitive environment. To achieve greater precision, different operational methods have been used in the past. These include reducing cycle times by improving techniques and technology or improving the organisation of production.

Improving the production process can lead to a reduction in the organisation and coordination of operations. For example, in the traditional method, each operation processes a complete batch of parts, and the next operation is only started after all the parts of the previous operation have been processed. This consistent combination of operations means that the machines are usually combined each time the next batch of parts is released.

In parallel tuning, parts are processed simultaneously in all operations. Mixed tuning, on the other hand, is the most common method used for the production of individual furniture pieces, where the duration of operations is not repeated.

When producing several complex individual pieces of furniture consisting of hundreds of individual parts, the organisation of the production process can be very complicated. Such production involves many groups of machines, and therefore compatibility of their performances must be assessed. For instance, some parts have short machining times and others long.

To tackle this, the full range of engineering expertise and special mathematical methods and models for data management, as well as business analysis software for problem-solving, must be used, including pricing individual pieces of furniture.

The price of individual products depends on production data, which makes the task of price discovery and forecasting complex. To solve this problem, intelligent analytical software based on machine learning methods is required.

To implement the third data management strategy (breakthrough innovation), products can be divided into groups such as individual products and mass-produced products. By assessing and forecasting prices within these groups, furniture manufacturers can reduce the complexity of the task and gradually change their approach to pricing. Solving such problems and accurately forecasting prices requires specific data management that depends on the amount and content of historical data. This brings the data closer to the requirements of big data.

Instant cost estimation is a crucial practical application of data mining. However, it is still complicated by various factors. Firstly, the customised production nature where unique product data is generated every time requires data mining technologies to handle unstructured, dispersed, repetitive and isolated data. Therefore, in practice, especially in low-technology industries such as the furniture industry, cost estimation still relies on and is mainly based on data about the materials used in the production process.

The production time is an equally important factor that significantly determines the final cost of a product. However, it is difficult to calculate the exact production time because only the time spent by the equipment or employee on a standardised procedure is recorded in the databases. Other time indicators are not taken into account. This is not solely a production issue, but a social phenomenon that is reflected in the wages. In such cases, the worker's qualifications, the presumed level of social welfare in the country, and other social factors should also be considered to determine the production time accurately.

To aid in understanding, a summary of these strategies is presented in Table 6.4.

In Conclusion

Based on the analysis of production data, it has been found that the data collected from the daily processes of manufacturing companies can provide valuable insights for complex decision-making and can be used for modelling and machine learning tasks. The abundance of data captured can help in automating and

Table 6.4. Data Usability Strategies for Furniture Manufacturing.

	Data usability strategies	Innovation level	Goals pursued/intended effect	Preferred methods	Resources required
1.	**Performance management** (to make decisions regarding process and output improvement)	incremental innovation	To achieve: Better efficiency Better process Cheaper production process Client relationship improvement	• Multi-decision utility theory • Operational research • Indexing • Process flow • Customer pathway	Market available tools Microsoft Excel Solver ERP (skills – mathematical evaluation) For immediate use
2.	**Go/no go strategy** (to make decisions regarding complicate new orders with limited knowledge)	incremental innovation	A rough assessment to decide whether the order is in line with the company's current strategy (production capacity, specificity of design, level of issuance that employees are willing to accept)	• Multi-decision utility theory • Operational research • Indexing • Process flow	Good preparation for analyses – to understand the level of their own capabilities, define write criteria and their output levelSkill analytical skills and time for company analyses
3.	**Early prices estimation** (to make decision with support of machine learning and experts estimation)	breakthrough innovation	To use machine learning and other estimation technique to minimise uncertainties and based price estimation not just on material prices only	• Artificial intelligence	Big data availability IT competences data clearing, machine learning interpretation)

optimising processes. Additionally, it has been observed that setting targets for new data is not necessary as the existing data is already complete enough. Despite the fact that companies accumulate a large amount of data, whether intentionally or unintentionally, data management remains a significant challenge for them. This data reflects various aspects of their economic and productive activities, as well as user profiles. Therefore, managing this data effectively requires companies to make a special effort.

The furniture industry involves a vast range of operations, making it necessary to establish a comprehensive and specific data capture system. To achieve this, business analysis programmes are utilised to minimise the cost of individual furniture production factors, by incorporating efficient action planning and production realisation methods, while also exploring all possible ways to reduce costs.

Unfortunately, the management of production data remains a challenge for small and medium-sized individual furniture manufacturing companies, as the data processing tools are not tailored to individual furniture processes, but are focussed on mass production trends. It is important to acknowledge that a shift needs to occur from cost-based to value and data-based pricing before data can be effectively utilised in customised production.

Adopting a wider range of data-application strategies can accelerate the shift by overcoming the behavioural norms that restrict thinking and practice.

The first data strategy is already widely used in business, and it's commonly referred to as performance management. This strategy involves appropriate data management for the production of customised furniture, with the primary function of informing.

The second strategy, known as go/no-go, involves applying data management to orders. With this strategy, the suitability of the order is automated, and the system assesses whether the project is suitable for the organisation. It also evaluates the complexity of the individual order that is acceptable to the furniture manufacturing company.

The third strategy, breakthrough innovation, involves identifying and predicting the future price of a custom piece of furniture.

Early Cost Estimation by Means of Machine Learning with Data Visualisation

Historical Production Data in Furniture Companies

In this section, we discuss information about the collection and preparation of historical data for machine learning applications for product price prediction. When collecting historical production data, we collaborated with businesses that participated in a qualitative study analysing the production process (further details about the characteristics of these companies are described in the Methodological section). We selected companies with extensive experience in customised production. These companies primarily deal with customised orders, often in large quantities, and operate on a business-to-business (B2B) basis. They focus on the international market, targeting the mid to high-priced segment and face intense international competition, adhering to high production quality standards. The growth of their production and order expansion is closely tied to their technical capabilities, including optimised production processes (CNC machinery) and extensive use of IT tools.

Initial evaluations by company executives regarding the value of IT tools in the production and pricing processes revealed concerns about the cost-benefit balance, data security, proper data collection, administration, utilisation, employee inertia, and potential resistance to innovation. Despite this initial scepticism, businesses participating in the project provided their historical production data. During data analysis, a data classification system was developed.

As these companies already use data management programmes in their production processes, they have accumulated over 10 years of historical production data, meeting the analysis requirement for a diverse range of products. Therefore, companies provided data not categorised by products but rather by production periods, which, in any case, encompassed the entire range of possible products, from complex to simple.

The companies submitted their data in the form of output files prepared from their business management systems in MS Excel format, along with invoices in PDF format. One company, however, modified the data to obscure price parameters. Unfortunately, they did not provide product prices, making their data unsuitable for price forecasting. Nonetheless, the data was suitable for examining production processes. For price forecasting prototyping, data from another company was sufficient.

Data Preparation and Transfer Process

To properly prepare data for machine learning and further develop machine learning algorithms, we followed a standard data science process (Donoho, 2017):

1. Data collection and exploration.
2. Data transformation.
3. Data analysis.
4. Data modelling (machine learning algorithm development).
5. Data visualisation and presentation.
6. Interpretation.

Data preparation, also known as pre-processing, involves the initial transformation and conversion of raw data to make it suitable for machine learning algorithms (the first two stages of the data science process), as very specific datasets are required.

Unprocessed data was received in '.xls' or '.xlsx' file formats. Some information was also provided in '.pdf' format, but due to the complexities of processing such files, this data was used sparingly in the project. The structure of these files often varied, even for information on orders spanning recent years. It's important to note that each company uses different tools for storing such data, so while document types may coincide, the organisation of data within them can vary significantly. A common format involved order information and production information being stored in separate documents. The documents containing order information were filled and edited by company employees without using any auxiliary applications, resulting in occasional grammatical or formatting errors that complicated the data processing process. Some datasets were missing values (most commonly in the price field), and some values were incorrect (often due to erroneous entries in operation fields). Missing data sometimes appeared as empty cells, strange values, or a specific symbol, such as a question mark. Anomalies also needed to be identified and handled.

To ensure accuracy and prevent incorrect predictions, significant attention was given to data cleansing processes. Even after selecting data following the cleansing process, further data shaping (e.g. aggregation or summarisation) was required. Additionally, many datasets lacked useful business context (e.g. poorly defined ID values), necessitating the enrichment of features (selecting significant business and production operations).

The data preparation process required substantial effort from IT specialists due to the specific data storage formats used by companies. However, this process was fully automated by developing a custom data parsing application for data collection, sorting, and data cleansing commands.

The data preparation system was built using the ***node.js*** framework. ***JavaScript*** was used for coding the system, and data was stored in a ***MySQL database***. Since the structure of incoming data constantly changed, significant attention was paid to potential future system expansion and the addition of algorithms for recognising new file formats during system development.

The data preparation process aimed to simplify the logic of the machine learning algorithm and avoid prediction problems. The following essential steps were taken with the data:

1. *Data presentation format*: Data from collaborating companies arrived in various formats. To simplify the algorithm's logic, all necessary data was selected and transferred to a common database with a strictly defined structure.
2. *Data type error*: Some data provided for analysis could have contained all the required values for calculations but pertained to the wrong process. For example, the data could describe the correction of errors in a previous order rather than the execution of a new order. Such data could distort the results of algorithm training and was filtered out.
3. *Missing data*: To perform as accurate training and predictions as possible, the machine learning algorithm requires certain data. Often, datasets provided for analysis lacked one or several mandatory data types. Such data were filtered out and not included in training or predictions.
4. *Interpretation errors*: Sometimes, the received data had all the required values for training, but they needed to be interpreted differently. For example, a numeric value describing the process was correct but was expressed in different units of measurement. In such cases, numeric values were converted into appropriate units of measurement during data preparation and subsequently used for algorithm training or predictions.

The data structure for machine learning was provided in Annex 4.

All data selected for machine learning training and testing, following data preparation, was transferred and stored in one of the database tables. The data selected for further analysis, such as order number, customer information, quantities, prices, materials used, operation durations, etc., is presented in Annex 4.

During data processing, almost 6,000 documents were read for system training, but fewer data points were collected. Over 1,500 documents had unique formats, making it impossible to find the required information or not containing information about new order production. The remaining files were successfully read, but some of them lacked critical order information or had specific issues, leading to their rejection. Ultimately, over 3,000 unique order records were collected and used for algorithm training.

Early Cost Estimation by Means of Machine Learning with Data Visualisation

In the literature, price evaluation methods are often based on the principles of similarity or analogy. The idea of such approaches is based on comparing a new product with previously produced products to identify similarities between them. Similarities help to incorporate previous data into the price assessment of the new product, thus avoiding the need to calculate the estimate from scratch each time. Thus, through the experience of cost estimation, previous design and production data provide useful information for estimating the cost of a new product. This type of cost estimation method can be divided into intuitive and analogy methods. Intuitive cost estimation methods are based on cost estimation experience. Expert knowledge can be stored in the form of rules, and decision trees in a specific database to facilitate the decision-making process and the preparation of cost estimates for new products. When analogy methods are used, the cost of a new product is estimated from the cost of similar known products. Here, similarity is usually assessed without any quantitative estimates. However, there may be cases where numerical estimates of product similarities are used, in which case analogue methods may be classified as quantitative. Quantitative pricing methods are based on a detailed analysis of the product design and manufacturing processes. Quantitative methods are further subdivided into analytical and parametric methods. Analytical methods require the product to be decomposed into parts, and then the cost of each part is calculated, the price being the sum of the costs of all the parts. Parametric methods (sometimes called statistical methods) assess historical data to establish the relationship between product characteristics and price. One of the most popular methods in this group is regression analysis. Artificial neural networks can also be successfully applied to the pricing problem, but they typically require large amounts of historical data.

Various methods have been developed for advanced price estimation, but they are not widely used in the furniture industry. This area of production is distinguished from others by the fact that products often range from small (and thus low-cost) pieces of furniture to large and expensive sets of furniture. Below is a more detailed description of how machine learning has been applied to forward pricing.

Proposal for a Machine Learning-based Forward Price Evaluation Method

Pre-estimating the price of furniture takes a lot of manual work and therefore time. When a new order is received, it is necessary to assess whether the product will be new to the company. If the product (or a very similar product) has been made before, the price will be known. Qualitative and survey research has shown that the cost of a product depends on the complexity of the furniture (furniture consists of many components, and parts are complex), the volume of production (a large production batch lowers the price per unit), the complexity of the new design (whether the new design requires additional designer time or simply adapts one of the existing ones), the production features (a lot of manual labour is used, expensive machines are used). However, if a new product is ordered, it is necessary to consider the components of the price: quantities and cost of materials, labour costs, packaging costs, etc. The overheads are added and the estimated price of the product is sent to the customer (Fig. 6.9a). Machine learning can significantly reduce the time of this process. Thus, the project activities proposed a preliminary price estimation method based on machine learning (Fig. 6.9b). Those figures and diagrams are related to the data discussed in our chapter (Kurasova et al., 2021).

In order to solve the problem of upfront price estimation, it is necessary to identify the relationship between the price and the characteristics of the components of the piece of furniture as well as the operating times. Thus, the price

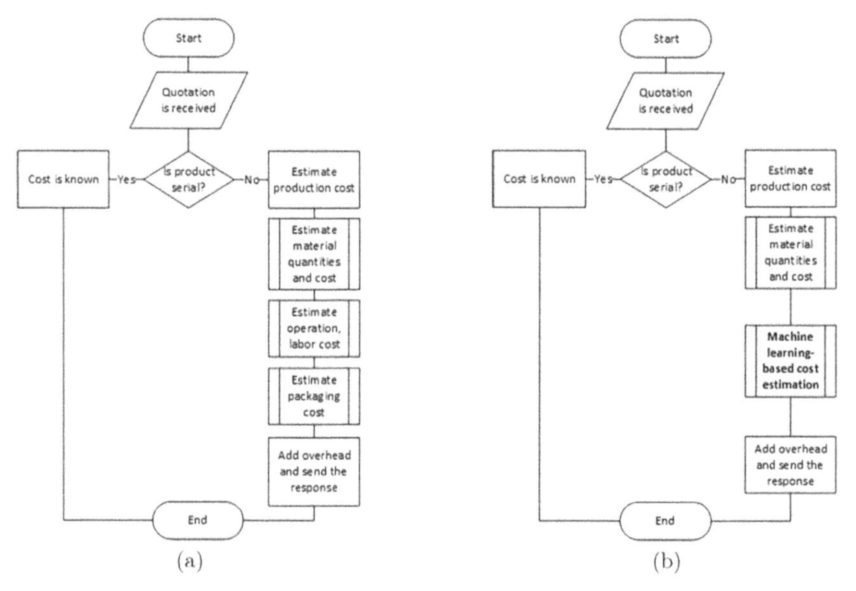

Fig. 6.9. Preliminary Cost Estimation: (a) Manual-intensive Process, (b) Proposed Machine Learning-based Approach (Kurasova et al., 2021).

estimation problem is formulated as a forecasting problem, for which various methods can be used:

1. Linear regression. This is the simplest regression model where the relationship between the dependent variable and one or more independent variables is linear.
2. Decision tree regression (decision tree, random forest, complementary tree, AdaBoost, gradient boosting). Here a tree-like structure is created.
3. K-neighbour regression. This is a non-parametric method used for regression where the input is the k nearest training examples, the output is the predicted value obtained by summing the average values of the k nearest neighbours.
4. Artificial neural networks. They are widely used as machine learning methods to solve complex data analysis problems. Artificial neural networks can be successfully used as a prediction method, but they require a large amount of historical data for training.

Experimental Study

The data sample includes the quantities of materials needed to produce the furniture and the operating times. The prices of these furniture items are also known, so that supervised learning algorithms can be applied. The dataset consists of the following attributes: y is the price (dependent variable to be predicted); $x_1, \ldots, x_n$ are the operating times (independent variables); $x_{n+1}, \ldots, x_m$ are the numerical properties of materials (independent variables):

$x_1 - x_5$ – metal processing times,
$x_6 - x_{10}$ – wood processing time,
x_{11} – quantity of materials in metres,
x_{12} – quantity of materials in square metres,
x_{13} – quantity of materials in units,
x_{14} – quantity of materials in kilograms,
x_{15} – quantity of parts,
x_{16} – quantity of different parts,
x_{17} – quantity of different materials,
x_{18} – order size.

The training process was cross-checked using 10 blocks. The resulting mean values (avg) and standard deviations (std) of the coefficient of determination R^2 and root mean square error (RMSE) are shown in Table 6.5. Two cases were analysed: when all attributes $x_1, \ldots, x_m$ are used for prediction and the five attributes that were identified as the most important ones using the essential attribute selection method. The highest R^2 value was obtained using the random forest (0.842) and gradient boosting (0.840) algorithms. The results show that attribute reduction has a negligible impact on the prediction results. Fig. 6.10 shows the results of the price estimation (prediction) using linear regression in graphical form. It can be seen that low prices are more accurately predicted. For high prices, the accuracy of the prediction should be increased.

Table 6.5. Forecasting Results.

Forecasting Algorithm	***R2* (avg)**	***R2* (std)**	**RMSE (avg)**	**RMSE (std)**
Linear regression (all attributes)	0.793	0.125	136.84	55.47
Linear regression (5 attributes)	0.739	0.110	155.97	50.52
Decision tree (all features)	0.726	0.118	159.50	42.18
Decision tree (5 attributes)	0.698	0.096	167.00	39.70
Random forest (all features)	0.842	0.079	122.15	44.64
Random forest (5 attributes)	0.806	0.094	134.14	47.24
Additional tree (all attributes)	0.738	0.098	157.87	44.99
Additional tree (5 attributes)	0.720	0.074	161.88	34.57
AdaBoost (all signs)	0.809	0.075	134.06	38.36
AdaBoost (5 signs)	0.775	0.097	144.78	46.08
Gradient boosting (all features)	0.840	0.074	123.02	40.48
Gradient boosting (5 features)	0.802	0.094	136.54	48.99
K Neighbours (all signs)	0.811	0.087	133.76	43.29
K neighbours (5 attributes)	0.799	0.099	138.61	52.51
DNT (all attributes)	0.805	0.077	133.97	36.46
DNT (5 attributes)	0.764	0.075	150.48	42.39

Visual Analysis of Production Data and Early Pricing for Customised Furniture

As previously noted, customised manufacturing is becoming increasingly popular in the furniture sector. Since each bespoke item requires a unique pricing calculation, businesses must rapidly generate a preliminary cost upon receiving an order. The challenge of delivering an accurate and prompt estimate can be addressed through a predictive approach, leveraging diverse machine learning methods. To improve the accuracy of price forecasting, a deeper analysis of the data and its nuances is necessary. Data visualisation techniques are well-suited for this purpose. It's important to keep in mind that product managers may not be experts in machine learning. Therefore, decision support systems should integrate simple and easy-to-understand visualisation methods as well as more advanced techniques like dimensionality reduction to reveal hidden data structures. The proposed data visualisation process in this chapter can provide valuable insights into customised furniture manufacturing data, leading to better price prediction models.

Data exploration can be greatly enhanced with the use of visualisation techniques. Humans are known to process visual information more effectively than other forms of data presentation. Visualisation enables data analysts to gain deeper insights into the data, allowing them to choose appropriate analysis

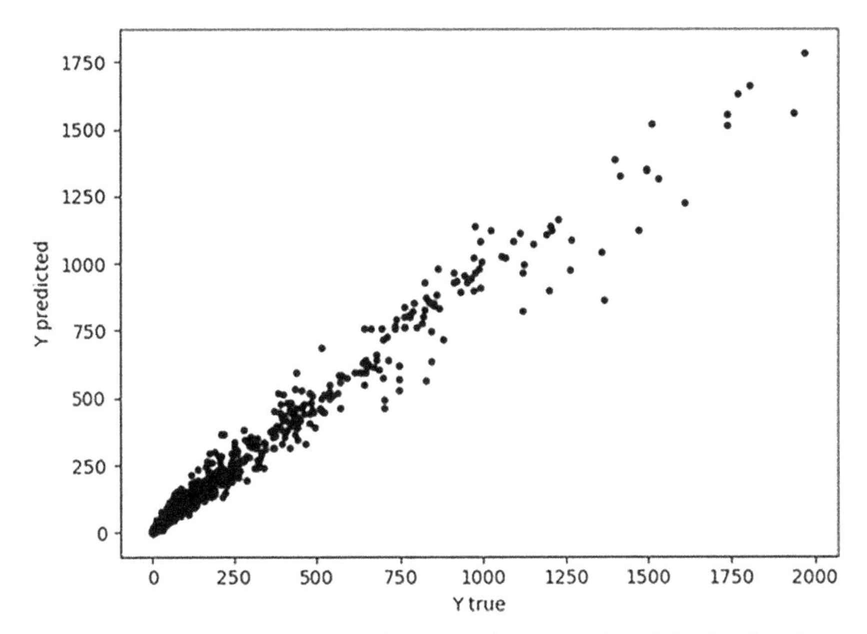

Fig. 6.10. Relationship Between the True Price (Y True) and the Predicted Price (Y Predicted) Obtained by Linear Regression (Kurasova et al., 2021).

methods for various tasks such as classification and prediction. However, real-world data can be complex and highly specific, necessitating the use of advanced visualisation techniques. To gain a comprehensive overview of multi-dimensional data, dimensionality reduction techniques are commonly employed. When integrating visualisation methods into decision support systems, it is important to ensure that the tools are user-friendly and easy to comprehend for those unfamiliar with data mining and machine learning. The challenge lies in balancing the need for sophistication with the need for simplicity. Furthermore, it is crucial to tailor the visualisation methods to capture the unique characteristics of the data and the domain being analysed.

When determining the price of furniture, there are typically two components to consider: the cost of production C_p and the profit margin P_p. The cost of production C_p includes direct costs like materials and labor, as well as overhead costs like manufacturing and administrative expenses. The price P_w is the sum of these components: $P_w = C_p + P_p$. Estimating these costs can be a challenging and time-consuming process, particularly in the early stages of furniture design when information is limited (Niazi et al., 2006). Once the cost has been estimated, a profit margin is added to arrive at the final price for the customer.

In order to use machine learning methods, it is necessary to have access to data within the relevant domain. In the case of analysing data related to the production of customised furniture, a data object would be a product (furniture) that is defined by a specific set of attributes. These attributes include product dimensions (such as length, height, width, and weight), material data (including the materials used for production and their associated costs), activity data (including a list of operations required and the time needed to complete the operation), labour data (which can include both manual labour and the use of expensive machine tools), production time (which is often based on the customer's requirements), batch size (as producing more of the same product can often reduce costs per unit), and production complexity (which is a qualitative parameter that reflects the product's uniqueness and the complexity of the work required to produce it). We denote these attributes by $x_1, x_2, \ldots, x_m$. In regression methods for price forecasting, independent variables are utilised, with price being the dependent variable denoted by y.

Estimating the price of non-standard furniture with machine learning is a challenging endeavour, resulting in less than optimal prediction accuracy. To address this, visualisation techniques can be employed to explore the data and provide external experts with additional insights. Through visual analysis, suitable data subsets and features can be identified to enhance the machine learning process and improve cost forecasting results.

Selected Methods for Multidimensional Data Visualisation

Data visualisation is crucial in the age of big data as it helps to comprehend the entirety of the analysed data, thus facilitating decision-making (Medvedev et al., 2017). Data visualisation serves a multitude of purposes, such as initial data discovery, exploratory data analysis, and interpreting machine learning results. This is especially crucial when the individual handling the data is not well-versed in data exploration and machine learning. As such, it's important to present the data in a format that is easily comprehensible. Fortunately, there are now a wide range of data visualisation techniques available that make the task of exploring and understanding complex data much simpler (Sorzano et al., 2014). These techniques can be broadly divided into two groups: direct visualisation and data dimensionality reduction-based visualisation. Direct visualisation methods present data in a visual format that is easily comprehensible to humans, such as scatter diagrams, bar charts, and histograms. Meanwhile, dimensionality reduction methods aim to represent multidimensional data in a lower dimensional space while still maintaining important structures like clusters or outliers (Sacha et al., 2017). Principal component analysis (PCA) is a widely used method to reduce data dimensionality. It identifies a linear subset that retains most of the variance in the data (Wang et al., 2016). Let $X_1, X_2, \ldots, X_d$ be the data to be analysed. Every object possesses specific attributes that define its characteristics $x_1, x_2, \ldots, x_m$. Assuming that all attribute values are numeric, a $d \times m$ data matrix $\mathbf{X} = x_{ij}$, $i = 1, \ldots, d$, $j = 1, \ldots, m$ can be constructed. The ith row of this matrix corresponds to the vector $X_i = (x_{ij})$, $j = 1, \ldots, m$, $i \in \{1, \ldots, d\}$, whose elements are the values of the attributes of the ith object. To find the principal components, we need to calculate the covariance matrix $C = (c_{kl})$, $k, l = 1, \ldots, m$ has to be computed at the beginning. The covariance matrix then needs to find the test vectors and the test values. The check value λk and its corresponding check vector E_k are the solutions of the equation $CE_k = \lambda_k E_k$. Given the check vectors E_k, $k = 1, \ldots, m$, they must be arranged in descending order of the corresponding check values ($\lambda_1 \geq \lambda_2 \geq \ldots \geq \lambda_m$). The matrix $A = (E_1, E_2, \ldots, E_m)$ is called the principal component matrix. The points X_i, $i = 1, \ldots, d$, are transformed into points Y_i, $i = 1, \ldots, d$, according to the formula $Y_i = (X_i - \bar{X})A, i = 1,\ldots,d$, where $\bar{X}$ is a vector consisting of the average of the values of the attributes $x_1, x_2, \ldots, x_m$. To reduce the dimensionality of the data, it is possible to take only the first l, $l < m$, instead of all principal components. Then, instead of the matrix A in the formula, the matrix $A_l = (E_1, E_2, \ldots, E_l)$ should be used.

PCA is a linear projection technique, while multidimensional scaling (MDS) is a non-linear approach. The objective of MDS is to identify points in a reduced dimension that maintain similar distances between them as in the original multidimensional space (Dzemyda et al., 2013). Suppose each m-dimensional vector $X_i \in R_m$, $i = 1, \ldots, d$, corresponds to a lower dimensional vector $Y_i \in R^l$, $l < n$. Let us denote the distance between vectors X_i and X_j by $d(X_i, X_j)$ and the distance between vectors Y_i and Y_j by $d(Y_i, Y_j)$. The MDS method attempts to approximate the distances $d(Y_i, Y_j)$ to the distances $d(X_i, X_j)$. If a quadratic error function is used, the minimised objective function is written as follows: $E_{MDS} = \sum_{i<j} \left(d(X_i, X_j) - d(Y_i, Y_j)\right)^2$. The result of this function minimisation problem yields lower dimensional vectors $Y_1, Y_2, \ldots, Y_d$.

The t-SNE method, which utilises non-convex optimisation, has emerged as the leading technique for visualising complex patterns across a broad range of fields. This algorithm generates low-dimensional representations that effectively capture intricate patterns from high-dimensional spaces, displaying them as distinct and separate point clusters. In addition, autoencoder neural networks have gained traction as another set of dimensionality reduction techniques, thanks to their ability to handle vast amounts of data through the use of deep learning strategies during training.

Due to the large number of data visualisation methods available, we are often faced with the challenge of choosing the most appropriate visualisation methods and how to organise the whole visualisation process Fig. 6.11 illustrates a proposed data visualisation process, which starts with the construction of a dataset. After that, the most relevant features of the data have to be selected.

There are two ways to accomplish this: manual selection or by utilising feature selection methods such as linear regression or feature correlation. Following this, data visualisation is the next step. To begin with, straightforward direct visualisation techniques such as scatter plots and histograms can be employed. As the analysis progresses, it may become necessary to use more advanced methods based on dimensionality reduction. Given that the data being analysed is typically multidimensional, dimensionality reduction visualisation methods can aid in uncovering any hidden structure within the data. Ultimately, the outcome is a collection of images.

The decision maker, who is responsible for analysing data and making informed decisions, can carefully examine and interpret the results to identify any hidden characteristics or relationships that may be useful in achieving the main

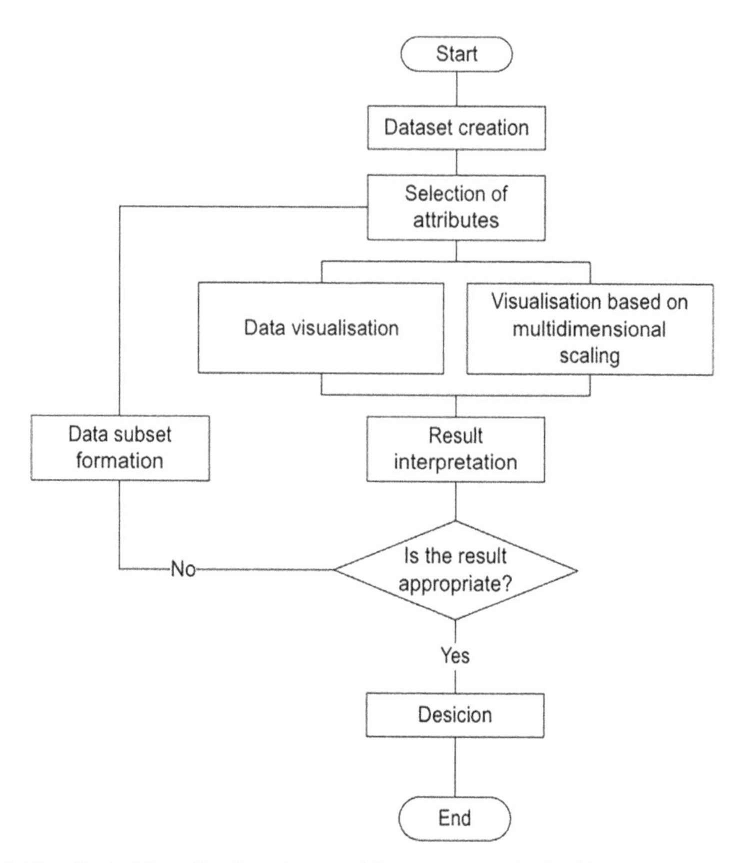

Fig. 6.11. Data Visualisation Process (Kurasova et al., 2021).

objectives of data analysis, such as classification or prediction. If the results meet the decision maker's expectations, a final decision can be reached and the process concluded. However, if the data appears to be clustered, it may be necessary to analyse individual clusters by selecting a subset of the dataset. In this case, a visual analysis of the data groups is conducted, and the process is repeated until an appropriate decision is made. The following figures and diagrams are directly related to the data discussed in our chapter (Kurasova et al., 2021).

The proposed visualisation process can be adapted to the field of customised furniture production. To collect historical data on previously manufactured products, we gathered information on metal and wood processing times, material properties, and the number of different components used. We then selected a subset of these attributes and utilised direct visualisation techniques like scatter diagrams and histograms to conduct initial exploratory analysis. To gain a comprehensive understanding of the data structure, we relied on well-known methods like principal component analysis and multidimensional scaling, as well as more sophisticated approaches like t-SNE and autoencoder neural networks. Additionally, we integrated cluster analysis into our visualisations to uncover deeper insights, employing the Louvain algorithm for this purpose (Traag et al., 2019). When the decision maker is not satisfied with the analysis performed on the entire dataset, a subset of the data should be selected and visually analysed.

Use Case for Visual Analysis

To showcase the suggested visualisation technique, we have utilised production data from a furniture manufacturing company based in Lithuania. This data comprises information on 1,007 products and has been gathered over the past five years. The pricing details for these products are available, and they range from small furniture pieces to elaborate furniture sets of varying sizes and complexity levels. Each product is characterised by certain attributes $x_1, x_2, \ldots, x_m$. The selected attributes are described in Table 6.6. Here the price is denoted by y. All attribute values in the data are numerical which means we can createto represent the data. Let us denote the feature vector X_i for the ith product. Then we have a $d \times m$ matrix $\mathbf{X} = (x_{ij})$, $i = 1, \ldots, d, j = 1, \ldots, m$. The ith row of this matrix corresponds to the vector $X_i = (x_{ij}), j = 1, \ldots, m, i \in \{1, \ldots, d\}$, whose elements are the values of the attributes of the ith product, X = 1,007, $m = 17$. The data visualisation was performed using the data mining software Orange (Demsar et al., 2013) (https://orangedatamining.com). The following python libraries were used to train the autocoder neural network and to visualise the results: *tensorflow, keras, scikit-learn, numpy, pandas, seaborn, matplotlib.*

Initial data analysis was carried out. Here, we limited ourselves to a visual analysis. It is appropriate to apply it in order to see the overall structure of the data. To achieve this goal, one can utilise histograms or other simple direct visualisation techniques. Fig. 6.12 shows a price histogram. It can be seen that there are many low-priced products in the dataset. Around 700 products are cheaper

Table 6.6. Attributes of Production Data.

xi	**Tag**	**Short Description**
$x_1, \ldots, x_5$	m10-m50	metalworking time
$x_6, \ldots, x_9$	w60-w90	woodworking time
x_{10}	M	total metres of material
x_{11}	m^2	total square metres of materials
x_{12}	Kg	total weight of materials
x_{13}	Qty	total quantity of materials
x_{14}	qty_ parts	quantity of parts
x_{15}	qty_ diff_parts	quantity of different parts
x_{16}	qty_ materials	quantity of different materials
x_{17}	qty_ order	order size
y	Price	price

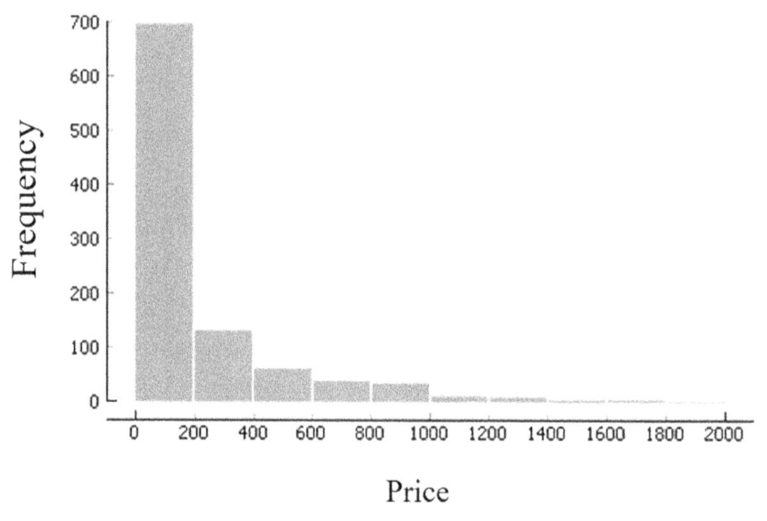

Fig. 6.12. Price Distribution (Kurasova et al., 2021).

than €200. Grouping the data into several price categories can be done since only a few expensive products exceed €1,000. It is also possible to explore histograms of other features of the data to see similarities and differences. Fig. 6.13 depicts the frequency of furniture parts. The number of components is limited to 20 in most cases, with only a few pieces having more parts.

Given the multidimensional nature of the data, it is possible to apply various dimensionality reduction techniques to visualise it. Among these methods, PCA is the most widely used. In order to represent lower dimensional data in a Cartesian coordinate system, only the first two principal components (PC1 and PC2) can be utilised (Fig. 6.14). Each point on the graph corresponds to a single product, or piece of furniture, and is color-coded according to its price value. The visualisation reveals that the less expensive products, indicated by dark grey dots, are clustered together, whereas the more expensive furniture, indicated by light grey dots, are spread out more widely. It's worth noting that in this case, the first two principal components account for only 53% of the variance. To achieve 80%, a total of seven principal components would be required, but this would render the data impossible to visualise.

If the data has a nonlinear nature, nonlinear dimensionality reduction techniques are more suitable. Therefore, in this case, the furniture manufacturing data was analysed using a non-linear MDS approach for visualisation purposes. In this context, the similarity between data points is measured using the Euclidean distance. In Fig. 6.15 displays both the clusters of points and outliers present in the data. The use of PCA resulted in the two light grey points being distant from the other points. On the other hand, when MDS was applied, the two light grey points remained as outliers, and additionally, the two dark grey points were also found to be distant from most other points.

Using the t-SNE method, we can visualise data and its unfolding, as shown in Fig. 6.16. It can be seen that the data representation is different from that shown in Figs. 6.15 and 6.16. In this figure, we can see several clusters of data. One large

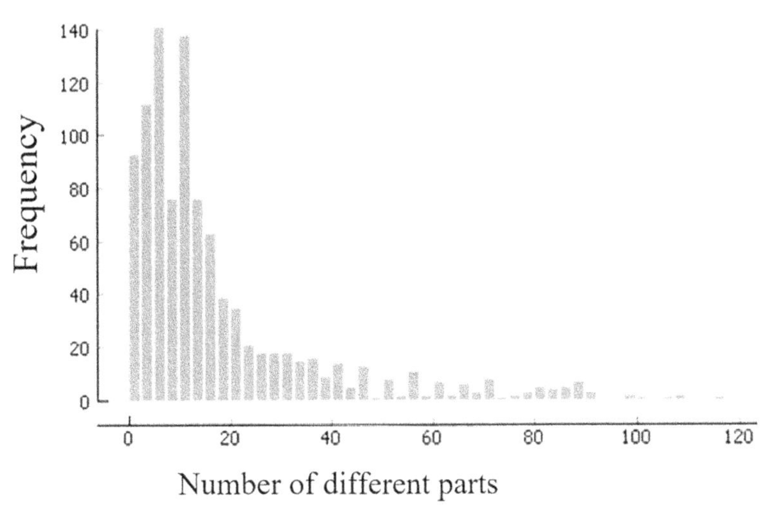

Fig. 6.13. Distribution of Different Parts (Kurasova et al., 2021).

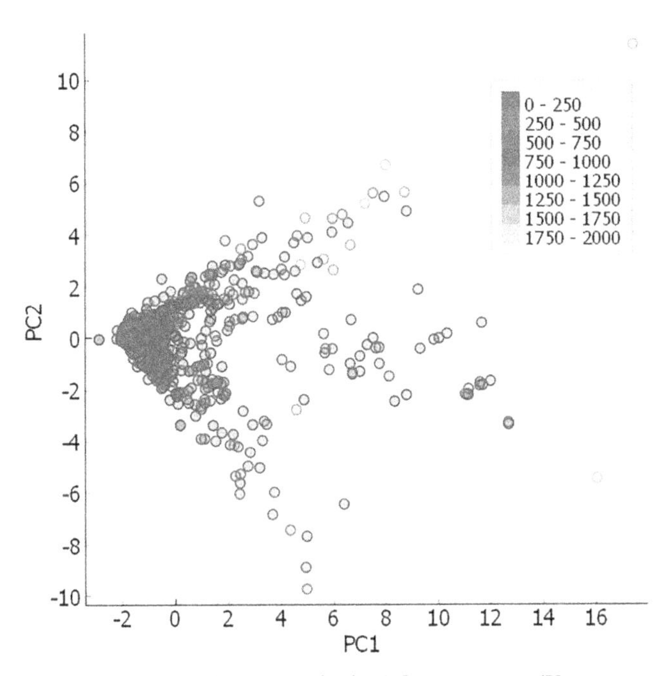

Fig. 6.14. Data Visualised by Two Principal Components (Kurasova et al., 2021).

group of points can be observed at the bottom, while a smaller cluster is present at the top. It is noteworthy that points representing expensive products (light grey) form specific clusters, unlike the PCA and MDS methods where no such clusters are formed. It is also important to mention that no clustering method

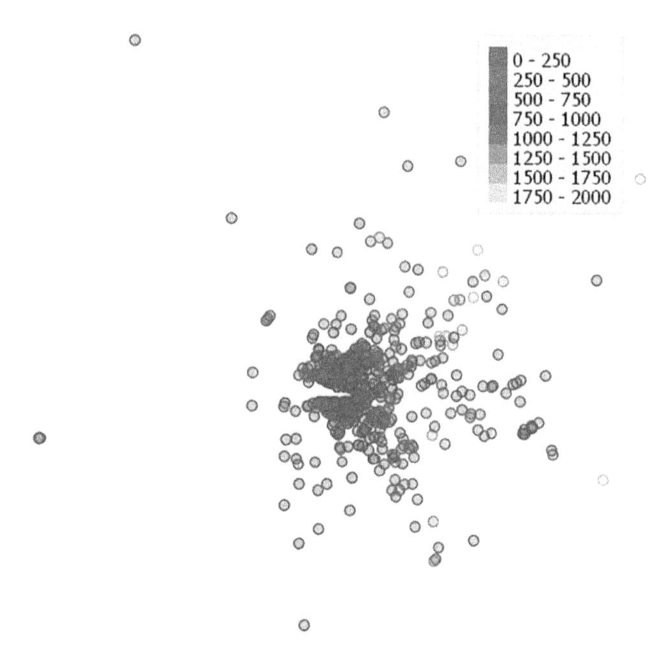

Fig. 6.15. Data Visualised Using the Multidimensional Scaling Method (Kurasova et al., 2021).

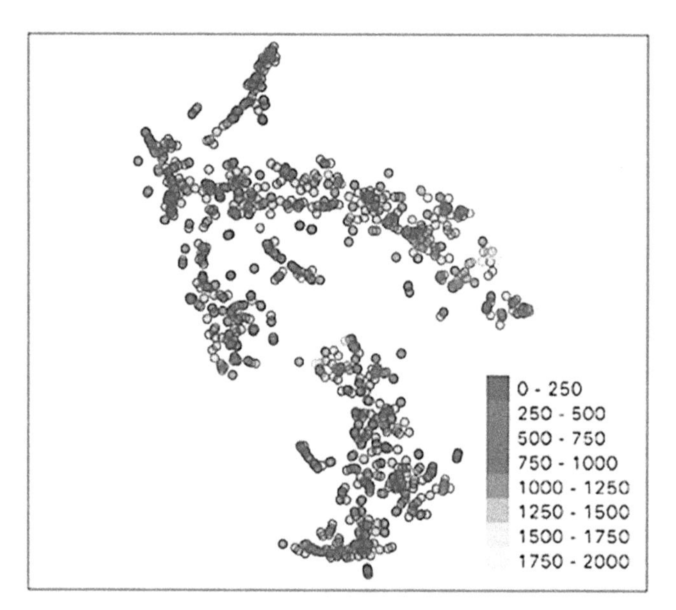

Fig. 6.16. Data Visualised by the t-SNE Method (Kurasova et al., 2021).

was used in this analysis. We only observe the formation of clusters by examining the resulting image.

Clustering methods are commonly used in visual analysis. The Louvain algorithm is a specific method used in this study.

Clustering methods are commonly used in visual analysis. The Louvain algorithm is a specific method used in this study (Traag et al., 2019). A clustering algorithm that operates hierarchically by merging communities into a single node is utilised for graph clustering. While primarily intended for community detection, it can also be used for other clustering purposes. The furniture data was clustered via Louvain's algorithm and subsequently visualised with the t-SNE method (Fig. 6.17). The data points have been colour-coded to correspond with their respective clusters, as determined by Louvain's algorithm, which has identified six distinct clusters. Examining the relationship between these clusters and the price point can be quite illuminating. For example, we could select the points belonging to cluster C6 (indicated by the colour purple) and plot them in a scatter diagram (Fig. 6.18). The points in the graph are coloured based on the price. We can observe that cluster C6 comprises of furniture items that are priced below €250. However, there are some points that do not belong to this cluster and it is important to investigate these products. The outliers may be due to errors in pricing, incorrect attribute values, or any other inaccuracies. Therefore, it is necessary to analyse these products and determine the cause of the outliers.

It can be observed that the cheaper products are closely clustered together, while the more expensive products are located farther apart. As mentioned earlier, large datasets are necessary to properly train an autoencoder neural network. Therefore, this method is not well suited for the data under study (Fig. 6.19).

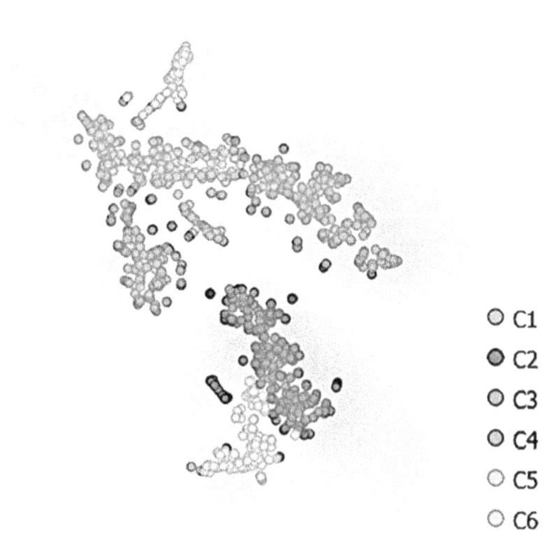

Fig. 6.17. Data Clustered by Louvain's Algorithm and Visualised by t-SNE (Kurasova et al., 2021).

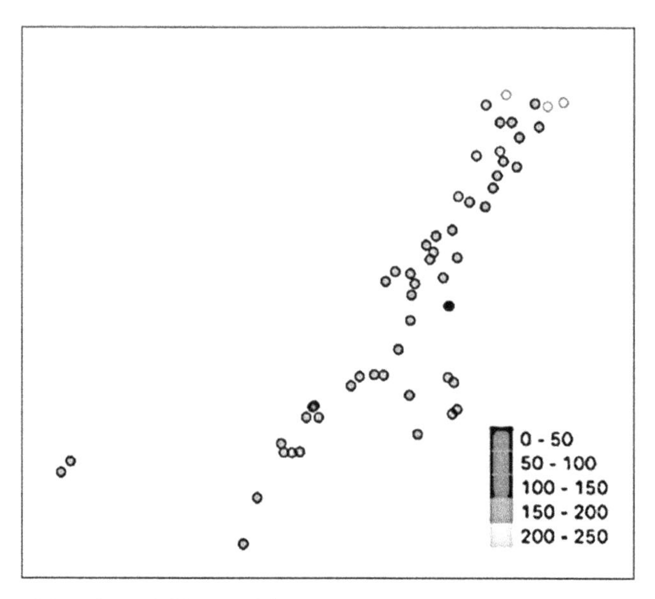

Fig. 6.18. Mapping of Cluster C6 (Kurasova et al., 2021).

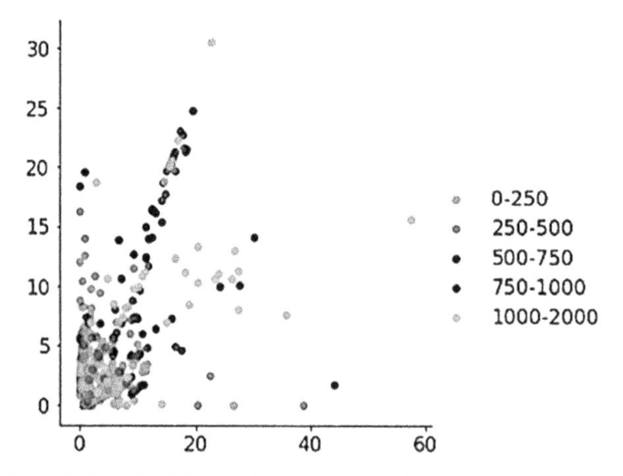

Fig. 6.19. Data Visualised by an Autoencoder Neural Network (Kurasova et al., 2021).

It is important to note that the products analysed in this study vary in size and complexity. The dataset includes both small individual pieces of furniture as well as large furniture sets. To better analyse the data, we have chosen a subset of products with prices ranging between EUR 100 and EUR 500. The data for these products has been visualised using the t-SNE method (Fig. 6.20). Here, the

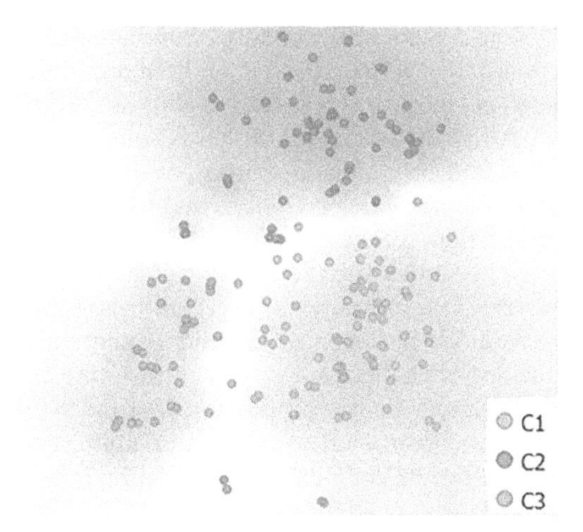

Fig. 6.20. Data Subset Clustered by Louvain's Algorithm and Visualised by t-SNE (Kurasova et al., 2021).

clusters are obtained using Louvain's algorithm. We observe the presence of three significant clusters.

In this chapter, we have highlighted the advantages of using visualisation in data mining tasks. We have specifically focussed on the production of customised furniture, where the main challenge is to predict the prices of products before the design phase. Our proposed integrated data visualisation process includes various visualisation techniques that can be useful for exploratory data analysis. These methods enable decision-makers to better understand the data, thereby allowing them to estimate and predict product prices more accurately. Moreover, the proposed process can be integrated into a decision support system that is intended for furniture company managers and designers. This system allows them to evaluate a custom furniture order quickly and conveniently and provide an initial price for the products. We have designed a visual analysis module to support this decision support system and ensure its usability and efficiency.

Empowering and Engaging Industrial Workers Using Structured Expert Judgment

In the previous sections, we discussed the challenges of customised manufacturing. These challenges include the rising global competition for customised solutions in the furniture manufacturing sector and the regional uniqueness of the Baltic States and Europe as a whole. Addressing these challenges requires a collaborative effort to stay informed with the latest digital and social knowledge (Colangelo et al., 2018).

Incorporating potentially new standards into a production process that allows for flexibility requires more resources in terms of both production time and cost.

To ensure this flexibility, avoid product quality errors, and reliably assess production procedures, costs, and deadlines, companies need decision support tools.

From a managerial perspective, it is crucial to address the difficulties and complexities involved in early cost estimation. It has become increasingly important to assess costs and lead times in a timely manner, as these factors also have their own cost component. The furniture manufacturing sector is not the only industry facing such challenges. Nanotechnology, aircraft manufacturing, and other sectors are also dealing with similar situations (Chou et al., 2010).

There are new trends in big data research (Elmaraghy et al., 2012) and decision support systems aided by machine learning (Caputo & Pelagagge, 2008) that can help us tackle the challenges we face. In furniture companies, we can use the concept of big data to analyse production data and extract useful knowledge from the overwhelming amount of production noise. This knowledge can then be used to make informed decisions and take action to improve production processes.

Sophisticated early price estimation systems, based solely on machine learning, may not be sufficient in a compact and ever-changing world. To ensure higher accuracy and greater flexibility of valuation systems, an element of employee involvement is proposed. This will allow for fine-tuning of digital solutions in real-time. In this discussion, we will explore the prospects for frugal innovation when expert involvement is used as a corrective tool for early cost estimation. Machine learning will be coupled with data visualisation to enable more efficient and accurate cost estimation.

Problems in decision-making and cost estimation often arise due to lack, insufficiency or inaccuracy of available data. Human knowledge often complements data driven estimation. However, when expert (human) knowledge is elicited, a structured elicitation should be employed. In addition to the challenges of incorporating the results from expert judgement sessions into an intelligent cost estimation system, there are also considerations of expert validation and establishing trust between tool users and experts. The cultural nuances further complicate the customisation process. The aim of this section is to explore the cultural dynamics of competence recognition in the furniture industry. With this understanding, we can recommend the most suitable structured expert judgement approach to authenticate expert input and integrate it with the cost estimation decision support tool. The idea of engaging and empowerment of employees has been refined over time in the social sciences, and is now being supported by new evidence of its substantial impact on performance, which adds value to the entire community, whether it is a policy process or a company with its own internal strategic objectives.

There is already compelling evidence that involving employees in decision-making, which includes decisions about various aspects of their work and the organisational environment, is increasingly valuable (Wikhamn et al., 2022). This is not only because it benefits the employees themselves, but also because they can contribute their knowledge and expertise to the decision-making process. Such values include organisational innovation, organisational performance, group organisational citizenship behaviour, individual job satisfaction.

Cost Estimation and Expert Based Knowledge

Cost Estimation

The simplest and most common technique for estimating complex, multifaceted costs, like the cost of manufacturing products is to partition the problem into smaller, constituent elements and then estimate the cost of these elements and the time necessary to produce them (Helbig et al., 2014). However simply adding up prices of individual components may also turn out to be a complex task, due to several uncertainties. The first source of uncertainty springs from the consequences of choosing between possible design solutions. Thus, the early design stage is very important in assuring the reliability of cost estimation models (Sjöberg & Jeppsson, 2017). A comprehensive approach to cost calculation involves unbundling elements by levels of uncertainty. Elements can be categorised as facts, estimated values, and application parameters (Helbig et al., 2014). We define facts as the precise measurements obtained at the end of the production process, while estimated values statistical observations or data collected through experiments (e.g. working hours).

Customised manufacturing often uses traditional cost estimation methods, which mostly depend on the level of complexity. However, attempts to estimate costs manually, particularly at the early design stage, have not yield accurate or timely results (Mikulskienė & Vedlūga, 2019).

Increasingly, a variety of industries – from manufacturing and software engineering to process engineering, construction, and scientific research - are turning to the power of artificial intelligence. As a result, cutting-edge techniques that leverage massive amounts of data are providing exciting new opportunities for customised manufacturing. Among the most sought-after analytical tools are those that rely on analogous cost estimation, bottom-up estimation techniques, and the fusion of computing technology with artificial intelligence. Cost estimation remains a complex problem, which continues to attract considerable research attention. According to Shane et al. (2009), accuracy and comprehensiveness in cost estimation are delicate issues and can be easily affected by many different parameters. In addition, each parameter must be properly addressed in order to maintain an acceptable level of accuracy during the process (Elfaki et al., 2014). Researchers have attempted different models (Leung & Fan, 2002). One such model based on artificial intelligence techniques, is detailed in Finnie and Wittig's work. They use artificial neural networks and case-based reasoning (CBR) in their evaluation efforts (Finnie & Wittig, 1996). The authors state that artificial neural networks can provide accurate estimates when there are complex relationships between variables and when the input data is distorted by high noise levels. Case-based reasoning solutions for new problems are based on solutions to old similar problems (Srinivasan & Fisher, 1995). However, the results from CBR were less encouraging than expected: in 73% of the cases, the estimates were within 50% of the actual effort, and for 53% of the cases, the estimates were within 25% of the actual effort (Leung & Fan, 2002). Srinivasan and Fisher used machine learning approaches based on regression trees and neural networks to estimate costs (Srinivasan & Fisher, 1995). A primary advantage of learning systems is that they are adaptable and nonparametric.

Another stream of literature in price estimation combines data with experts' knowledge. Briand, El Eman and Bomarius proposed a hybrid cost modelling method called COBRA. COBRA stands for Cost estimation, Benchmarking and Risk Analysis (Briand et al., 1998). This method is based on expert knowledge and quantitative project data from a small number of projects.

The current methods have been using artificial intelligence techniques in a fairly improvised manner, such that they can easily estimate the cost of a project even where limited data is available. More automation and improvements of this process led to the emergence of stronger methodologies (Chirra & Reza, 2019). Artificial intelligence is largely understood as machine learning, which can be considered as a data analysis method used to automate model building and quantification. The basic idea is that machine learning software can learn patterns and models from data and use previous experience to predict future outcomes. Machine learning algorithms used in general and in particular for cost estimation, can refer to neural networks, fuzzy logic, genetic algorithms, Bayesian networks, support vector regression or expert-based knowledge (Chirra & Reza, 2019).

Expert-based Knowledge

The estimated cost of a new customised order does not achieve enough accuracy despite the various methods and multiple methodologies involved in the estimation, due to its highly variable nature. Moreover, the tight deadlines require quick correction or solid justification of the preliminary numbers, aided by additional sources of knowledge. One possible source of valuable information is expert advice. Expert contribution is often utilised in areas where there is no explicit conceptual framework or where data are scarce (Scapolo & Miles, 2006) However, before considering how to integrate expert knowledge in to price estimation process, we must first address the issue of expert recognition. Continuous debates have been about who can be considered an expert and how to select the most suitable expert in the field. How can one know whose judgement is valid (or more reliable) than another? Previous studies assure us that we cannot assess that priority, and the only way of knowing is by assessing prior performance on similar tasks. Recently, expert judgment has been increasingly recognised as another type of scientific data. Recognition of expert-based knowledge is almost inseparable from some cultural issues underlying understanding the value added of professional knowledge. There is a certain tension between professional knowledge and other knowledge, which begins to emerge when knowledge uses overlap and comprehensiveness can only be achieved when any type of knowledge is used in combination. As research progresses, it becomes increasingly clear that professional expertise holds more weight than the practical knowledge of clients (de Graaff et al., 2019). There is a gap in the literature regarding product customisation and its impact on price estimation. When addressing how expert knowledge can be used in a price estimation task, the problems associated with the difficulties of integrating expert knowledge into a universal knowledge management system cannot be ignored. Knowledge management systems depend not only on the source of the knowledge (in this case, the expert), but also on the knowledge accumulation strategies. There are three most common knowledge preservation strategies: technology-based, interaction-based (capturing the process and practices) and culture-based (best practices with interactions of professionals) (Levallet & Chan, 2019). One of the significant aspects of a technology-based approach is its ability to gather and retain factual data and information (Blankenship & Bruck, 2008). One effective way of accumulating knowledge is through interaction and learning from past experiences. Another approach is to gain specific experience in a particular field by immersing oneself in its culture. Both of these methods can lead to valuable insights and growth (Cooke & Goossens, 1999). The last two are most relevant when using expert knowledge. Thus, the accuracy of pricing is highly dependent on the established knowledge accumulation strategy, which has a direct link with production practices and expert recognition and interaction with factual data strategies. Indeed, culture-based strategies vary from sector to sector, depending on the size of the sectors, the profitability and the degree of sector customisation. Typically, trust in experts is usually limited by the formal skills of the expert, recognising only specialists with a narrow specialisation. The inclusion of other related professionals in price estimation can be seen as a cultural process and can be analysed as an integral part of pricing. It is rational to use an expert-based approach when information is limited and the criteria to be assessed are undefined, for instance: furniture complexity, unique specifications, changing market variables, market size of specific furniture, market risk. Thus, areas where uncertainties are common are more confident in using expert knowledge. However, some businesses are often still cautious and more often avoid expert input, rather than using it in calculating costs (Cooke & Goossens, 1999).

Structured Expert Judgment (SEJ)

A wide variety of techniques are considered to be under the expert judgment umbrella. A range of techniques are available, each varying in their degree of rigour and scope, from individual opinions to think-tanks with external validation. There are frequently uncertainties around the parameters required for decision-making frameworks or modelling physical behaviour. Experts in different fields may possess the necessary expertise regarding models and applicable parameters within their specific area of interest. The quantitative assessment and combination of expert opinions can be an invaluable contribution to decision-making, potentially resulting in an optimal quantification of model parameters. These assessments can be simply point estimates of unknown quantities or parameters, or point estimates accompanied by a quantification of the uncertainty surrounding these estimates. This quantification of uncertainty is usually elicited in the form of probability distributions (or certain percentiles of such distributions). Studies have revealed that an individual's capacity to express assessments in terms of probabilities fluctuates based on their level of exposure and experience (French, 2011).

Whatever we chose to elicit as experts' opinions, by agreeing that we are going to use a panel of experts, we must assume that the elicitation outcome will consist of a set of estimates, distributions, etc., that will need to be aggregated for later use. There are two main ways in which experts' opinions are pooled (Clemen & Winkler, 1999): using *behavioural aggregation*, which involves striving for consensus via discussion (e.g. O'Hagan et al., 2006) or using *mathematical aggregation* which provides a more explicit, auditable, and objective approach to aggregation (e.g. Cooke & Goossens, 2008). Since mathematical aggregation methods allow little interaction between experts, the discussion is usually limited to ensuring that there is a common understanding of the questions among the experts prior to the elicitation.

A weighted linear combination of opinions is one example of such aggregation. Equal weighting is often used mostly because of its simplicity (since no justification for weights is required). Evidence also shows that the equal weighting scheme frequently performs quite well relative to more sophisticated aggregation methods (e.g. Clemen & Winkler, 1999), but not always (e.g. Cooke & Goossens, 2008).

Nonetheless, it is worth noting that an expert whose assessments deviate greatly from those of their colleagues can exert a considerable influence on the final decision. This can be a significant disadvantage if there is no valid justification for the expert's evaluations. Furthermore, as the number of experts involved in the study increases, the aggregated distribution, weighted evenly, often becomes quite uncertain (Cooke & Goossens, 1999).

It is also important to stress that differential weighting based on anything other than prior expert performance on similar tasks should be avoided. A few examples from the literature support this recommendation: for example, the authors of (Burgman et al., 2011; Cooke et al., 2008; Woudenberg, 1991) used self-ratings, peer-ratings, and citation indices to formulate weights and found that the performance of an aggregated opinion using such weights performs poorly in terms of statistical accuracy and informativeness. Probably the most well-known and widely used version of a differential weighting scheme is the Classical Model or Cooke's model (CM) for SEJ (Burgman et al., 2011; Cooke et al., 2008; Woudenberg, 1991) which uses calibration or seed variables to derive performance-based weights proportional to how calibrated and informative the experts' estimates are. Seed variables are variables taken from the experts' domain for which the true values are known (to the analyst but not to the experts), or will become known, within the time frame of the study.

Expert judgment is used in a wide range of areas, including nuclear safety (Pulkkinen et al., 2002; Simola et al., 2002); aircraft engineering (Peng et al., 2011); air traffic control (Nunes & Kirlik, 2005) and software production (Jørgensen, 2004). Two recent books on SEJ include applications in food security, health care decision making; nuclear threat risks; risk assessment in a pulp and paper manufacturer in the Nordic countries (Dias & Morton, 2018); and medicine policy and management, supply chain cyber risk management, geopolitical risks, terrorism, and the risks facing businesses looking to internationalise (Hanea et al., 2021). A database maintained by Professor Roger Cooke reports over 100 uses internationally, including from the nuclear sector; chemical and gas industries: groundwater/water pollution/dike ring/barriers; aerospace sector: space debris/aviation; occupational sector: ladders/buildings; health sector: bovine/

chicken (Campylobacter)/SARS; banking sector: options/rent/operational risk; and volcanoes/dams. In the context of this chapter expert judgement is also used extensively for cost estimation (Rush & Roy, 2001), where experts have to make assumptions and judgments about the cost of a new product.

The standard steps used in any structured protocol for eliciting experts' opinion start with the formation of an expert group, formulation of seed (if used) and target questions for experts, expert briefing and training, followed by the individual expert assessments of variables [e.g. Cooke & Goossens, 1999). The classical model for SEJ is no exception, and these steps are followed by calculating the calibration and informativeness scores for each expert, converting of these scores into weights and finally using these weights to form the most calibrated and informative mathematical aggregation of judgements. These last steps are implemented in the EXCALIBUR software for CM (Aspinall, 2008).

Methodology for SEJ Application

To analyse the cultural attitudes surrounding expert recognition and evaluate the statistical precision of expert estimates in price discovery, a mixed-methods research approach was employed, as outlined in Chapter 4. The mixed methods strategy included a qualitative study, a quantitative study, and an SEJ experiment. A qualitative study was carried out to examine the pricing stages utilised by companies and their employment of expert knowledge to validate price predictions. This investigation provides valuable insights for:

- determining the approach to inviting experts to the SEJ experiment;
- identifying the pertinent questions (related to prices) to ask; and
- establishing the anticipated level of accuracy in price prediction.

A quantitative survey was carried out to gather information on the use of expert knowledge in pricing and the cultural norms surrounding competence recognition in the furniture industry. The third study focussed on SEJ elicitation to assess experts' ability to forecast prices. By using both methods, the research aims to identify any behavioural patterns that may hinder the implementation of SEJ in a furniture company. Additionally, the study seeks to develop an effective SEJ strategy to validate expert input in the decision support tool for cost estimation.

Structured expert judgement. The research is based on a structured approach to expert judgement, namely the Classical model (Burgman et al., 2011; Cooke et al., 2008; Woudenberg, 1991). CM uses seed questions to identify the most appropriate combination of experts as measured by the performance of the aggregation. This process identifies the most accurate experts within the company who are best suited for cost estimation. The selection of the panel is crucial, as demonstrated by a study conducted at a Lithuanian furniture manufacturing company. Ten employees were nominated as experts, including the Managing Director, Design Engineer, Chief Accountant, Chief Financial Officer, Head of Finance, IT Manager, Designer, Senior Project Administrator, Project Administrator, and Senior Product Manager. The survey evaluated two dimensions of expert performance: statistical accuracy (calibration) and informativeness. The calibration questionnaire included general questions about the expert and initial inquiries. Prior to the expert elicitation, a set of 20 questions about their experience were answered by each expert. The experiment was conducted in two stages, with a list of ten products given for price estimation. Experts were tasked with suggesting possible prices based on available data about the product: metal preparation time, metal assembly time, metal painting time, metal packing time, wood preparation time, wood assembly time, wood collection time, wood colouring time, wood packing time, material quantity in meters (all what is measured by the amount in meters, for example cords, etc.), the amount of materials in square meters (everything is measured by the amount in square meters, for example plates, etc.), the amount of materials in units (everything is counted in units, for example bolts, etc.), preliminary price of the product.

After the first experiment, a second experiment was performed. In this second experiment, a list of 32 products was provided for the price calculation. Eight of these products were among the most expensive products, nine where from the cheapest range of products, and 15 products were from the average priced products. In terms of size, seven products were among the largest and 12 were among the smallest products. For this second experiment, more detailed product information than in the first experiment was provided. That is: metal preparation time, metal assembly time, metal painting time, metal packing time, wood preparation time, wood collection time, wood collection time, wood colouring time, wood packing time, material amount in meters (everything in terms of quantity in meters, examples of cords, etc.), quantity of materials in square meters (everything in quantity in square meters, for example plates, etc.), quantity of materials in units (all in units such as bolts, etc.), quantity of different materials, quantity of parts (how many parts are made of product), quantity of different parts.

The EXCALIBUR software package (Ababei, 2019) was used for compiling the answers and for calculating the two separate scores: calibration and informativeness. Each expert's overall score is determined by multiplying their individual scores, which is then used to assign a weight to each expert (Cooke & Goossens, 2008). In order to calculate the scores, experts are asked to provide their best estimate along with the lowest and highest plausible values within a certain tolerance.

In order to calculate these scores, experts are asked to provide their best estimates along with the lowest and highest plausible values within a certain tolerance. These are modelled as percentiles of the experts' subjective distribution characterising the uncertainty around the cost to be estimated. The best estimate is considered to be the median, that is, the 50th percentile, the lowest bound is taken to be the fifth percentile, and the upper bound is the 95th percentile of the distribution.

From Experiments to SEJ Benefits of Employees' Engagement for Early Cost Estimation

Structural expert judgment. A couple of SEJ elicitation exercises were undertook in one company which employs about 50 people. Of these 50, 15 are in management and finance, including two supply managers, and four of the employees are constructors. The aim was to explore whether the previously mentioned behavioural observations, indicating that companies rely heavily on designers, constructors, and product managers during the cost estimation process, and the three theories derived from the observed patterns, hold merit and can be supported to some extent through precise forecasting of trusted personnel.

Ten employees of this company kindly agreed to participate in the first elicitation, eight in the second, with four of them being active in both elicitations. The experts answered questions individually, on paper questioners for the first elicitation and electronically in the second. Experts answered 20 general questions about furniture sector, economics, and company demographics during the first elicitation exercise. Then they answered questions about the price of 10 randomly selected products. To ease this second task, the experts were presented with details about twelve aspects of the products: metal preparation, assembly, painting and packing time, wood preparation, assembly, collection, colouring and packing time, material quantity (m), amount of material (m^2), amount of different materials (units), and the preliminary price of the product.

For the second experiment, 32 products were selected such that they cover the price ranges: eight most expensive products, nine of the cheapest products, and 15 of the average priced products. More detailed product information was provided, for example, the quantity of different materials, the quantity of different parts, etc.

When true answers are known for the questions asked (i.e. when seed questions are available), they can be checked against the experts' answers. A good expert is one whose assessments capture the true values consistently in the long run (statistically accurate), with bounds that are as narrow as possible (informative). Informativeness is gauged by 'how far apart the percentiles are' relative to an appropriate background (e.g. the uniform distribution on the potential range which can be chosen in absence of information/knowledge). Measuring statistical accuracy requires the true values for a set of assessments. If the expert's assessments are statistically accurate, or calibrated, then in the long run, 5% of their answers should fall below the 5th percentile, 90% of the answers should fall between the fifth and the 95th percentiles, and 5% of their answers should fall above the 95th percentile. In gauging overall performance, calibration is more important than informativeness. Non-informative but calibrated assessments are useful, as they sensitise us to how large the uncertainties may be; highly informative but uncalibrated assessments are not useful.

The calibration is a number between 0 and 1, with 1 representing perfect calibration. Informativeness is a positive number with larger numbers representing good informativeness scores.

SEJ experiment 1. The experts' assessments from the first SEJ elicitation were saved in the EXCALIBUR software and calibration and informativeness scores were calculated based on the set of 20 seed variables. Fig. 6.21a shows the scores of all experts (calibration scores in column 3 and informativeness in columns 4 and 5) and the weights they would obtain if we were to look for the best-weighted linear aggregation of opinions. Only two experts out of ten were

Results of scoring experts
Bayesian Updates: no Weights: global DM Optimisation: yes
Significance Level: 1,673E-005 Calibration Power: 1

Nr.	Id	Calibr.	Mean relative total	Mean relative realization	Numb real	UnNormalized weight	Normaliz.weigl without DM	Normaliz.weig with DM
1	ZG	1,022E-008	1,724	1,724	19	0	0	0
2	LA	3,139E-011	1,449	1,449	20	0	0	0
3	DM	7,213E-010	1,896	1,896	20	0	0	0
4	NA	1,215E-007	1,165	1,165	20	0	0	0
5	MG	1,673E-005	1,336	1,336	20	2,236E-005	0,3489	0,0006074
6	MK	2,099E-007	1,714	1,714	20	0	0	0
7	DB	3,134E-007	1,405	1,405	19	0	0	0
8	LG	8,33E-013	2,049	2,049	20	0	0	0
9	VK	2,534E-005	1,646	1,646	20	4,172E-005	0,6511	0,001133
10	JM	2,506E-009	1,696	1,696	20	0	0	0

(a)

Results of scoring experts
Bayesian Updates: no Weights: global DM Optimisation: yes
Significance Level: 0,01811 Calibration Power: 1

Nr.	Id	Calibr.	Mean relative total	Mean relative realization	Numb real	UnNormalized weight	Normaliz.weigl without DM	Normaliz.weig with DM
1	ZG	0,002841	1,549	1,549	11	0	0	0
2	LA	3,651E-005	1,719	1,719	11	0	0	0
3	DM	2,614E-006	1,322	1,322	11	0	0	0
4	NA	0,002126	1,024	1,024	11	0	0	0
5	MG	0,01364	1,538	1,538	11	0	0	0
6	MK	0,01811	1,223	1,223	11	0,02216	0,3178	0,03204
7	DB	9,874E-005	1,097	1,097	10	0	0	0
8	LG	3,82E-005	1,73	1,73	11	0	0	0
9	VK	0,037	1,286	1,286	11	0,04756	0,6822	0,06879
10	JM	8,967E-006	1,192	1,192	11	0	0	0

(b)

Fig. 6.21. Results of Scoring Experts on (a) 20 Seed Questions, and on (b) 11 Seed Questions.

given non-zero weight, which suggests that they were the two experts with the best performance on the seed variables. These were the chief financial officer and the senior project administrator. Given that calculating performance based on very general (sometimes called 'almanac') questions may give unreliable results, an additional analysis was carried out. In this, the scores were recalculated based on only the questions about internal processes in the company (11 seed questions). The results indicate that the IT manager and chief product manager received the highest weights from the experts (Fig. 6.21b). It is worth noting that while the calibration scores changed dramatically (by orders of magnitude) the informativeness shifted just a little. This behaviour is explained by several factors such as: the calibration is a very fast changing function, very sensitive to the number of seed variables, whereas informativeness is a slow function which fluctuates very little. Moreover, the 11 seed questions left are more representative the knowledge base of our experts, hence their performance improved when the questions spanned their knowledge space better. The calibration is basically the *p*-value of a statistical test whose null hypothesis is that experts are well calibrated. When more seed variables are used the power of the statistical test increases, and the values of the calibration score decrease accordingly.

It is interesting to notice that top managers were not rank among the best performers in none of the two analyses.

Ideally, a decision-making strategy is informed both by experts and data, when available. If sufficient representative data can be obtained, machine learning algorithms can be used to model it. To advance research, authentic data from a Lithuanian furniture manufacturing company was utilised to estimate costs via machine learning algorithms, including actual product costs obtained through surveying experts (Kurasova et al., 2021). The experiment involved six experts from within the company, including the chief accountant, IT manager, two designers, a project administrator, and a chief product manager. The cost evaluation was based on data regarding material costs and the duration of manufacturing procedures. Earlier research had identified the chief accountant, IT manager, and chief product manager as the top-performing experts. The results of this experiment showed that the cost estimates provided by the chief accountant were very close to the actual product price and those generated by machine learning techniques. However, it is noteworthy that machine learning techniques require a significant amount of historical data and identification of essential data features to produce accurate results, while experts can provide cost estimates without historical data.

SEJ experiment 2. The experts' assessments from the second SEJ elicitation were also saved in the EXCALIBUR software and calibration and informativeness scores were calculated again. From the 32 products selected for the second elicitation, 23 prices were known with certainty. Results of cost estimation for four products demonstrated in Fig. 6.22. This equates with having 23 seed variables. However, some of 23 questions were very similar, and would skew the scores towards values which may not reflect realistic scores. We have selected 14 of these questions, making sure that we still cover the entire range of products with high, medium, and small prices, to recalculate scores (Fig. 6.23).

The constructor who is the person in charge of calculating preliminary prices and the chief product manager were ranked as the two best performers on this set of questions (Fig. 6.24). Extra scores are calculated for two 'virtual' experts: EW corresponding to an equally weighted combination of the eight experts' distributions (expert 17), and PW corresponding to a differentially (performance based) weighted combination of the eight experts' distributions (expert 16). EW seems to obtain the best calibration possible but is less informative than PW. A smaller informativeness score means a wider distribution of prices, reflecting more uncertainty.

The experts seemed to have difficulties with estimating the larger prices, and when all seven questions about these are removed, both the equally weighted

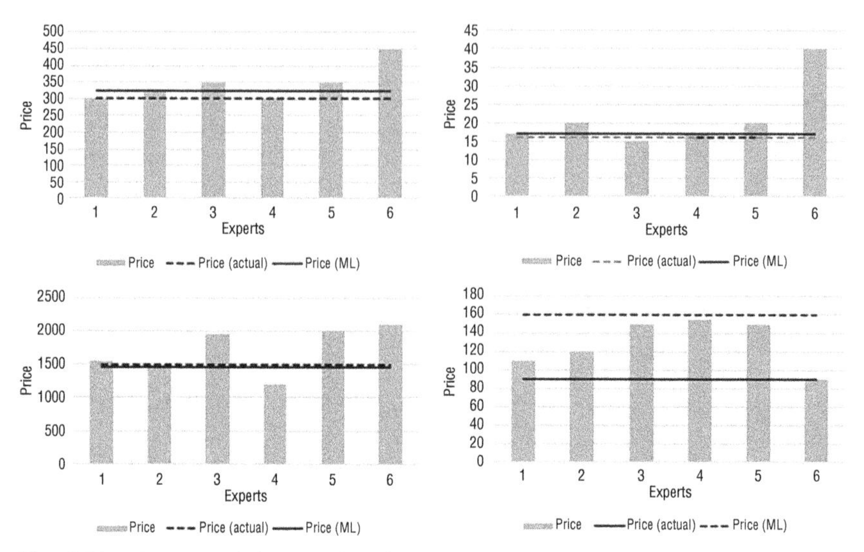

Fig. 6.22. Results of Cost Estimation for Four Products (Solid Line – the Real Cost of the Product, DOTTED Line – Machine Learned Cost Estimation).

Nr.	Id	Calibr.	Mean relati total	Mean relati realizatioo	Numb real	UnNormalize weight
1	AL	3.347E-007	1.735	1.814	14	6.073E-007
2	DB	0.05606	0.6795	0.767	14	0.04299
3	L	1.25E-005	1.787	1.806	14	2.258E-005
4	LG	1.051E-005	2.074	2.21	14	2.321E-005
5	NA	1.682E-005	1.428	1.419	14	2.387E-005
6	RS	8.19E-005	1.529	1.803	14	0.0001477
7	RU	1.866E-008	1.853	1.994	14	3.719E-008
8	VK	0.0001825	1.249	1.343	11	0.0002451
9	EW	0.2048	0.24	0.2666	14	0.05461
10	PW	0.05606	0.6204	0.7059	14	0.03957

Fig. 6.23. Results of Scoring Experts on 14 Seed Questions and the Equally and Differentially Weighted Combinations of Experts.

combination (EW_noB, expert 14) and the performance based differentially weighted combination (PW_noB, expert 13) achieved good performance. The performance of PW_noB improved even more (see expert 12, PW_noB_rp) when we decrease the power of the test by making it equivalent with that of a test using ten seed questions (which is the standard number of seed questions used in most of the CM applications). Other subsets of seed questions (no small prices, or no large prices and less medium prices, half powered test for all questions, etc., see experts 9, 10, 11, and 15 as numbered in the first column of Fig. 6.24 were investigated and most resulted in better performance than any one individual expert.

Nr.	Id	Calibr.	Mean relati total	Mean relati realizatioo	Numb real	UnNormalize weight
1	AL	3.884E-016	1.841	1.98	15	7.689E-016
2	DB	4.239E-005	0.7303	0.8423	15	3.571E-005
3	L	6.576E-011	1.933	2.039	15	1.341E-010
4	LG	2.361E-011	2.321	2.596	15	6.127E-011
5	NA	6.404E-007	1.444	1.444	15	9.246E-007
6	RS	3.019E-009	1.546	1.812	15	5.469E-009
7	RU	6.019E-015	1.959	2.153	15	1.296E-014
8	VK	2.248E-005	1.188	1.221	15	2.745E-005
9	EW_noS2M	0.003899	0.2407	0.266	15	0.001037
10	PW_noB2M	0.1393	0.6593	0.7391	14	0.103
11	EW_noB2M	0.4318	0.2474	0.2786	14	0.1203
12	PW_noB_rp	0.1813	0.5978	0.664	16	0.1204
13	PW_noB	0.07309	0.6712	0.7455	16	0.05449
14	EW_noB	0.3506	0.244	0.2694	16	0.09447
15	PW_all_hp	0.01614	0.5849	0.6282	23	0.01014
16	PW_all	0.0001267	0.6623	0.7089	23	8.984E-005
17	EW_all	0.009447	0.2318	0.2448	23	0.002313

Fig. 6.24. Results of Scoring Experts on Different Subsets of Seed Questions & and the Corresponding Equally and Differentially Weighted Combinations of Experts.

These analyses suggest that having a group of experts' opinions may serve as a reliable solution when trying to quantify the uncertainty around the cost estimation problem. The formation of the group is a crucial matter to consider. While it is true that key insights and valuable viewpoints can sometimes come from non-experts (Burgman et al., 2011), it is still necessary for the selected participants to possess the requisite knowledge to comprehend the technical, organisational, and financial aspects of cost estimation. An external expert may also be a valuable addition to the group for bringing a fresh perspective and alternative opinions. Having the possibility to create 'virtual' experts by combining the judgements of the experts in the group offers the huge advantage of always having a reliable estimation available, even when one or two experts are not present for the elicitation. The group of remaining experts can compensate for this absence.

Another valuable element of a structured protocol is feedback and discussion. Sometimes experts draw on different knowledge spaces and reason based on different mental models. If they would be able share reasons and discuss differences of opinions, new info may inform their estimates. Below (Fig. 6.25) there are three examples of the spread of answers. Often experts disagree to the extent where their intervals do not overlap. Sometimes, the majority has the estimates tilted towards the right direction (see the top questions), but every so often, it is the outlier who drives the combination towards the right answer (see the bottom question).

The CM for SEJ can be also integrated into a different structured protocol, called the IDEA protocol. IDEA offers the possibility of incorporating feedback and discussion into the elicitation, in a controlled and facilitated fashion and may be considered as an extension in the future. IDEA stands for Investigate, Discuss, Estimate and Aggregate and more details about this protocol can be found in (Hanea et al., 2016; Hemming et al., 2020).

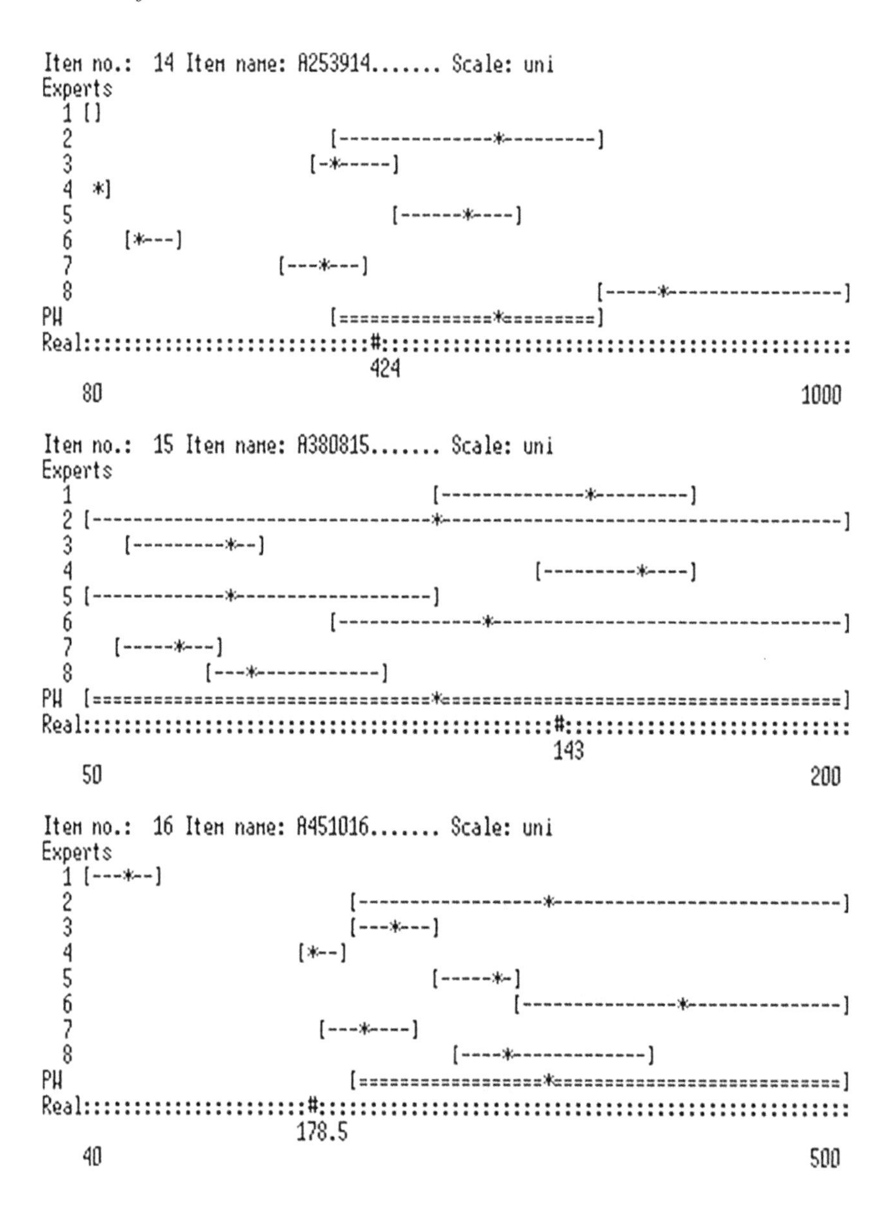

Fig. 6.25. Example of 3 Seed Questions and the Experts Estimates. PW is the Performance-Based Aggregation Based on the 23 Seeds, Calculated with Power Reduced to Half. The '['Represents the Fifth Percentiles, ']' Represents the 95th Percentile and the '*' Represents the Median. The '#' Represents the True Value.

In Conclusion

The main aim of this chapter was to explore whether there are any cultural patterns in the furniture industry that affect how employees' competences are recognised. It also investigates how these patterns may limit the integration of different methodologies, such as machine learning and expert judgement, in making decisions and estimating costs.

Cost estimation at an early design stage turns out to be very important not only because it allows cost optimisation, but also because it contributes to the development of customised production processes, where employees can use their competences to improve the machine learning result. The principles that companies would expect from an on-demand price evaluation system were reviewed, highlighting the importance of material pricing systems and the scepticism towards systems based on past evaluations. The survey conducted on companies' dependence on experts revealed several trends. In the furniture industry, only a handful of specialists are responsible for determining the right price. There seems to be a lack of trust in both the expert methodology and the participation of employees, particularly those who are not directly involved in the production process.

The structured approach to expert evaluation that we have chosen has enabled us to extract the knowledge available to the employees in the best possible way to realise a specific task, and thus to create a form of involvement in the decision that becomes a bridge between the neutral digital solution and the actualised experience of the employee.

Based on a structured experiment on expert elicitation, it was found that experts selected from the company's workforce performed better than any others. Even those who were culturally considered to be good price assessors did not perform as well as the company's own experts. However, these experts cannot be identified prior to calibration, which requires the participation of a well-formed and diverse group of experts.

The study found no evidence to support the idea that top managers can accurately forecast prices, which may be due to the clear division of responsibilities among managers. However, in the furniture sector, top managers and the best engineers are still considered experts. This should be taken into account when developing methodological guidance on the implementation of structured expert assessment, especially in the furniture sector where the cultural model recommends a slightly different approach.

We have proposed a tool for SEJ employee involvement that can aid in making rational decisions. This tool can be customised by establishing cultural boundaries for identifying experts in a particular organisation, sector, or region. The research we have conducted in this chapter is based on the scientific principles of SEJ. We have utilised qualitative analysis to experiment with SEJ and have also explored the use of machine learning methods to validate and augment SEJ.

The previous usage of the tool in many SEJ applications has demonstrated its versatility (Dias & Morton, 2018; Hanea et al., 2021). Its applicability is crucial as the proposed methods follow a structured protocol consisting of a clear sequence of steps. The protocol contains a quantitative part and an acknowledgement of uncertainty, which allows for integration with other data-driven assessment methods (such as machine learning methods).

The price assessment tools discussed in this chapter can be easily transferred to other manufacturing sectors, such as glass manufacturing and automotive manufacturing, provided similar steps are taken and sector-specific business process data is collected.

References

Ababei, D. (2019). *Excalibur*. http://www.lighttwist.net/wp/excalibur

Aspinall, W. (2008). Expert judgment elicitation using the Classical Model and EXCALIBUR. In Briefing notes for Seventh Session of the Statistics and Risk Assessment Section's International Expert Advisory Group on Risk Modeling: Iterative Risk Assessment Processes for Policy Development Under Conditions of Uncertainty/Emerging Infectious Diseases: Round IV. http://dutiosc.twi.tudelft.nl/~risk/extrafiles/EJcourse/Sheets/Aspinall%20Briefing%20Notes.pdf

Azevedo, A. L., & Sousa, J. P. (2000). A component-based approach to support order planning in a distributed manufacturing enterprise. *Journal of Materials Processing Technology*, *107*(1–3), 431–438.

Azvine, B., Cui Z., Nauck, D. D., & Majeed, B. (2006). Real time business intelligence for the adaptive enterprise. In E-commerce technology, 2006. The 8th IEEE International Conference on and Enterprise Computing, E-Commerce, and E-Services, The 3rd *IEEE International Conference On* (p. 29). *IEEE*. https://doi.org/10.1109/CECEEE.2006.73.

Balakrishnan, A., Kumara, S. R., & Sundaresan, S. (1999). Manufacturing in the digital age: Exploiting information technologies for product realization. *Information Systems Frontiers*, *1*(1), 25–50.

Blankenship, L., & Bruck, T. (2008). Planning for knowledge retention now saves valuable organizational resources later. *Journal – American Water Works Association*, *100*(8), 57–61. https://doi.org/10.1002/j.1551-8833.2008.tb09699.x

Boothroyd, G., & Fairfield, M. C. (1991). Assembly of large products. *CIRP Annals – Manufacturing Technology*, *40*(1), 1–4.

Briand, L. C., El Emam, K., & Bomarius, F. (1998). COBRA: A hybrid method for software cost estimation, benchmarking, and risk assessment. In *Proceedings of the 20th international conference on software engineering*, Science Council of Japan, Kyoto (pp. 390–399). https://doi.org/10.1109/ICSE

Brinch, M., Stentoft, J., Jensen, J. K., & Rajkumar, C. (2018). Practitioners understanding of big data and its applications in supply chain management. *The International Journal of Logistics Management*, *29*(2), 555–574. https://doi.org/10.1108/IJLM-05-2017-0115

Brynjolfsson, E., Hitt, L. M., & Kim, H. H. K. (2011). Strength in numbers: How does data-driven decision-making affect firm performance?*Available at SSRN 1819486*. https://doi.org/10.2139/ssrn.1819486

Brynjolfsson, E., Jin, W., & Wang, X. (2023). *Information technology, firm size, and industrial concentration (No. w31065)*. National Bureau of Economic Research.

Burgman, M., Carr, A., Godden, L., Gregory, R., Mcbride, M., Flander, L., & Maguire, L. (2011). Redefining expertise and improving ecological judgment. *Conservation Letters*, *4*, 81–87. https://doi.org/10.1111/j.1755-263X.2011.00165.x

Burgman, M. A., Mcbride, M., Ashton, R., Speirs-bridge, A., Flander, L., Fidler, F., & Twardy, C. (2011). Expert status and performance. *PLoS ONE*, *6*(7), 1–7. https://doi.org/10.1371/journal.pone.0022998

Caputo, A. C., & Pelagagge, P. M. (2008). Parametric and neural methods for cost estimation of process vessels. *Journal of Production Economics*, *112*, 934–954.
Chen, H., Chiang, R. H., Storey, V. C. (2012). Business intelligence and analytics: From big data to big impact. *MIS Quarterly*, *36*(4), 1165–1188.
Chesbrough, H. W. (2003). *Open innovation: The new imperative for creating and profiting from technology*. Harvard Business Press.
Chirra, S. M. R. C., & Reza, H. (2019). A survey on software cost estimation techniques. *Journal of Software Engineering and Applications*, *12*, 226–248. https://doi.org/10.4236/jsea.2019.126014
Chou, J. S., Tai, Y., & Chang, L. J. (2010). Predicting the development cost of TFT-LCD manufacturing equipment with artificial intelligence models. *International Journal of Production Economics*, *128*(1), 339–350. https://doi.org/10.1016/j.ijpe.2010.07.031
Clemen, R. T., & Winkler, R. L. (1999). Combining probability distributions from experts in risk analysis. *Risk Analysis*, *19*(2), 187–203.
Colangelo, E., Kröger, T., & Bauernhansl, T. (2018). Substitution and complementation of production management functions with data analytics. *Procedia CIRP*, *72*, 191–196. https://doi.org/10.1016/j.procir.2018.03.145
Cooke, R. M., ElSaadany, S., & Huanga, X. (2008). On the performance of social network and likelihood-based expert weighting schemes. *Reliability Engineering and System Safety*, *93*(5), 745–756. https://doi.org/10.1016/j.ress.2007.03.017
Cooke, R. M., & Goossens, L. J. H. (1999). *Procedures guide for structured expert judgment*. European Commission.
Cooke, R. M., & Goossens, L. L. (2008). TU Delft expert judgment data base. *Reliability Engineering & System Safety*, *93*(5), 657–674. https://doi.org/10.1016/j.ress.2007.03.005
de Graaff, M. B., Stoopendaal, A., & Leistikow, I. (2019). Transforming clients into experts-by-experience: A pilot in client participation in Dutch long-term elderly care homes inspectorate supervision. *Health Policy*, *123*(3), 275–280. https://doi.org/10.1016/j.healthpol.2018.11.006
Dean, P. R., Xue, D., & Tu, Y. L. (2009). Prediction of manufacturing resource requirements from customer demands in mass-customisation production. *International Journal of Production Research*, *47*(5), 1245–1268.
Demsar, J., Curk, T., Erjavec, A., Gorup, C., Hocevar, T., Milutinovic, M., & Zupan, B. (2013). Orange: Data mining toolbox in Python. *Journal of Machine Learning Research*, *14*, 2349–2353.
Dhaniawaty, R. P., Fadillah, A. P., & Lubi, D. (2020). Design of furniture production monitoring information system. In *IOP Conference Series: Materials Science and Engineering*, *879*(1), 012044. https://doi.org/10.1088/1757-899X/879/1/012044
Dias, L. C., & Morton, A. (2018). *Elicitation: The science and art of structuring judgement*. Springer Cham. https://doi.org/10.1007/978-3-319-65052-4
Donoho, D. (2017). 50 years of data science. *Journal of Computational and Graphical Statistics*, *26*(4), 745–766. https://doi.org/10.1080/10618600.2017.1384734
Dzemyda, G., Kurasova, O., & Žilinskas, J. (2013). *Multidimensional data visualization : methods and applications*. Springer. https://doi.org/10.1007/978-1-4419-0236-8
Elfaki, A. O., Alatawi, S., & Abushandi, E. (2014). Using intelligent techniques in construction project cost estimation: 10-year survey. *Advances in Civil Engineering*, *1*, 1–11. https://doi.org/10.1155/2014/107926
Elfving, J. A., Tommelein, I. D., & Ballard, G. (2005). Consequences of competitive bidding in project-based production. *Journal of Purchasing and Supply Management*, *11*(4), 173–181.
Elmaghraby, W., & Keskinocak, P. (2003). Dynamic pricing in the presence of inventory considerations: Research overview, current practices, and future directions. *Management Science*, *49*(10), 1287–1309. https://doi.org/10.1287/mnsc.49.10.1287.17315

Elmaraghy, W., Elmaraghy, H., Tomiyama, T., & Monostori, L. (2012). Manufacturing technology complexity in engineering design and manufacturing. *CIRP Annals – Manufacturing Technology*, *61*(2), 793–814. https://doi.org/10.1016/j.cirp.2012.05.001

Europos ekonomikos ir socialinių reikalų komitetas. (2020). *Pasiūlymas dėl Europos Parlamento ir Tarybos reglamento dėl Europos duomenų valdymo (Duomenų valdymo aktas)* [COM(2020) 767 final].

Fayyad, U., Piatetsky-Shapiro, G., & Smyth, P. (1996). From data mining to knowledge discovery in databases. *AI Magazine*, *17*(3), 37. https://doi.org/10.1609/aimag.v17i3.1230

Finnie, G. R., & Wittig, G. E. (1996). AI tools for software development effort estimation. In *Proceedings 1996 International Conference Software Engineering: Education and Practice* (pp. 346–353).

French, S. (2011). Aggregating expert judgement. *Revista de la Real Academia de Ciencias Exactas, Fisicas y Naturales. Serie A. Matematicas*, *105*(November 2010), 181–206. https://doi.org/10.1007/s13398-011-0018-6

Grabenstetter, D. H., & Usher, J. M. (2015). Sequencing jobs in an engineer-to-order engineering environment. *Production & Manufacturing Research*, *3*(1), 201–217.

Gunasekaran, A., Yusuf, Y. Y., Adeleye, E. O., & Papadopoulos, T. (2018). Agile manufacturing practices: The role of big data and business analytics with multiple case studies. *International Journal of Production Research*, *56*(1–2), 385–397. https://doi.org/10.1080/00207543.2017.1395488

Hanea, A., McBride, M. F., Burgman, M. A., Wintle, B. C., Fidler, F., Flander, L., & Mascaro, S. (2016). Investigate discuss estimate aggregate for structured expert judgement. *International Journal of Forecasting*, *33*(1), 267–279. https://doi.org/10.1016/j.ijforecast.2016.02.008

Hanea, A., Nane, G., Bedford, T., & French, S. (2021). *Expert judgement in risk and decision analysis*. Springer.

Harris, F., McCaffer, R., Baldwin, A., & Edim-Dorwe, F. (2021). *Modern construction management* (8th ed.). Wiley Blackwell.

Hegedus, M. G., & Hopp, W. J. (2001). Due date setting with supply constraints in systems using MRP. *Computers & Industrial Engineering*, *39*(3–4), 293–305.

Helbig, T., Hoos, J., & Westkämper, E. (2014). A method for estimating and evaluating life cycle costs of decentralized component-based automation solutions. *Procedia CIRP*, *17*, 332–337.

Hemming, V., Hanea, A., Walshe, T., & Burgman, M. (2020). Weighting and aggregating expert ecological judgments. *Ecological Applications*, *30*(4), e02075. https://doi.org/10.1002/eap.2075

Hofmann, E. (2017). Big data and supply chain decisions: The impact of volume, variety and velocity properties on the bullwhip effect. *International Journal of Production Research*, *55*(17), 5108–5126. https://doi.org/10.1080/00207543.2015.1061222

Jäger, J., Schöllhammer, O., Lickefett, M., & Bauernhansl, T. (2016). Advanced complexity management strategic recommendations of handling. *Procedia CIRP*, *57*, 116–121. https://doi.org/10.1016/j.procir.2016.11.021

Jørgensen, M. (2004). A review of studies on expert estimation of software development effort. *Journal of Systems and Software*, *70*(1–2), 37–60.

Kang, H. S., Lee, J. Y., Choi, S., Kim, H., Park, J. H., Son, J. Y., & Noh, S. D. (2016). Smart manufacturing: Past research, present findings, and future directions. *International Journal of Precision Engineering and Manufacturing-Green Technology*, *3*(1), 111–128. https://doi.org/10.1007/s40684-016-0015-5

Kedmey, D. (2014). This is how uber's 'surge pricing' works. *Time*. https://goo.gl/Wo5xSB

Kienzler, M., (2017). Does managerial personality influence pricing practices under uncertainty? *Journal of Product & Brand Management*, *26*(7), 771–784.

Kingsman, B., Hendry, L., Mercer, A., & Souza, A. D. (1996). Responding to customer enquiries in make-to-order companies: Problems and solutions. *Production*, *6*(2), 195–207.

Kingsman, B. G., & de Souza, A. A. (1997). A knowledge-based decision support system for cost estimation and pricing decisions in versatile manufacturing companies. *International Journal of Production Economics*, *53*(2), 119–139.

Kluth, A., Jäger, J., Schatz, A., & Bauernhansl, T. (2014). Evaluation of complexity management systems – Systematical and maturity-based approach. *Procedia CIRP*, *17*, 224–229. https://doi.org/10.1016/j.procir.2014.01.083

Kurasova, O., Marcinkevičius, V., & Mikulskienė, B. (2021). Enhanced visualization of customized manufacturing data. *Computer Science Research Notes*, *4617*, 109–114. https://doi.org/10.24132/CSRN.2021.3101.12

Kurasova, O., Marcinkevičius, V., Medvedev, V., & Mikulskienė, B. (2021). Early cost estimation in customized furniture manufacturing using machine learning. *International Journal of Machine Learning and Computing*, *11*(1), 28–33.

Kusiak, A. (2018). Smart manufacturing. *International Journal of Production Research*, *56*(1–2), 508–517.

Leung, H., & Fan, Z. (2002). Software Cost Estimation. In S. K. Chang (Ed.), *Handbook of Software Engineering* (pp. 1-14). Hong Kong Polytechnic University. https://doi.org/10.1142/9789812389701_0014

Levallet, N., & Chan, Y. E. (2019). Organizational knowledge retention and knowledge loss. *Journal of Knowledge Management*, *23*(1), 176–199. https://doi.org/10.1108/JKM-08-2017-0358

Liozu, S. M., & Hinterhuber, A. (2012). Industrial product pricing: A value-based approach. *Journal of Business Strategy*, *33*(4), 28–39.

Liozu, S. M., & Hinterhuber, A. (2013). Pricing orientation, pricing capabilities, and firm performance. *Management Decision*, *51*(3), 594–614.

Liozu, S. M., Hinterhuber, A., Boland, R., & Perelli, S. (2012). The conceptualization of value-based pricing in industrial firms. *Journal of Revenue and Pricing Management*, *11*(1), 12–34.

Lu, Y. (2017). Industry 4.0: A survey on technologies, applications and open research issues. *Journal of Industrial Information Integration*, *6*, 1–10. https://doi.org/10.1016/j.jii.2017.04.005

Marti, M. (2007). *Complexity management: Optimizing product architecture of industrial products*. Deutscher Universitäts-Verlag Wiesbaden.

Medvedev, V., Kurasova, O., Bernatavičienė, J., Treigys, P., Marcinkevičius, V., & Dzemyda, G. (2017). A new web-based solution for modelling data mining processes. *Simulation Modelling Practice and Theory*, *76*, 34–46. https://doi.org/doi.org/10.1016/j.simpat.2017.03.001

Menezes, S., Creado, S., & Zhong, R. Y. (2018). Smart manufacturing execution systems for small and medium-sized enterprises. *Procedia 51st CIRP Conference on Manufacturing Systems, CIRP*, *72*, 1009–1014.

Mikulskienė, B., Vedlūga, T., Navickienė, O., & Medvedev, V. (2019). Behaviour patterns in expert recognition by means of structured expert judgement in price estimation in customized furniture manufacturing. In A. Nitin, S. Leonidas, & W. Gerhard-Wilhelm (Eds.), *Modeling and simulation of social-behavioral phenomena in creative societies: First international EURO mini conference, MSBC 2019 Vilnius, Lithuania, September 18–20* (pp. 112–125). Springer. https://doi.org/10.1007/978-3-030-29862-3

Mikulskienė, B., & Vedluga, T. (2019). Strategies for complexity management coping with cost estimation. The case of customized furniture manufacturing. In *Proceedings of 8th international conference on industrial technology and management ICITM 2019 March 2–4* (pp. 212–217). Institute of Electrical and Electronics Engineers. https://doi.org/10.1109/ICITM.2019.8710725

Navickienė, O., & Mikulskienė, B. (2019). Orientation of Lithuanian furniture sector to customized production: Impact on corporate governance and knowledge management. In *2nd sustainable solutions for growth conference: Book of abstracts: September 16–17* (pp. 29–30). Wojciech Budzianowski Consulting Services.

Niazi, A., Dai, J. S., Balabani, S., & Seneviratne, L. (2006). Product cost estimation: Technique classification and methodology review. *Journal of Manufacturing Science and Engineering*, *128*(2). https://doi.org/10.1115/1.2137750

Nunes, A., & Kirlik, A. (2005). An empirical study of calibration in air traffic control expert judgment. In *Proceedings of the 49th annual meeting of human factors and ergonomics society*(Vol. 49, pp. 422–426). SAGE Publications

O'Hagan, A., Buck, C., Daneshkhah, A., Eiser, J., Garthwaite, P., Jenkinson, D., & Rakow, T. (2006). *Uncertain judgements: Eliciting experts' probabilities*. Wiley.

Omar, Y. M., Minoufekr, M., & Plapper, P. (2019). Business analytics in manufacturing: Current trends, challenges and pathway to market leadership. *Operations Research Perspectives*, *6*, 100127.

Peng, W., Zan, M. A., & Yi, T. (2011). Application of expert judgment method in the aircraft wiring risk assessment. *Procedia Engineering*, *17*, 440–445. https://doi.org/10.1016/j.proeng.2011.10.053

Piller, F. T., Vossen, A., Ihl, C. (2012). From social media to social product development: The impact of social media on co-creation of innovation. *Die Unternehmung*, *66*(1), 7–27. https://doi.org/10.5771/0042-059X-2012-1-7

Pugh, P. G. (1992). Working top-down: Cost estimating before development begins. *Proceedings of the Institution of Mechanical Engineers. Part G: Journal of Aerospace Engineering*, *206*(2), 143–151.

Simola, K., Pulkkinen, U., Taija, H., Saarenheimno, A., Karjalainen-Roikonen, P. (2002). *Comparative Study of Approaches to Estimate Pipe Break Frequencies, NKS-79. General Purpose Reliability Tool.* https://inis.iaea.org/collection/NCLCollectionStore/_Public/34/022/34022481.pdf

Reddy, B. R., Sujith, A. (2018). A comprehensive literature review on data analytics in IIOT (industrial internet of things). *HELIX*, *8*(1), 2757–2764. https://doi.org/10.29042/2018-2757-2764

Ren, S., Zhang, Y., Liu, Y., Sakao, T., Huisingh, D., & Almeida, C. M. V. B. (2019). A comprehensive review of big data analytics throughout product lifecycle to support sustainable smart manufacturing: A framework, challenges and future research directions. *Journal of Cleaner Production*, *210*, 1343–1365. https://doi.org/10.1016/j.jclepro.2018.11.025

Reuter, C., Brambring, F., Hempel, T., & Kopp, P. (2017). Benefit oriented production data acquisition for the production planning and control. In S. Takata, Y. Umeda, & S. Kondoh (Eds.), *Procedia CIRP: The 24th CIRP conference* (pp. 487–492).

Rimpau, C., & Reinhart, G. (2010). Knowledge-based risk evaluation during the offer calculation of customised products. *Production Engineering*, *4*(5), 515–524.

Rush, C., & Roy, R. (2001). Expert judgement in cost estimating: Modelling the reasoning process. *Concurrent Engineering*, *9*, 271–284. https://doi.org/10.1177/1063293X01009004

Sacha, D., Zhang, L., Sedlmair, M., Lee, J. A., Peltonen, J., Weiskopf, D., & Keim, D. A. (2017). Visual interaction with dimensionality reduction: A structured literature analysis. *IEEE Transactions on Visualization and Computer Graphics*, *23*(Section 6), 241–250. https://doi.org/10.1109/TVCG.2016.2598495

Santoro, G., Fiano, F., Bertoldi, B., & Ciampi, F. (2019). Big data for business management in the retail industry. *Management Decision*, *57*(8), 1980–1992. https://doi.org/10.1108/MD-07-2018-0829

Santos, B. P., Charrua-Santos, F., & Lima, T. M. (2018). Industry 4.0: An overview. In *Proceedings of the world congress on engineering* (Vol. 2, pp. 4–6). IAEN.

Scapolo, F., & Miles, I. (2006). Eliciting experts' knowledge: A comparison of two methods. *Technological Forecasting and Social Change*, *73*(6), 679–704.
Shane, J. S., Asce, A. M., Molenaar, K. R., Asce, M., Anderson, S., Asce, M., & Asce, D. M. (2009). Construction project cost escalation factors. *Journal of Management in Engineering*, *25*(4), 221–230.
Singh, A. (2018). Scope of Business Analytics in Manufacturing, the value of manufacturing analytic. https://alkajmc.medium.com/scope-of-business-analytics-in-manufacturing-dbd9f43f191d
Song, D. P., Hicks, C., & Earl, C. F. (2002). Product due date assignment for complex assemblies. *International Journal of Production Economics*, *76*(3), 243–256.
Sorzano, C. Sánchez O., Vargas J. & Pascual-Montano A. (2014). A survey of dimensionality reduction techniques. *ArXiv* abs/1403.2877 (2014): n. pag.
Srinivasan, K., & Fisher, D. (1995). Machine learning approaches to estimating software development effort. *E Transactions on Software Engineering*, *21*(2), 126–137. https://doi.org/10.1109/32.345828
Svensson, D., & Malmqvist, J. (2002). Strategies for product structure management at manufacturing firms. *Journal of Computing and Information Science in Engineering*, *2*(1), 50–58.
Thoben, K. D., Wiesner, S., & Wuest, T. (2017). "Industrie 4.0" and smart manufacturing – A review of research issues and application examples. *International Journal of Automation Technology*, *11*(1), 4–16.
Traag, V. A., Waltman, L., & Van Eck, N. J. (2019). From Louvain to Leiden: Guaranteeing well-connected communities. *Scientific Reports*, *9*(1), 1–12.
Trost, S. M., & Oberlender. G. D. (2003). Predicting accuracy of early cost estimates using factor analysis and multivariate regression. *Journal of Construction Engineering and Management*, *129*(2), 198–204.
Umeda, Y., Fukushige, S., Tonoike, K., & Kondoh, S. (2008). Product modularity for life cycle design. *CIRP Annals*, *57*(1), 13–16. https://doi.org/10.1016/j.cirp.2008.03.115
Versli Lietuva. (2019). Business Sector. Furniture industry. Retrieved 29 June, 2020, from https://www.enterpriselithuania.com/en/business-sectors/furniture-industry/
Vogel, W., & Lasch, R. (2016). Complexity drivers in manufacturing companies : A literature review. *Logistics Research*, *9*(1), 1–66.
Wang, D., Fang, S. C., & Hodgson, T. J. (1998). A fuzzy due-date bar gainer for the make-to-order manufacturing systems. *IEEE Transactions on Systems, Man, and Cybernetics. Part C (Applications and Reviews)*, *28*(3), 492–497.
Wang, T., Berthet, Q., & Samworth, R. J. (2016). Statistical and Computational Trade-Offs in Estimation of Sparse Principal Components. *The Annals of Statistics*, *44*(5), 1896–1930. http://www.jstor.org/stable/43974703
Wikhamn, W., Wikhamn, B. R., & Fasth, J. (2022). Employee participation and job satisfaction in SMEs: Investigating strategic exploitation and exploration as moderators. *The International Journal of Human Resource Management*, *33*(16), 3197–3223. https://doi.org/10.1080/09585192.2021.1910537
Woudenberg, F. (1991). An evaluation of Delphi. *Technological Forecasting and Social Change*, *40*, 131–150. https://doi.org/10.1016/0040-1625(91)90002-W
Yadavalli, V. S., Darbari, J. D., Bhayana, N., Jha, P., & Agarwal, V. (2019). An integrated optimization model for selection of sustainable suppliers based on customers expectations. *Operations Research Perspectives*, *6*, 100113. https://doi.org/10.1016/j.orp.2019.100113
Zach, O., & Olsen, D. H. (2011). ERP System Implementation in Make-to-Order SMEs: An Exploratory Case Study. In *44th Hawaii International Conference on System Sciences*, Kauai, HI, USA, 2011 (pp. 1-10). doi: 10.1109/HICSS.2011.190.

Zennaro, I., Finco, S., Battini, D., & Persona, A. (2019). Big size highly customised product manufacturing systems: A literature review and future research agenda. *International Journal of Production Research*, *57*(15–16), 5362–5385.

Zolas, N., Kroff, Z., Brynjolfsson, E., McElheran, K., Beede, D. N., Buffington, C., & Dinlersoz, E. (2021). *Advanced technologies adoption and use by us firms: Evidence from the annual business survey. No. w28290*. National Bureau of Economic Research.

Zorzini, M., Corti, D., & Pozzetti, A. (2008). Due date (DD) quotation and capacity planning in make-to-order companies: Results from an empirical analysis. *International Journal of Production Economics*, *112*(2), 919–933.

Chapter 7

Conceptional Knowledge Management Tool for Early Furniture Cost Estimation (Integrated Early Price Assessment System)

Birutė Mockevičienė

Mykolas Romeris University, Lithuania

Abstract

This chapter is dedicated to presenting the essential features and technical capabilities of the prototype 'Integrated early price assessment system', with an overview of how the machine learning system interacts with an expert evaluation system. The functional structure of the prototype's operation is presented by means of an overview of the prototype's communication flow diagram.

Keywords: Prototype; integrated early price assessment system; communication flow diagram; knowledge management; user interface

Communication Flow Diagram for the Operation of the Prototype

The purpose of creating a prototype of the system is to show that the algorithm used is not only consistent with the theoretical framework but is also technically feasible.

To develop the prototype, which we call an integrated early price assessment system, we begin with a conceptual framework for knowledge management, which is then realised as a basic prototype in the form of a communication flow diagram. This diagram visually represents the information flows that keep the system

Participation Based Intelligent Manufacturing:
Customisation, Costs, and Engagement, 239–262

Published under exclusive licence by Emerald Publishing Limited
doi:10.1108/978-1-83797-362-020241007

running and captures the transformational processes as information moves from input to output. The communication flow visualisation diagram allows users to examine the system specifications using visual modules easily (Al Ashry, 2017).

The communication flow diagram is needed prior to prototype development to justify the strategy of the prototype and to prepare a list of functionalities for the prototype programming.

The prototype is based on the principles of knowledge management and the stages of knowledge creation (prior knowledge, new knowledge and knowledge of agreement). We must ensure the system's functionality, as depicted in the flow diagram, in order to create a user-friendly system.

The prototype of an *integrated early price assessment system* has to operate like a conventional knowledge-based IT system – a coherent logical structure of reasoning, cognition and practice that prevails among the members of one specific community. This knowledge can be the equivalent of the external world (thought patterns) or certain beliefs about the external world.

Types of Knowledge

Any IT system must be able to handle any type of knowledge. Several types of knowledge dominate any innovation process. One typology divides knowledge into two main categories: 'local knowledge' and scientific knowledge. Individual experience (not expert or other information) is classified as local knowledge. Scientific knowledge is more related to systematically collected information (in this case, e.g., machine learning-based furniture price forecasting is the knowledge that has emerged from scientific hints). Due to the vagueness of scientific knowledge, it is subject to different evaluation requirements, as it is usually validated and generated using principles and validated processes for collecting data and knowledge that are quite specific to the scientific field. It often has to be borne in mind that the level of local and scientific knowledge of stakeholders (e.g. furniture manufacturers) can vary dramatically, which invariably poses a challenge for the translation of knowledge into furniture production practices.

Another important typology of knowledge is the distinction between expressive and non-expressive knowledge. Expressive knowledge is the knowledge that people carry around in their minds, with little or no sharing. Initially, all knowledge appears implicit and only over time, through a long process of knowledge management, tests and mistakes, is this knowledge perceived, understood and recorded. Explicit knowledge exists in the form of words, sentences, documents and organised data. In this typology, knowledge management is the process of transforming non-explicit knowledge into explicit knowledge to be realised by the prototype under development.

Knowledge Management Cycle

In order to develop a user-friendly instrument for the furniture industry, the system must be able to handle any type of knowledge, standardising the knowledge management cycle. The classical knowledge management process includes the following stages: knowledge acquisition, creation, development, storage, transfer, sharing and use (Bang et al., 2019; Levallet & Chan, 2019).

Each knowledge management cycle from explicit to knowledge (Grandinetti, 2016) passes through a series of stages of knowledge storage, accumulation and verbalisation. These stages become quite complex. Based on the knowledge management paradigm, a knowledge management cycle can be developed (Fig. 7.1):

- Pre-knowledge. Historical production data is provided to the system.
- Associated knowledge. The manipulation of historical data (data classification, sorting, evaluation) creates a unified view of the accumulated data, which is continuously updated. The calibration of expert knowledge is also knowledge that is linked to the object under consideration (furniture manufacturing).
- New knowledge. The application of selected machine learning algorithms to evaluate the data and to calculate the parameters for new orders. Expert opinion on a specific new product is captured and aggregated.
- Consensus knowledge. Consensus or agreement is the fusion of machine learning and expert opinion data.
- Output knowledge. Knowledge for use and decision-making. Knowledge derived from peer-to-peer actions becomes the source of a continuous learning process. It is the cost of an order/new product that is used by the decision-maker in his/her activities.

In addition to the principles of the knowledge cleansing cycle, the development of the prototype aimed to integrate decision-making approaches based on the principles of creativity, rationality and interaction (employee involvement). Firstly, the creative formalisation of knowledge (machine learning and evaluation) is facilitated, while in the next phase, the aim is to integrate the structured evaluation by experts and the interaction is realised by merging the result of the evaluation of a new product achieved in two ways, where the expert can adjust his/her decision by observing the prediction made by the machine learning.

A communication flow diagram of the prototype's operation is given in Fig. 7.2. The flow diagram is presented as a two-dimensional diagram with the

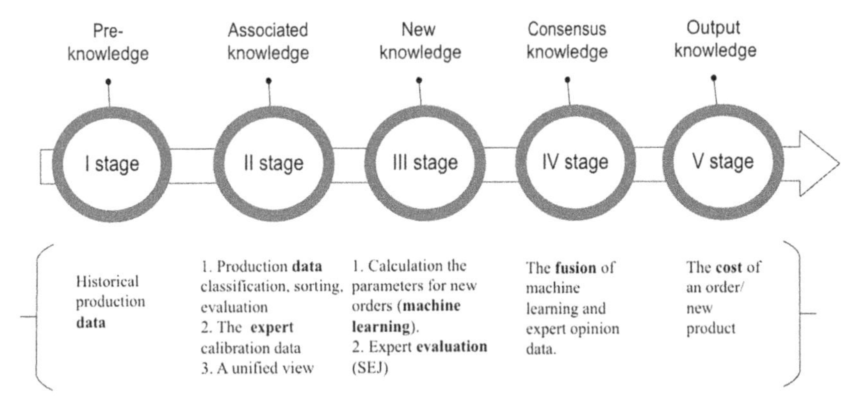

Fig. 7.1. The Mapping of the Knowledge Management Process to the Operational Phases of the Cost Estimation/Prognosis Prototype.

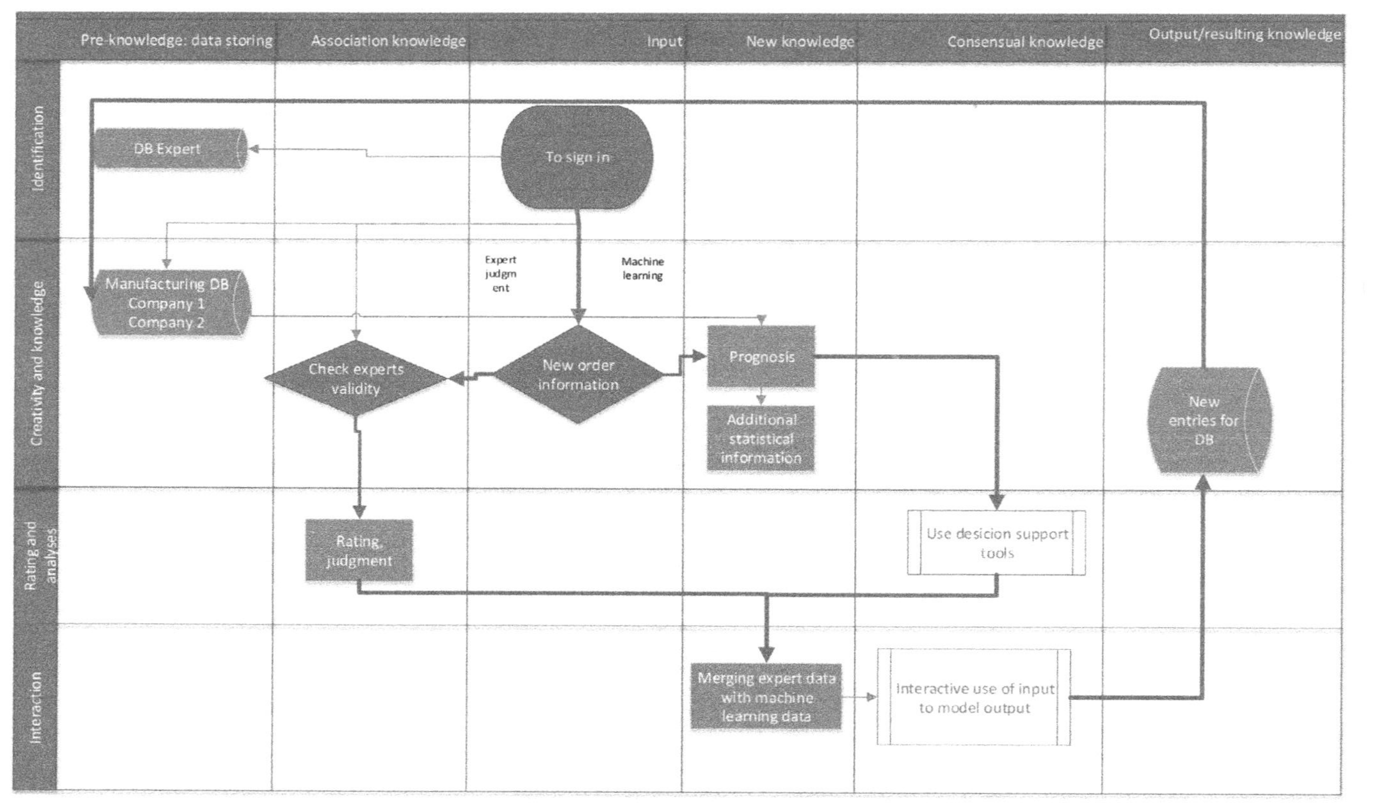

Fig. 7.2. Communication Flow Diagram for Prototyping.

principles of knowledge management on one axis and the principles of actor or user interaction on the other.

On the horizontal axis, the processes are arranged in a way that is consistent with the principles of the knowledge management paradigm, starting with the accumulation of existing knowledge and ending with the generation and storage of new knowledge (prior knowledge, associated knowledge, input, new knowledge, knowledge of agreement and knowledge of inference or use).

In the vertical axis, the flow of activity is controlled by the required functionalities of the prototype:

- Identifiability (assigning the right roles to potential users of the system [order manager and expert evaluator]).
- Creativity and knowledge creation:
  - Creativity function (machine learning and forecasting, statistical analysis, adding historical data to the results of a new order).
  - Knowledge function (building up historical data, providing data for a new product order).
- Rating functionality (validating experts by setting confidence limits on their predictions, other decision support methods to facilitate the manager's decisions on machine learning and data provided by experts, for example, visualisation.
- Interaction functionality – an interactive solution fusing expert and machine learning results.

Prototype Functionality and Structure

Functionality

After defining the communication flow diagram and the prototype specification, it's always useful to define the functionalities of the system to help users determine how useful it is in the production process (Lyly-Yrjänäinen et al., 2019). Two levels of functionality can be distinguished. The first level is in line with the initial objective of prototyping, which is to predict the early cost of a customised product. Secondary functionalities can also be defined to make the system even more valuable and attractive to companies in the furniture sector. Two secondary functionalities can be identified: (1) A multi-sided platform that matches data providers (for sharing or for buying) - this provides the possibility to collect and share data not only within the company, but also to exchange it with similar companies, collect data from several companies and use it to train their own system for better forecasting accuracy. (2) A database of experts with estimates of the quality of their forecasts, the scope of the forecast, the specialisation of the products in terms of their complexity and so on.

To enable the secondary functionalities, the prototype developers must meet certain conditions and commit to the security principles of operation. Historical company data, when multiple companies use the prototype simultaneously, is stored on a secure, isolated server. Confidentiality is guaranteed by allowing users to choose their preferred degree of confidentiality. Data anonymity is ensured by impersonalising it, preventing any possibility of tracing company names.

Structure

Taking into account the theoretically developed flow diagram of the prototype, the structured expert evaluation methodology and the principles of machine learning, the most important and necessary functionalities of the prototype were identified:

- User registration – the ability to securely register and identify oneself in the system with a specified role.
- Data storage securely – to ensure the physical and programmatic security of the data stored (uploaded and generated by the prototype).
- Data input – to allow the user to upload data and datasets in a user-friendly way. Automated and manual input is required. In addition, bulk and single order data entry is required.
- Price evaluation/forecasting – possibility to activate machine learning algorithms, structured expert evaluation methodology.
- Display of data – possibility to present the result in a user-friendly way.

In order to develop a cost and process evaluation tool for the furniture industry, the theoretical principles of standardisation of furniture production processes were selected and combined with the requirements of customised furniture production in terms of process management. After studying the theoretical process frameworks of the manufacturing industry and taking into account the methodology developed in the theoretical study and the data collected on the processes of the selected companies, an operational prototype for the evaluation of the integrated early price assessment system was developed. The Integrated early price assessment system prototype is designed to evaluate the forward prices of a furniture production order and to compare the data obtained with the results obtained by conventional evaluation methods. The estimation of the upfront price of furniture is a manual and time-consuming exercise, as it requires the evaluation of price components such as material quantities and costs, labour costs, packaging costs, production processes, overheads, etc. If the furniture has been produced in the past for the same (or a very similar) product, the estimation of the price is not a complex process and will usually be known.

Therefore, when a new order is received, it must be assessed whether the product will be new to the company, that is, whether it is a customised order that requires innovation in both design and construction. The Integrated early price assessment system prototype, based on machine learning, employee inputs and a structured expert evaluation methodology is in place, so that each new custom order solution improves the accuracy of the prototype in evaluating the production processes. The prototype can be used in a wide range of production areas and it can be an effective and accurate tool for the evaluation of furniture forward prices.

The prototype consists of front end (user interface) and back end. The whole prototype architecture as prototype operating diagram is presented in Fig. 7.3.

Back End of the Prototype

The back end of the Prototype is constructed based on three components:

1. System training.
2. System for prediction.
3. Databases.

Consider each element separately:

1. *System training* is the model, which process of collecting and understanding historical production data. The whole process is based on the principles of data mining architecture (Saini, 2021). Data mining is a process of using software to find patterns in massive amounts of data (Pendharkar & Rodger, 2000). It is a process that extracts interesting knowledge from large amounts of data. In most cases, the data to be worked with is stored in companies' internal systems in various data forms or sources. These may be file systems or Excel tables, structured databases, unstructured data repositories. Therefore, at this stage, before preparing the data for training the system to predict the cost, it will be necessary to carry out data mining procedures, which are implemented according to the principles of the already established data mining architecture, and to realise them, it is necessary to provide a data source, a data warehouse server, a data mining engine and a knowledge base. In this phase, it is necessary to standardise the data, then perform automated (or, in the case of agent is new, initial data cognitive analysis) machine learning procedures and build a supporting learning base on top of them. This function of the system integrates a data mining engine consisting of a number of modules selected to be most relevant to the cost of the furnishing, which performs data clustering, linking, classification, characterisation, clustering, etc.
2. *System for prediction* is organised in two independent modules (*machine learning-based prediction and expert passed prediction*). One of them performs the prediction of the price of a new order, using the information it receives from the system training module. The second module performs the inclusion function when the price is predicted by an expert. This module is responsible for validating the predictive ability of the new experts and making a specific prediction based on the SEJ. The System prediction component is therefore like the core of the whole system, integrating the actions performed by all the other modules. Namely, it uses the information stored in the databases, transmits its output result to be stored in the databases, interacting with the data mining engine.
3. *Databases*. A database is instrumental throughout the data mining process. We use it to guide the search patterns of the results. The database contains both user-generated information and system-generated information, including the data mining product. The data mining engine (i.e. the system training

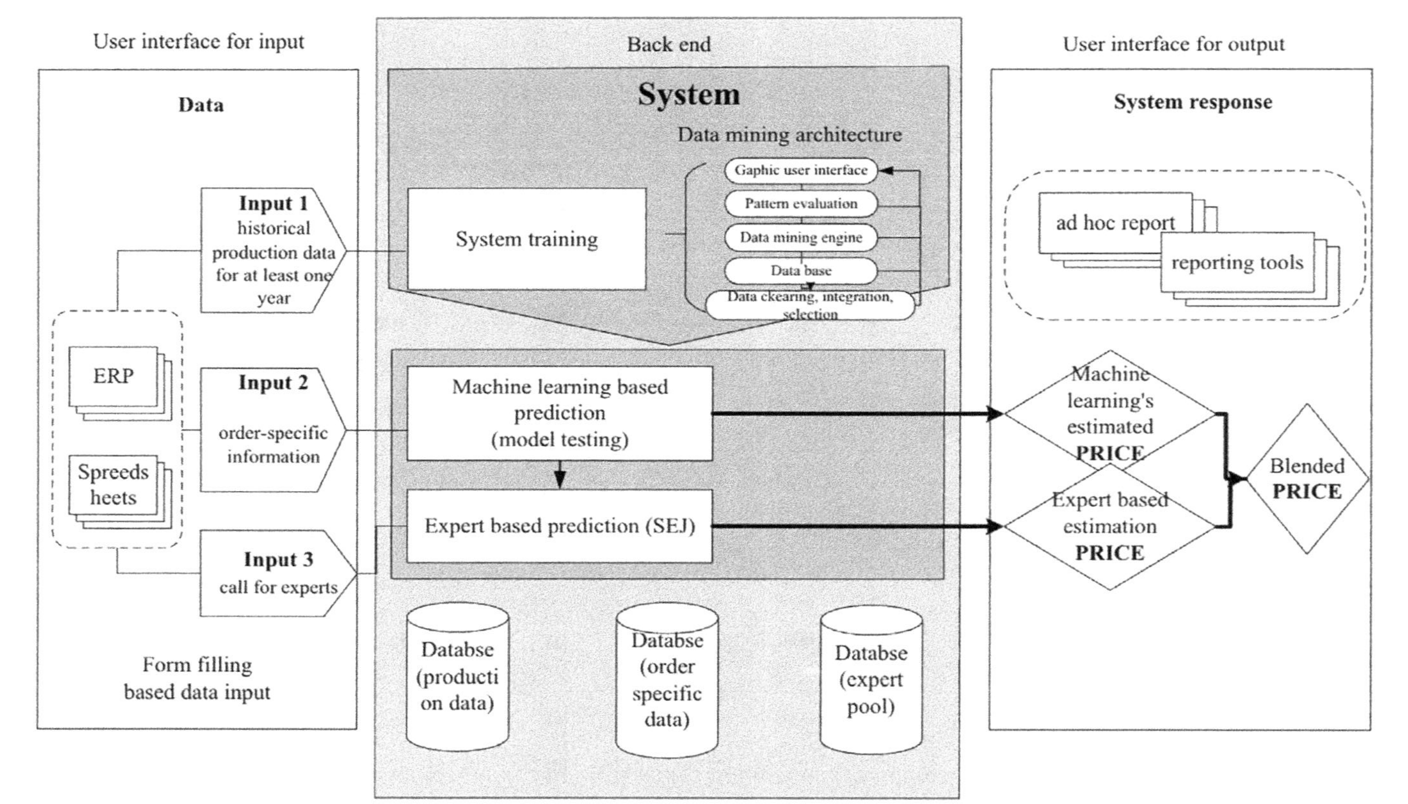

Fig. 7.3. Integrated Early Price Assessment System Prototype Operating Diagram (Prognostic Approach and Interface Diagram).

module) is provided with input data from the database via bi-directional communications in order to receive the input data and also to renew it. This prototype generates three types of databases according to the intended use of the data. These are the real data sources. A database server contains the factual data that is ready to be processed. Thus, the server handles the retrieval of relevant data. This is done on the basis of a user's data search query.

- One of their databases is designed to store production data, both historical and current.
- The second database stores order data. Since not all orders are processed into a production item, the data stored here does not reflect the production itself, but rather the needs of the customer and the ability to map them to future production processes. The first and second database are linked by a unique order number.
- The third database is design for the collection and storage of expert data. It not only contains personal data on the potential evaluator as an individual, but also his/her involvement in previous price forecasting cases.

Front End

The purpose of the user interface is to simplify the user experience, making it simple and intuitive. Although the prototype being developed breaks with the established practice among production professionals with different backgrounds (accountant, engineer, designer and quality manager), it is essential to reduce the effort the user has to put in to get the maximum desired result (preferably as intuitive as possible).

Research suggests that user interfaces are best built with layers of interaction that are appealing to the human senses – touch, sight, hearing, etc. To describe user interfaces in detail, it is necessary to talk about input devices, software interface and output devices. These include input devices such as keyboards, mice, fingerprint scanners, touch screens, microphones, cameras, electronic pens, etc., and output devices such as speakers, monitors and printers. Unfortunately, due to the limitations of the research in developing a prototype without the intention of commercialising it at this stage, a minimal impact user interface structure has been adapted to meet the necessary requirements for user acquisition and retention. Therefore, the focus was on input usability and saving user time, and consequently rich multimedia was not used. The graphical user interface is used to communicate between the user and the data mining system. The *front end* for users has two key input windows. These are for data input when the application is started and for data output when the forecasting task is completed. The interface is based on the form filling principle in order to make it easy and efficient for the user to use the system without knowing the actual complexity of the process. The template for the forms is optional.

Prototype Users

The interface was built with the consideration that there are two types of users with completely different roles. The first type of user (associated with a company and henceforth referred to as company) is an employee of the company that is querying the cost of a new order. This may be an order manager, a designer, a senior manager, a pricing officer or even the CEO of the company (this is relevant for small companies where the decision to accept a new order is localised at a higher management level). The second type of user is the expert, who will draw on his/her knowledge and experience to provide an accurate translation of the machine learning result. Although in some cases the two users may be the same person, we recommend separating them in the prediction due to the cost of the functions.

Structural Elements

The user interface has the following structural elements:

1. *Home window*: It displays the most relevant information structured for the user, for example, graphs, statistics, etc. The main window allows the registration of the company (User type I) and/or the expert (User type II) and/or the login if the company and/or the expert already has an account. The identification of users is secure and complies with all security requirements. Summary results of studies, discoveries, solutions and tests are also provided.
2. *Input window*.
   (a) Historical data/Orders. A list of the orders generated is displayed with their function buttons.
   (b) Four options for data entry. Import template, upload historical data, import order and generate order. The prototype allows you to upload a form with data into the system for training, download a blank form to fill in or a sample of an already filled in form, import your own new inquiry (order) or enter an order manually.
   (c) Expert dialog. The company representative can select the experts from the existing list in the database, but can also call in the required experts from outside (the system). You can invite an unlimited number of people to become experts. The experts have a dedicated space on the front page for their activities.
3. *Expert interface*. The expert is presented with a window containing all the options he/she has been given to evaluate the price. In another window, he is given information about the order and the price given by machine learning and the form to present his input as his prognostic price.

User Interface

A form-based interface (with a basic menu option) was chosen from all the most popular types of user interfaces (e.g. graphical user interfaces [GUI], command line interfaces [CLI], form-based interfaces, menu-based interfaces and natural language interfaces).

As our proposed system, especially for system training, requires a lot of information to be uploaded (preferably of big data type), which will be mostly done by specialists delegated by the company, it is important to keep the format of the upload as simple as possible. For such needs, form-based software interfaces are widely used. Inputs to the system are generally more easily accepted by the user when they are predictable and are also linked to the format of the ERP output used by the participating company, so form-based interfaces tend to have some common features and are quite intuitive to understand. At this point, we prefer the principle of automatic generation to drop-down selection elements, which are often the preferred types of response fields.

It is true that forms can be filled in the system to provide the order information, as in this case a piece of limited information (single order) is already filled in, which can be adjusted according to the information available from the customer during the forecasting task.

The form automatically moves the cursor from one field to another. This is done in order to guide the user logically through the form and to ensure that all relevant information is provided.

The data entered is validated. This is done to ensure that the data entered is meaningful and that unsubstantiated data is rejected by the system. This process helps to ensure that the data entered into the system is consistent. In this case, the validation is done by synchronising the template mapping with the ERP output of the enter price resource planning instrument used internally for the company. Several methods can be used to validate the data.

Design of User Interface

User interface design is the process of taking wireframes and turning them into polished graphical user interfaces. This makes the product more user-friendly and can even create an emotional connection between the user and the product.

User interface design focusses on the design of individual screens or interfaces in the user's path. It is essential in the last stage of commercialisation when bringing a product to the market. So, in this applied research we have deliberately omitted this stage.

Algorithm for Verification of Performance Under Laboratory Conditions (Combining ML with SEJ)

This algorithm has been developed to test the functionality of the prototype.

Prototype Operating Principle for Estimating the Cost of a Piece of Furniture

In the prototype developed for furniture price estimation, the price of furniture is predicted using machine learning algorithms. The price can also be predicted by a panel of experts, whose prediction is validated. A methodology for combining the estimated prices, that is, combining the price predicted by the machine learning algorithms with the expert estimate, is developed, given a two-way estimate of the price of a new product to be ordered. Therefore, the price estimation is carried out in the following steps:

1. Price prediction by machine learning algorithms.
2. Price estimation by expert evaluation.
3. Combining two estimates: Estimation of the final price by combining machine learning results and expert judgement.

We will discuss each of these separately below.

1. *Price prediction by machine learning algorithms*:

The task of estimating the early price can be formulated as a forecasting task. For this purpose, supervised learning algorithms are applied (see Section 2.1.3 for details). The training (machine learning) dataset consists of the attributes (characteristics) and price of the products already produced. The objective of the prediction task is to determine the price of a new product based on the same attributes.

2. *Price estimation by expert evaluation*:

Price estimation is proceeded using SEJ (structured expert judgement) assessment methodology.

Since several experts may evaluate the price of the same product, it is necessary to establish expert estimates that show the level of their expertise in the field under study. To do this, they must first answer a questionnaire which does not yet deal with the pricing of specific products, but rather with general questions whose answers should indicate the level of competence of the expert. Experts are assessed on two measures:

- The calibration estimate is the probability that the distribution of the experts' answers corresponds to the target results, based on a chi-square test.
- The information estimate is a measure of the informativeness of the expert's judgements.

The notations used below and brief descriptions of them are given in Table 7.1.
The expert evaluation methodology is structured as follows:

Step 1. Determining the expert calibration estimate.
Each expert is asked to provide his/her best guess (median) and the 5% and 95% confidence limits, respectively, for each *i-th* question.
Intervals are constructed for each expert's response $I_{1,i,e}$, $I_{2,i,e}$, $I_{3,i,e}$, $I_{4,i,e}$.

Step 2. Determine which ranges the answers of each expert fall into;

Table 7.1. The List of Notions.

Notation	**Description**
E	Number of experts.
E	*Expert e, e* = 1, …, *E.*
N	Number of questions used to calibrate experts.
x_i	*the correct value of the answer to question (i) used to calibrate experts, i* = 1, …, *N*.
P	Distribution of correct answers, $p = \{p_1, p_2, p_3, p_4\} = \{0.05, 0.45, 0.45, 0.05\}$.
$\left(x_{5,i,e},\ x_{50,i,e},\ x_{95,i,e}\right)$	the most likely value of the answer to question i of the *e*-expert's question ($x_{50,i,e}$) and 5% and 95% confidence interval ($x_{5,i,e}$, $x_{95,i,e}$).
$I_{1,i,e} = \left[m_i,\ x_{5,i,e}\right]$ $I_{2,i,e} = \left(x_{5,i,e},\ x_{50,i,e}\right]$ $I_{3,i,e} = \left(x_{50,i,e},\ x_{95,i,e}\right]$ $I_{4,i,e} = \left(x_{95,i,e},\ M_i\right]$	Intervals into which the 5, 50 and 95 percentiles divide the x_i interval. m_i and M_i are the respective minimum and maximum values of the answer to question *i* among all answers to that question.
$s_j(e)$	The probability that the value of x_i falls within the interval $I_{j,i,e}$, j = 1, 2, 3, 4.
$\Theta(e)$	The *e*-calibration estimate of the expert.
k	This is an arbitrary value that allows the boundaries of the interval $[m_i, M_i]$ to be extended; if *k=0,* the boundaries will not be extended.
L_i	Lower bound of the extended interval $[m_i, M_i]$.
U_i	Upper limit of the extended interval $[m_i, M_i]$.
$I(f_{i,e}\|g_i)$	Estimate of the e-th expert's information for the i-th question.
$\Lambda(e)$	The average information estimates of the e-^{third} expert.
α	This is a freely selectable value that evaluates which expert's results will be included in the final product assessment, based on the calibration results of that expert. This value can be used to adjust whether only the best calibrated experts are included in the final evaluation or whether those with low calibration scores are also included.
$h(e,l)$	The price of the *l*th product, as determined by the *e*-expert.

Table 7.1. (*Continued*)

Notation	Description
$Price_{SEJ}(l)$	Total expert price for the lth product.
$Price_{ML}(l)$	The price predicted by machine learning algorithms for product l.
$Price_{final}(l)$	The final price of the lth product
s_1	Weighting the price predicted by machine learning algorithms.
s_2	Weight of the total expert price.

if correct answer from the *e-expert* $x_i \in I_{1,i,e}$, tai $s_{i,1}(e)=1$;

if $x_i \in I_{2,i,e}$, tai $s_{i,2}(e)=1$;

if $x_i \in I_{3,i,e}$, tai $s_{i,3}(e)=1$;

if $x_i \in I_{4,i,e}$, tai $s_{i,4}(e)=1$.

Step 3. Calculated averages: $s_j(e)=\frac{1}{N}\sum_{i=1}^{N} s_{i,j}$, $j=1,\ 2,\ 3,\ 4$.

Step 4. The estimate of the expert calibration $\Theta(e)$ is calculated, being the p value of the statistical test H_e that the expert is well calibrated:

$$\Theta(e)=P(\chi_3^2>\chi^2(e)\mid H_e),$$

here $\chi^2(e)=2n\sum_{j=1}^{4} s_j(e)\ln\frac{s_j(e)}{p_j}$.

Step 5. Calculating the information score.

The value of k is selected.

For each question i, the lower and upper range is calculated:

$$L_i=m_i-\frac{k}{100}(M_i-m_i) \text{ ir } U_i=M_i+\frac{k}{100}(M_i-m_i).$$

The values of the *e*-expert's score for question i shall be calculated according to the following formula:

$$I(f_{i,e}\mid g_i)=\ln(U_i-L_i)+0.05\times\ln\left(\frac{0.05}{x_{5,i,e}-L_i}\right)+0.45\times\ln\left(\frac{0.45}{x_{50,i,e}-x_{5,i,e}}\right)$$
$$+0.45\times\ln\left(\frac{0.45}{x_{95,i,e}-x_{50,i,e}}\right)+0.05\times\ln\left(\frac{0.05}{U_i-x_{95,i,e}}\right)$$

here $i = 1,\ldots,N$, $e = 1,\ldots,E$; f and g are represented by two probability density functions.

The average score of the information estimate of the *e*-expert is calculated according to the formula:

$$\Lambda(e) = \frac{1}{N}\sum_{i=1}^{N} I(f_{i,e} \mid g_i)$$

Step 6. Calculating expert weights.

With the calibration and information scores of all the experts, the weights need to be calculated.

Select the value of the coefficient α.

For each expert, the calculation shall be $1_\alpha(\Theta(e))$ value: if $\alpha > \Theta(e)$, then $1_\alpha(\Theta(e)) = 1$; otherwise, $1_\alpha(\Theta(e)) = 0$.

The weights for each expert shall be calculated according to the following formula: $w_\alpha(e) = 1_\alpha(\Theta(e)) \times \Theta(e) \times \Lambda(e)$.

Step 7. Calculating the aggregate price by expert.

Several experts may have different results when assessing the price of the same product. Therefore, after taking into account the weights of the experts obtained in step 3, it is necessary to aggregate the results and determine the total price.

Each expert is presented with a *l*th product and asked to determine its price *h*(*e*, *l*).

With the weights of all the experts, the aggregated price of the product is estimated according to the formula:

$$Price_{SEJ}(l) = \frac{\sum_{e=1}^{E} w_\alpha(e) \times h(e,l)}{\sum_{e=1}^{E} w_\alpha(e)}$$

Step 8. Estimating the final price for the product.

Taking the price $Price_{ML}(l)$ predicted by the machine learning algorithms and the experts' aggregated $Price_{SEJ}(l)$, is necessary to estimate the final price of the *l*th product. This can be calculated according to the formula:

$$Price_{final}(l) = s_1 \times Price_{ML}(l) + s_2 \times Price_{SEJ}(l)$$

Here s_1 – weight of the price predicted by machine learning algorithms, s_2 – weight of the total expert price, $s_1 + s_2 = 1$. Šiuos svorius gali parinkti pats sprendimų priėmėjas.

A Case Study on the Application of the Price Evaluation Methodology

The case of the use of an expert evaluation methodology is presented below. For simplicity, let us assume for the sake of simplicity that there are three experts

(E = 3) and the questionnaire consists of two questions (N = 2). The correct answer for the first question is 110 and for the second 20 (see Table 7.4).

Step 1. Each expert has to provide the best guess he/she thinks is appropriate. The results are presented in column - 50. Each expert must also indicate the confidence limits (columns 5 and 95 respectively), that is, the possible deviation from his best guess. The minimum (min) and maximum (max) values are then calculated. In the last rows of Table 7.2, the minimum and maximum values of all the values for each answer are bolded. We can see that for the first answer, the minimum value is 70 and the maximum is 130. For the second answer, the values are 10 and 35 respectively.

Step 2. Determine how many correct answers fell into the four ranges:

$$I_{1,i,e},\ I_{2,i,e},\ I_{3,i,e},\ I_{4,i,e}.$$

Step 3. The results of each expert's predictions for each question are divided into the following four intervals (see Table 7.3). If the value of a particular expert's prediction falls within a certain range, the s_i_j part of the table is set to 1, otherwise to 0.

Step 4. Using the results of the two questions from all three experts, the predictions are averaged into the ranges $s_j(e)$, j = 1, 2, 3, 4 (see Table 7.4).

Step 5. The expert calibration estimate must be calculated. For this purpose, the following shall be calculated $\chi^2(e) = 2n\sum_{j=1}^{4} s_j(e)\ln\frac{s_j(e)}{p_j}$ values, which are given in '2N x SUM' of Table 7.5, taking $p = \{p_1, p_2, p_3, p_4\} = \{0.05, 0.45, 0.45, 0.05\}$ and $s_j(e)$ values are shown in Table 7.4.

It is now necessary to calculate the values of $\Theta(e)$. To simulate the case study in MS Excel, the CHIDIST function can be used. The results are shown in the column 'TETA' in Table 7.5. The closer this value is to zero, the stronger the calibration of the expert. In the case considered, experts 1 and 2 are fairly well calibrated, while the third is very poorly calibrated.

Step 6. Select k=10.

Step 7. For each question, calculate the lower and upper values of L_i and U_i (see Table 7.6).

Step 8. For each question and for each expert, the information estimation values $S_{i,e}$ are calculated (see Table 7.7).

Step 9. Average values for each expert's information are calculated (see Table 7.8).

Step 9. Choose a coefficient value α=0.1.

Step 10. Calculate $1_{⊠}$ values for each expert (see Table 7.9, column '1_alpha').

Step 11. For each expert, the weights $w_{⊠}(e)$ are calculated (see Table 7.9, column 'w').

Step 12. Each expert is presented with the lth product and asked to determine its price $h(e, l)$ (see Table 7.10).

*The final step***.** Considering the price $Price_{ML}$ (l) predicted by the machine learning algorithms and the experts' total price $Price_{SEJ}(l)$, the final price of the

Table 7.2. Answers to Questions for Expert Calibration (Expert was Asked to Give His/Her Best Guess (The Median) and the 5% and 95% Confidence Bounds and, Respectively, for Each of the Variables).

True value	**110**			**20**		
	5	50	95	5	50	95
	***i*=1**			***i*=2**		
***e*=1**	90	100	110	20	25	30
***e*=2**	110	120	130	10	20	30
***e*=3**	70	90	110	15	25	35
min	**70**	90	110	**10**	20	30
max	110	120	**130**	20	25	**35**

Table 7.3. Breaking Down the Experts' Predictions into Ranges.

	5	50	95	
	x_i<=5	5<x_i<=50	50<x_i<=95	x_i>95
***i*=1**				
e=1	<=90	(90;100]	(100;110]	>110
e=2	<=110	(110;120]	(120;130]	>130
e=3	<=70	(70;90]	(90;110]	>110
	*s*1	*s*2	*s*3	*s*4
e=1	0	0	1	0
e=2	1	0	0	0
e=3	0	0	1	0
***i*=2**				
e=1	<=10	(20;25]	(25;30]	>30
e=2	<=10	(10;20)	[20;30]	>30
e=3	<=15	(15;25]	[25;35)	>35
	*s*1	*s*2	*s*3	*s*4
e=1	1	0	0	0
e=2	0	0	1	0
e=3	0	1	0	0

Table 7.4. Averages of the Experts' Predictions Falling Within the Specified Ranges.

	Average			
	*s*1	*s*2	*s*3	*s*4
e=1	0,5	0	0,5	0
e=2	0,5	0	0,5	0
e=3	0	0,5	0,5	0

Table 7.5. Calculation of the Calibration Score.

p	0,05	0,45	0,45	0,05			
		*s*2	*s*3	*s*4	SUM	2N x SUM	TETA
e=1	1,15	0,00	0,05	0,00	1,20	**4,816**	**0,52**
e=2	1,15	0,00	0,05	0,00	1,20	**4,816**	**0,52**
e=3	0,00	0,05	0,05	0,00	0,11	**0,421**	**0,96**

Table 7.6. Calculation of Lower and Upper Limits.

	k=	10.00
***i*=1**	L1	64.00
	U1	136
***i*=2**	L2	7.50
	U2	37.5

Table 7.7. Calculation of the Information Estimate (Per Expert, Per Question).

***i*=1**						
	(U-L)	**0,05**	**0,45**	**0,45**	**0,05**	**I(f|g)**
***e*=1**	4,28	−0,31	−1,40	−1,40	−0,31	0,86
***e*=2**	4,28	−0,34	−1,40	−1,40	−0.24	0,91
***e*=3**	4,28	−0,24	−1,71	−1,71	−0.31	0,31
***i*=2**						
	(U-L)	**0,05**	**0,45**	**0,45**	**0,05**	**I(f|g)**
***e*=1**	3,40	−0,28	−1,08	−1,08	−0,25	0,71
***e*=2**	3,40	−0,20	−1,40	−1,40	−0,25	0,16
***e*=3**	3,40	−0,25	−1,40	−1,40	−0,20	0,16

Table 7.8. Calculation of the Average Information Values.

	LAMBDA
e=1	**0,78**
e=2	**0,53**
e=3	**0,24**

Table 7.9. Calculation of the Weights of the Experts.

	alpha	0.1		
	1_alpha	**TETA**	**LAMBDA**	**W**
e=1	1,00	0,52	0,78	0,406
e=2	1,00	0,52	0,53	0,277
e=3	1,00	0,96	0,24	0,227
			suma	**0,910**

Table 7.10. Estimation of the Final Expert Price.

	What is the price (h)?		
e=1	500	203,18	
e=2	550	152,42	
e=3	600	136,17	
		491,78	SUM
		540,15	**Price (SEJ)**

*l*th product is estimated. Suppose that the price predicted by the machine learning algorithms for a certain product was 620 €. Table 7.11 shows the results for the five options. In the first case (Row 1), the machine learning algorithms' predicted price and the experts' total price have the same weights (s_1 = s_2 = 0.5). In the second case (Row 2), only the price predicted by the machine learning algorithms is taken into account (s_1 = 1, s_2 = 0), and in the third case (Row 3), only the experts' total price is taken into account (s_1 = 0, s_2 = 1). In the fourth case (Row 4), a higher weight is given to the experts' total price (s_1 = 0.2, s_2 = 0.8), and in the fifth case (Row 5), a higher weight is given to the price predicted by the machine learning algorithms (s_1 = 0.8, s_2 = 0.2). In this way, the decision-maker can determine for himself how much the final price will be influenced by the price predicted by the machine learning algorithms and the price determined by the experts.

Table 7.11. Estimation of the Blended Price.

No	*s_1*	*Price_ML*	*s_2*	*Price_SEJ*	**Blended Price**
1	0.5	620	0.5	558.15	589.08
2	1	620	1	558.15	620.00
3	0	620	0	558.15	558.15
4	0.2	620	0.8	558.15	570.52
5	0.8	620	0.2	558.15	607.63

Prototype Testing Conditions and Challenges

Preconditions

As the success of the commercialisation of a project depends on a successful testing process, it is important that companies are willing to share their production data and understand the added value of the prototype to their business.

For prototype testing, this process is valuable to the extent that its essential functionalities can be tested (Vestad & Steinert, 2023). The difficulties arising from prototype testing are not only due to the incomplete representation of the product through the prototype, but also due to the incomplete matching of the working environment during the test to simulate real conditions.

Prototype testing is a vital step in the development of a final product. During this stage, customers are usually involved to identify their needs. The testing process primarily focusses on the aesthetic and ergonomic aspects of the product or subsystems of complex products (Campbell et al., 2007). The main objective of prototype testing is to facilitate customer interaction and gather valuable feedback to enhance the final product. This testing ensures that the product meets the customers' requirements and expectations.

This type of prototype testing is characterised by the following conditions, which are important to maintain. Requirements for the type of company:

1. The company must have an ERP system in place to collect historical production data.
2. The company must be working with customised production practices in the area of production.

Other desirable conditions:

- Be willing to invest staff time to test the prototype.
- Share a research-based culture to the problem-solving.
- Accept difficulties during testing that will delay employees' daily work until they become comfortable with the prototype.

Success criteria required for prototype testing:

First, to have historical production data available (this reduces the frustrations of testing a new product, which will still be there no matter how perfect the prototype).

The second prerequisite is the sector's attitude towards competitiveness - if a company realises that the success of such a prototype can give them the necessary competitive advantage, then the testing process is taken seriously. Otherwise, any process that interferes with the company's internal processes and interrupts the company's day-to-day operations can be seen as a disruption.

The third prerequisite is the sector's attitude towards confidence in scientists.

It is acknowledged that it has been quite difficult to find companies willing to work together on the tests. These markets are low-tech, dominated by small companies and therefore have limited experience of collaborating with researchers, have limited capacity and may not have sufficient production quantities for testing. For these reasons, the search for experimental infrastructures is a challenging exercise.

Selecting the Company

Knowing that these furniture markets are low-tech, that trust in scientists in the sector is uncertain and that competition in the market is quite aggressive, it could be seen as dangerous to openly share sensitive production data, including the price component, we have taken quite active steps to publicise the idea (press releases on the progress of the prototype, popular publications in the specialised press and popular media, interviews on business-oriented radio, a 1-hour innovative programme on a YouTube channel for business on collaboration with business. In total, there were 14 publicity events in three years. Proactive steps were also taken to contact companies and present the prototype under development.

Although this was a communication activity, it revealed quite interesting trends in the market. After the first press releases, two large companies got in touch and offered to cooperate. Representatives of three small companies also came forward on their own initiative, just to get a better understanding of the functionality of the prototype. Further individual discussions and the resolution of their specific problems with the companies showed that the issue of early product pricing is a major concern. The problem of pricing a new product is perceived as a problem by the companies in all departments (they had allocated a budget for this or recruited new people to deal with this issue only, usually IT specialists to program the tool). One company contacted us themselves and during 4 meetings the possible working principle of the prototype was discussed, presenting the cost requirement for forecasting. Unfortunately, the discussion on cooperation broke down when the prototype was presented, and the companies heard that the prototype needed their production data linked to the real price of the final product. Price as a competitive differentiator proved to be too high a value. Without hiding their fears, companies did not dare to take the decision to open up the space for testing. So, the range of testing possibilities was limited to one company.

The experiment was carried out in a furniture company, the same one that provided the data for the prototype, in order to test the prototype under real conditions.

The prototype required testing, which necessitated corporate preparation. As soon as it arrived, the Project Manager, acting as the Deputy Director, was presented with the prototype. During the visit, both the manager and potential experts were trained in how to use it. Since the 'live' process is limited to ordering and designing a new product, and the cost of the product is only realised after it has been manufactured and sold, testing of all the features of the prototype is only possible after a certain period of time has elapsed (1–3 months, depending on the complexity of the product and the ordering timescale). The testing experiments were therefore divided into two. The Type I experiment aimed to test the performance and functionality of the prototype. It focussed on determining which features make sense to practitioners, understanding how the system works and monitoring the skills and willingness of workers to use similar types of instruments. The experiment also monitored which features or query moments surprised and puzzled the employees. The Type I experiment was not intended to test how close the real price of the product is to the price presented by the prototype. As mentioned above, a project manager from the company was delegated to test the prototype, who was the main tester and was also the person in charge of the testing process within the company. The Type II experiment was designed to test the evaluation function of the prototype. Instead of using a live environment, the company selected its production data, for which the sales price from the last six months was already fixed, by selecting orders/products at random. The latest production data was shared with us, which we automatically prepared for uploading into the system using the data preparation, formatting and cleaning method described in section.

This data was not used as historical data to train the system but as data for new orders.

Data for Testing

The data used to test the prototype were mainly obtained from the collaborating companies. This data, like the previous ones, was obtained in xlsx file format. Although the number of files amounts to several thousand, most of the information was only invoices, without information on the work done and the materials used. A large part of the data is also related to corrections or additions to the work carried out previously. All the remaining relevant data files were scanned using the same file reading system and the information required for the calculations was triggered in the database.

From this database, 100 new entries were selected at random. As the system accepts project data in xlsx files (the format of this file was described earlier), these records were triggered in one such file, where each line represents a new record. The resulting file was uploaded to the system and the results were saved.

Challenges of data loading technology. The system was designed so that a user could upload one project at a time in the form of an xlsx file. Although this is the most likely way of uploading projects when the system is used by a standard user, it is not sufficient for this test case. Uploading a hundred projects into the system one by one, and storing each result separately, would be a lengthy process, and could lead to a number of errors and distort the test results.

Refining the Prototype

No changes were needed to the core strategy, but additional communication statements explaining the system's functionality were found to be important. Furthermore, the need for data security was noted. To ensure data security, public access to the database on the server was prohibited. Internal queries on the server reduce the risk of data theft but make it more difficult for developers to access the data. Although access is possible by logging in to the server, it is a more complex process than the standard secure public access. Adding a facility for easier access in the future could benefit developers and administrators. The user registration interface has been revised, and communication errors that caused users to lose their login session have been fixed. During testing, it was found that users would lose the ability to log in after clicking the browser's 'Back' button due to incorrect connections between the application's pages. The bug was resolved by properly configuring the links between the pages. The user interface for communication with experts has been revised to include additional explanatory sentences. Experts missed certain data because the principles of expert and machine learning evaluation are different. While the prototype provides the same data for both, experts require additional data. Therefore, the explanation texts for experts have been extended. During a live presentation to a collaborating company, it was observed that experts extract more criteria from the project than the AI algorithm. One of the most important criteria for experts is the Project Drawing, but the current system only requires the data needed for the algorithm's calculations, making it difficult for experts to evaluate a project. The company representatives recommended adding the option to upload additional data that will only be used by the experts.

References

Al Ashry, M. (2017). The importance of data flow diagrams and entity relationship diagrams for the design of data structure systems in C++ a practical example. *Journal of Strategy and Management*, *8*(4), 51–61.

Bang, S. H., Ak, R., Narayanan, A., Lee, Y. T., & Cho, H. (2019). A survey on knowledge transfer for manufacturing data analytics. *Computers in Industry*, *104*, 116–130. https://doi.org/10.1016/j.compind.2018.07.001

Campbell, R. I., De Beer, D. J., Barnard, L. J., Booysen, G. J., Truscott, M., Cain, R., Burton, M., Gyi, D., & Hague, R. (2007). Design evolution through customer interaction with functional prototypes. *Journal of Engineering Design*, *18*(6), 617–635. https://doi.org/10.1080/09544820601178507

Grandinetti, R. (2016). Absorptive capacity and knowledge management in small and medium enterprises. *Knowledge Management Research & Practice*, *14*(2), 159–168. https://doi.org/10.1057/kmrp.2016.2

Levallet, N., & Chan, Y. E. (2019). Organizational knowledge retention and knowledge loss. *Journal of Knowledge Management*, *23*(1), 176–199. https://doi.org/10.1108/JKM-08-2017-0358

Lyly-Yrjänäinen, J., Aarikka-Stenroos, L., & Laine, T. (2019). Mock-ups as a tool for assessing customer value early in the development process. *Measuring Business Excellence*, *23*(1), 15–23. https://doi.org/10.1108/MBE-11-2018-0096

Pendharkar, P. C., & Rodger, J. A. (2000). Data mining using client/server systems. *Journal of Systems and Information Technology*, *4*(2), 72–82. https://doi.org/dx.doi.org/10.1108/13287260080000756

Saini, A. (2021). Data Mining Architecture – Data Mining Types and Techniques. *International Journal of Innovative Research in Technology*, *8*(5), 310–318.

Vestad, H., & Steinert, M. (2023). Creating your Own Tools : Prototyping Environments for Prototype Testing. *Procedia CIRP, 84*(March), 707–712. https://doi.org/10.1016/j.procir.2019.04.225

Chapter 8

Epilogue

Birutė Mockevičienė

Mykolas Romeris University, Lithuania

Customised Furniture Manufacturing in the Era of Industry 4.0

Industry 4.0 presents immense potential for creating customised products with more flexibility and personalisation options. However, implementing changes in production requires a significant transformational capacity, which can be perceived as a challenge in the furniture sector. As customised furniture production becomes increasingly specialised and individualised, it is also associated with significant uncertainties.

Furniture companies that specialise in customised production is changing the traditional principles of manufacturing. As they produce unique, one-of-a-kind products, they need to adopt different organisational forms, management solutions, and product quality assurance methods to reduce uncertainty in production. This shift also leads to a greater focus on customer needs and additional investment in IT tools. However, the furniture industry is dominated by small and medium-sized enterprises with limited financial resources and a flat organisational structure, which hinders their ability to invest in operational optimisation and digital solutions related to price evaluation systems. As a result, they struggle to catch up with new technologies and compete in the market.

Regionalisation Issues

The furniture industry in the Baltic region has been growing steadily since 2010, with an increase in the number of new start-ups, the growth and consolidation of older companies, a rise in profitability, and an increase in order intake. While there have been some fluctuations in production volumes over the past 14 years, the overall trend indicates a strengthening of the regional furniture business. Mixed production (a combination of customised and mass production), which requires appropriate specialised knowledge and, therefore, management tools,

Participation Based Intelligent Manufacturing:
Customisation, Costs, and Engagement, 263–268

doi:10.1108/978-1-83797-362-020241008

is steadily gaining ground in the portfolios of growing companies. Companies are now focussing on customised production, which suggests that it is becoming more cost-effective to organise the production of certain products (furniture) in regional locations closer to potential consumers. This approach helps businesses better understand the needs of regional consumers and pursue sustainable production practices such as minimising transport costs, using local resources, and avoiding cultural differences that could lead to constraints. The concept of frugal innovation embraces and realises the potential of digitisation through Industry 4.0.

Attitudes Towards Digitalisation and Data Usability of Furniture's Companies

Furniture manufacturers in Lithuania and other Baltic countries have been facing challenges in adopting smart technologies in their production processes. This has been particularly evident in the context of Industry 4.0, where the development of new business models and system designs has been inadequate, automation levels are low, and organisations are struggling to manage data in a high-uncertainty context.

Through our research on the use of digital tools in furniture manufacturing, we have discovered that manufacturers are facing challenges in integrating IT into their pricing processes as well. They are finding it difficult to automate their processes and are therefore avoiding the deeper penetration of technology in the pricing process. Introducing digital tools can lead to new uncertainties in quality and process management.

The interrelationships of factors have been identified, leading to a clearer definition of incentives for manufacturers to pursue process technologisation and digitisation. Manufacturers in the furniture sector tend to adopt digital tools more readily when the tools have simpler functions that do not significantly alter their routines. However, they are less likely to adopt tools that have more complex functions or are surrounded by uncertainty. The implementation of data collection and management technologies is hindered by the lack of interest among managers in adopting technological advancements. They are reluctant to invest in smart manufacturing, which indicates that there is insufficient motivation to move into the more advanced era of big data.

Attitudes Towards Pricing Strategies of Furniture's Companies

The furniture companies in the region still use traditional costing methods for their costing system. This is evident from the principles they expect from a customised costing system. This highlights the significance of material-based pricing systems and the scepticism towards other systems, especially when they require a paradigm shift.

It would initially seem that costing in companies aspires to be based on accurate data analysis. However, after analysing historical data collected by companies, it was evident that although the material-related data were accurate, the time variable was not being used to its full potential in costing. Estimated production times were often fragmented and not reflective of the true cost of labour time, which was predetermined by management. Manufacturers typically predetermine lead times and do not consider other factors, such as construction time and actual production time, which makes it difficult to calculate prices from the available data. Controlling production time is a key factor in managing wages and total costs. By optimising production time, companies can effectively reduce their costs, making it a critical lever for cost control. This may explain why companies resist the data analytics of pricing investments, as they may view it as a disproportionate share of the expectation of this competitive advantage.

Another element that is important for pricing and that companies consider as significant depth is the fact that the final price is influenced by a range of additional factors, such as information about the customer's brand, the customer's expected order volume and the potential long-term benefits of working with the customer. Unfortunately, such information is not captured, systematised or analysed by any digital tools. On the contrary, senior managers have a monopoly on this information.

The hybrid pricing approach, which combines cost-based and value-based elements, is more appropriate for businesses that produce unique products, sell directly or through intermediaries, participate in a global market, and have access to information and data processing capacities. This approach allows businesses to reflect the specificity of their production, competitive environment, and business strategy in their pricing decisions.

Although small businesses seem to have limited potential to adopt new data management techniques, larger companies are increasingly investing in business analytics tools. This can be attributed to a growing understanding of how important data analytics is for business success. Companies are looking to gain deeper insights into phenomena and leverage data to improve their operations. In this regard, effective cost management plays a crucial role in optimising production processes while simultaneously controlling expenses. By assessing costs early on, businesses can better align production and cost management, leading to more efficient and profitable operations.

Preconditions for Making the Pricing Paradigm Shift Happen

The seemingly narrow task of early price estimation and forecasting reveals the complexity of customised manufacturing processes, which arise from the product development process, product requirements, product manufacturing and business models.

The increasing complexity of production and the challenge of pricing in early product development stages push companies to adapt their organisational structure. In other words, customised production encourages companies to seek more flexible organisational structures. The price estimation task in the early design phase becomes a strategy for managing complexity. Otherwise, compensating managerial strategies to reduce complexity start to dominate: that is, avoiding customised orders, improving standardised forms of operation (standardised operating times), increasing hierarchy (decoupling of pricing), and modularising individual activities (e.g. optimisation of warehouse balances). Unfortunately, even this reduction in complexity does not guarantee long-term competitiveness. Therefore, managers are beginning to realise that other approaches are needed to manage complexity, and they must accept the inevitability of a paradigm shift.

Uniqueness of Solution of Proposed Integrated Early Price Assessment System

Our desire to develop an early price estimation algorithm was born out of a desire to address the challenges that custom furniture companies face in estimating the cost of developing.

Typically, these companies rely on detailed cost models and a time-consuming early price evaluation process that can take up to a month. However, this process can be prone to errors in price evaluation, production technology, process planning, and quality management. To overcome these issues, custom furniture businesses require an innovative solution that can shorten the process of early price evaluation.

Our team developed the prototype we call the *Integrated early price assessment system*, which is based on the main findings using our own created methodology. This study has successfully developed a unique and practical methodology for early price assessment. This methodology is not only valuable in terms of selecting methods but also in terms of methodological sequencing and data extraction and utilisation. Its strength lies in its interdisciplinary approach, which links seemingly distant fields of science, such as social science through employee involvement, mathematics and computer science through machine learning algorithms and mathematical formalisation of expert opinions, and engineering through the compatibility of furniture manufacturing processes.

The integrated early price assessment system explored findings that the price estimation problem can be solved as a prediction problem using machine learning methods and interactive expert evaluation. The study has shown that the price of customised furniture can be estimated quite accurately when only data on material quantities are used, with a coefficient of determination of the results obtained of not less than 0.8. The use of machine learning techniques can reduce the time needed to estimate costs in the early design phase and speed up the time to market.

The integrated early price assessment system uses the advantages of visualisation techniques. The process proposed for visualising multidimensional data includes both conventional direct visualisation and methods that rely on dimensionality reduction and artificial neural networks. This process can be extremely helpful in analysing data in the custom furniture production industry. By gaining a better understanding of the data, more accurate price prediction models can be developed. Additionally, the process of visualising data in this way can be adapted for use in other manufacturing industries.

The proposed **the integrated early price assessment system** for pricing approaches not only provides price estimation and forecasting information but also addresses the company's own quality and order management issues. It can also be used as a self-learning tool to better identify future orders and sometimes even serves as the only prerequisite for informing the manufacturer of a customised product about the consumer's wishes.

The integrated early price assessment system replaces the traditional accounting and engineering cost model with a parametric one based on AI and expert involvement. The prototype has a high potential for commercialisation, but it temporarily faces some difficulties due to the furniture sector's scepticism towards data science and the fear of price competition in the market.

The integrated early price assessment system is equipped with expert input. By using expert decision-making methodologies, it has been proven that employee input on pricing decisions can complement machine learning-based price predictions, resulting in better overall pricing and decision-making outcomes. After exploring the acceptability of expert judgement in the furniture sector and its fit with IT tools, it was found that a properly constituted panel of experts could contribute to a more accurate estimation of contingent costs. According to the research, a diverse group of experts, including employees such as an accountant, a sourcing specialist, and a designer, can price products more accurately than individual recognised experts in a company or sector. Although managers and top engineers in the sector are still considered experts, a diversified team has the advantage of bringing different perspectives and skills to the table, which can lead to better predictions. This team could predict prices more effectively.

During the prototyping process in 2020, an external expert elicitation tool was introduced to anticipate that the company's employees might not have sufficient knowledge and evaluation capacity to predict the price of a new product. This approach was considered innovative at the time, but it is now becoming popular and widely used in other systems that rely on experts' opinions, even in nonmanufacturing systems. As a result, a large number of users are adopting this approach.

Transferability of Findings

This study conducted a comprehensive analysis of the difficulties faced by the furniture industry. However, the findings from this research can be applied to other industries that provide tailor-made products to the market on a business-to-consumer (B2C) basis and going quality challenges related to price forecasting. These industries include the automotive and construction industries. The automotive industry is constantly introducing technological advancements, and the proposed struggle with the metal consistency approach in this study can be beneficial in evaluating their feasibility. The proposed model can potentially apply to other activities such as moulding assembly technology, car body production, corrosion protection, painting, and final assembly. Construction companies that work on large orders involving many non-standard structural components face similar issues. The proposed model pricing. The price is also suitable for impact testing, predicting the cutting edge of glass together with the strength of selecting a fracture model for thermally prestressed glass, and so on. However, it is essential to note that when transferring the prototype algorithm to other manufacturing sectors, experimental development studies are required to determine which historical data is most suitable for machine training.

Future Research

Further research would be useful to complement the pricing algorithm with machine learning for image recognition. Image recognition research is well advanced. While the task of image recognition is well known to AI developers, the link between image recognition and price is a completely new area of research. The furniture industry may need a flexible recognition system for manufactured furniture that can continuously build on its own capabilities and adapt to its individual needs. Such additional information would further shorten the price forecasting process by providing more visualisation for experts and more opportunities for the user to be involved in the design process.

Annex 1

Diagnostic Study on the Furniture Manufacturing Sector: Questionnaire

Hello,

My name is _______ and I'm calling you from the market research agency RAIT. We are currently conducting a survey of furniture manufacturing companies. We would like to ask you to take up to 20 minutes to answer the questions below. We guarantee complete anonymity of the data you provide.

COMPLETED BY THE INTERVIEWER.

Specify the contact ID ________________________ Enter a number (question type 'Number')

Ask all

1. Your position in the company.
   a. Owner
   b. Director
   c. Production Manager
   d. Project Manager
   e. Head of Unit or Section (write which one): Other (Write In[Required])
   f. Deputy Director
   g. Engineer
   h. Designer
   i. Designer
   j. Constructor
   k. Technologist
   l. Manager
   m. Other (Write In[Required])

Ask all

1. Enter the year the company was founded: Enter a number (question type 'Number') ______________________________

Ask all

2. How many people work for your company? SINGLE
   a. 1–9
   b. 10–49
   c. 50–249
   d. 250 and more

Questions about the structure of the organisation:
Ask all

3. What is your company structure? SINGLE
    a. Manager and other staff
    b. Head, deputies and other staff
    c. General Manager, Heads of Units, Executors and other staff
    d. General Manager, Heads of Units, Heads of Departments, Executives and other staff
    e. President, Chief Executive Officer, Heads of Units, Heads of Departments, Executives and other staff
    f. Other (Write In[Required])

Ask all

4. Which departments/divisions are in your company? MULTIPLE
    a. No units/departments
    b. Woodworking shop
    c. Metalworking shop
    d. Marketing or marketing
    e. Accounting or finance
    f. Planning or design
    g. Staff
    h. Other (Write In[Required])

Ask all

5. What kind of positions do you have in your company? MULTIPLE
    a. Engineer
    b. Manager
    c. Designer
    d. Artist
    e. Project Manager
    f. Production Manager
    g. Marketing or marketing professional
    h. Financier or accountant
    i. Mechanic
    j. Head of Unit or Division
    k. Deputy Director
    l. Designer
    m. Constructor
    n. Technologist
    o. Warehouseman
    p. Other (Write In[Required])

Questions about process management:

Ask all

6. Do you have precise operating times for production processes? SINGLE
    a. Yes
    b. No
    c. For some

IF 8 = a

7. How did you set the operating times? SINGLE
   a. The Director has determined
   b. Experimenting with machine tools and other equipment
   c. Established over a long period of time working with machine tools and other equipment
   d. Other (Write In[Required])

Ask all

8. How do you calculate the price of an order? MULTIPLE
   a. We calculate accurate material prices, operating times, and make construction drawings
   b. Calculate average prices and add a mark-up
   c. Deciding 'by eye' – no exact calculations
   d. Other (Write In[Required])

Ask all

9. How long do you aim to calculate the cost of an order? (please specify in working hours, if you want to calculate per day – 8 hours, 2 days – 16 hours, etc.)
   ____________________ working hours Enter a number (question type 'Number')

Ask all

10. Who is involved in calculating the price of an order? MULTIPLE
    a. Owner
    b. Director
    c. Production Manager
    d. Project Manager
    e. Head of Unit or Division
    f. Deputy Director
    g. Engineer
    h. Designer
    i. Designer
    j. Constructor
    k. Technologist
    l. Manager
    m. Accountant
    n. Other (Write In[Required])

Ask all

11. Are there cases where the calculated price of an order, after the order has been completed, is not the same as the original calculation, that is, it is higher or lower? SINGLE
    a. Yes, quite often
    b. Yes, sometimes
    c. No
    d. We do not calculate the price for repeat orders
    e. I don't know

If 13 = a or b

12. What is the most common error in the price calculation? SINGLE
    a. Errors up to 10%
    b. The margin of error is between 10 and 20%
    c. The error is more than 20%

Ask all

13. Which specialists/staff are most important in deciding on the price of the order (whose judgements are most important)? MULTIPLE
    a. Production Manager
    b. Project Manager
    c. Head of Unit or Division
    d. Deputy Director
    e. Engineer
    f. Designer
    g. Designer
    h. Constructor
    i. Technologist
    j. Manager
    k. Accountant
    o. Other (Write In[Required])

Ask all

14. Who makes the final decision on the price offer to the customer? MULTIPLE
    a. Owner/Director
    b. Deputy Director
    c. Production Manager
    d. Project Manager
    e. Head of Unit or Division
    f. Engineer/Designer/Constructor
    g. Designer/Technologist
    h. Manager
    i. Accountant
    j. Other (Write In[Required])

Ask all

15. Which do you think is the most difficult/problematic process in furniture production for your company? MULTIPLE
    a. Supplies
    b. Logistics (Outgoing)
    c. Customer Relationship Management
    d. Calculating the price of orders
    e. Construction drawings for orders
    f. Production process
    g. Storage of orders, products
    h. Warranty service
    i. Other (Write In[Required])

Ask all

16. Do you buy services from other companies for furniture production? SINGLE
    a. Yes, quite often
    b. Yes, sometimes
    c. No

If 18 = a

17. What services do you buy? SINGLE
    a. Metalworking
    b. Woodworking
    c. Glass processing
    d. Plastic and rubber moulding
    e. Advertising
    f. IT services
    g. Bookkeeping
    h. Design
    i. Other (Write In[Required])

Ask all

18. Which employee do you think has the biggest impact on product competitiveness? SINGLE
    a. Production Manager
    b. Project Manager
    c. Head of Unit or Division
    d. Deputy Director
    e. Engineer
    f. Designer
    g. Designer
    h. Constructor
    i. Technologist
    j. Manager
    k. Accountant
    l. Other (Write In[Required])

Ask all

19. What are the different production functions/operations you can perform with the equipment you have? MULTIPLE
    a. Cutting
    b. Drilling
    c. Milling
    d. Sewing
    e. Lamination
    f. Bending
    g. Gluing
    h. Painting
    i. Turning

j. Stamping
k. Engraving
l. Grinding
m. Welding
n. Clapping
o. Preparation of the shell
p. Other (Write In[Required])

Ask all

20. Do you use CNC machines in production? SINGLE
    a. Yes
    b. No
    c. I don't know

Ask all

21. What specialised IT process management tools does your company use? MULTIPLE
    a. We are installing/setting up an original/specialised information system, please specify: Other (Write In[Required])
    b. We are installing/implemented an *ERP system,* please specify which one: other (Write In[Required])
    c. Completing approved MS Word/MS Excel tables
    d. Each department/employee creates and uses tools that are convenient for them, name them: Other (Write In[Required])
    e. Other (Write In[Required])

Ask all

22. In which processes do you use specialised IT process management tools? MULTIPLE
    a. Customer management
    b. Warehouse management
    c. Managing customer offers
    d. Order management
    e. Production planning and management
    f. Quality management
    g. Logistics management
    h. Bookkeeping
    i. Human resources management
    j. Analytics, report management
    k. Other (Write In[Required])

Ask all

23. What specialised IT tools are most needed in your company? MULTIPLE
    a. For faster information circulation between staff/departments
    b. Get aggregated information for decision-making
    c. Coordination of information/data from different departments

d. Manage production ‘interruptions’
e. Accumulate/systematise data
f. Analyse information
g. Perform calculations
h. Gather/process information faster
i. Other (Write In[Required])

Ask all

24. Do the specialised IT tools/systems used in your company meet your needs/expectations? SINGLE
    a. Yes, absolutely
    b. Partly, because only in some areas do we have the right IT tools/systems
    c. Partially, because not all relevant information/data is neatly placed in IT tools/systems
    d. Partly, because it requires a lot of extra work
    e. No, because it doesn’t produce the expected result
    f. No, because it costs too much and the results are insufficient
    g. Difficult to assess, still developing IT tools/systems
    h. Other (Write In[Required])

Ask all

25. In which processes do you feel there is a lack of information technology in your company? MULTIPLE
    a. Customer management
    b. Warehouse management
    c. Managing customer offers
    d. Order management
    e. Production planning and management
    f. Quality management
    g. Logistics management
    h. Bookkeeping
    i. Human resources management
    j. Analytics, report management
    k. Other (Write In[Required])

Ask all

26. What are the advantages of information technology in your company? MULTIPLE
    a. Making the production process more manageable
    b. Better assessment of untapped potential
    c. Reduced production errors
    d. Reduced turnaround time
    e. More efficient resource management
    f. More transparent labour accounting
    g. Increased production efficiency
    h. More flexible production management

i. More accurate inventory management
j. Optimised production
k. Other (Write In[Required])

Ask all

27. Please assess how these indicators have evolved over the last year compared to the previous year: have they decreased, stayed the same or increased? ONE IN A ROW

		Decreasing	No Change	Increased
1	Number of orders			
2	Profit margin (net margin)			
3	Revenue			
4	Number of employees			
5	Number of sales			
6	Number of equipment or technologies used			

Ask all

28. What was your company's turnover in 2017? SINGLE
    1. Up to EUR 50,000
    2. 50,000–250,000
    3. 250,000–500,000
    4. 500,000–1 million
    5. 1–2 million
    6. 2–3 million
    7. 3–4 million
    8. 4–5 million
    9. Over 5 million
    10. Refused or don't know **(DO NOT READ)**

Ask all

28. What is the predominant type of production in your company? SINGLE
    a. Personalised
    b. Small-scale
    c. Serial
    d. Stambiaserial
    e. Bulk
    f. Mishri
    g. Other (Write In[Required])

Ask all

29. What does your company produce? MULTIPLE
    a. Manufacture of furniture for offices and commercial enterprises (shops)
    b. Manufacture of kitchen furniture
    c. Manufacture of mattresses
    d. Manufacture of other furniture
    e. Other (Write In[Required])

Ask all

31. Do you produce for export? SINGLE
    a. Yes
    b. No

If 33 = a

30. Share of orders for export in total orders: SINGLE
    a. Up to 10%
    b. Up to 25%
    c. Up to 50%
    d. Up to 75%
    e. 75% or more

Ask all

31. Have you received/are you receiving orders from companies representing international brands? SINGLE
    a. Yes, regularly
    b. Yes, sometimes
    c. Yes, rarely
    d. No, never

These were all questions. Thank you for your time!

Annex 2

Interview Questionnaire

The aim of the study is to develop a rapid, operational system for complex production order estimation based on employee involvement and machine learning.

During data processing, the respondent's personal information (name, company name, and other personally identifiable data) will be encrypted and not published.

The questions are indicative and their location and wording may vary depending on the interview process.

Interview and informant data	Interview location/date Job Duties Work experience in an organisation Code of the informant
Demographic characteristics	1. How big is the company? 2. What is your product? 3. Which product/service are you proud of? 4. Who are your customers (average customer portrait)? 5. Who are your competitors? 6. What percentage of total production does customised manufacturing represent?
Manufacturing and cost estimation	1. Which production phase or non-production activity do you consider to be the most important one that determines the success of the whole project/order? Why? 2. Which production phase or non-production area creates the most challenges/tension/difficulties? Which ones? Which is the most complex and/or risky? Why? 3. Costing: Who makes that initial calculation? From your point of view, how important is it to make this calculation as accurate as possible? How do you manage to do it? What are the reasons if you think it is not being done accurately enough?

(*Continued*)

IT and ERP	1. How would you rate the information systems used or currently being developed/implemented in your company to manage your company's operations/production? Which of the systems you are familiar with was the most suitable? Which was the least? Why? How often do you update? 2. Which module or modules/feature of such systems do you find most useful or helpful in your day-to-day work? 3. To what extent are staff involved in the development/improvement of such information systems? 4. What is the biggest obstacle to the information systems you have/are developing being of real value to you? What could/should be done differently in the development/development of such systems? How?
Involvement/role of employees	1. How does cooperation work between each other? What are the problems and challenges you face when dealing with staff from other units or other positions? 2. Which specialists are most important in costing? What could/should be the contribution of employees on these issues? Which staff members' evaluations/opinions should be taken into account most? Is this currently the case? 3. How often are you briefed on innovations planned in the company? 4. What opportunities do you see or could you suggest to make employees feel sufficiently involved in the management of the company or the improvement of production processes?

Annex 3

OBJECTIVE

To determine the qualifications of the experts according to a set of selection criteria. The Structured Expert Judgement (SEJ) model developed by Roger M. Cooke is used to identify experts. It is a mathematical and performance-based method for evaluating, disaggregating and combining expert judgements. SEJ evaluates each expert on the basis of its statistical accuracy and informativeness by measuring performance against a set of calibration variables.

DEMOGRAPHIC DATA

First name Last name ______________________________

Position

____

Age: ☐ <18 m. ☐ 18–29 m. ☐ 30–45 m. ☐ 46 – 65 m. ☐ > 65 m

Years of experience in the field: ☐ <1 m. ☐ 1–5 m. ☐ 5–10 m. ☐ >10 m

Name of workplace: ______________________________

Education: ☐ School ☐ College ☐ Higher non University ☐ University ☐ PhD ☐ Prof.

QUESTIONS	Minimum possible value	Correct guess	Maximum possible value
Historical/competitive environment			
What year did IKEA open in Vilnius?			
In what year was the first furniture workshop established in Kaunas?			
How many furniture manufacturing companies are there in Lithuania?			

(*Continued*)

QUESTIONS	Minimum possible value	Correct guess	Maximum possible value
Production volumes			
What was your company's biggest order (in price)?			
What proportion of your furniture is sold on the domestic market?			
Accounting issues			
What is the average salary in your company?			
How many employees work in your company?			
How old is the oldest employee in your company?			
How old is the youngest employee in your company?			
What is the average age of employees in your company?			
What is the average length of service per employee in your company?			
How many employees are responsible for calculating the cost of a piece of furniture?			
Operating times			
What is the average time it takes you to complete an order?			
What is the average design time for a piece of furniture?			
In the context of Lithuania			
What percentage of GDP does furniture production represent in Lithuania?			
What percentage by value does seating furniture and its parts account for of the total furniture production in Lithuania?			

(*Continued*)

QUESTIONS	Minimum possible value	Correct guess	Maximum possible value
What percentage of value added is generated by furniture production in the Lithuanian manufacturing industry?			
What is the monetary value of the furniture industry per person per year?			
What percentage of production is exported?			
What percentage of production is exported to Sweden?			

Annex 4

Data Structure for a Machine Learning Algorithm

All the data selected for use in training and testing the machine learning algorithm is transferred and stored in one of the database tables after data preparation. The following data were selected for further analysis:

- Order number.
- Customer information (description in words).
- Quantity of products.
- Order price.
- Quantity of materials used to prepare the order.
- Quantity of different types of materials used to prepare the order.
- Number of parts ordered.
- Number of different parts in the order.
- Class number indicating the product type.
- Number of parts of the product measured in units.
- Number of parts of the product measured in metres.
- Number of parts of the product measured in square metres.
- Number of parts of the product measured in conical metres.
- Gross volume of parts of the product measured in conical metres.
- Number of parts of the product measured in kilograms.
- Gross volume of parts of the product measured in litres.
- Total metal working time in seconds.
- Total metal assembly time in seconds.
- Total metal painting time in seconds.
- Total metal packaging time in seconds.
- Total duration of metalworking operation (operation code 'm50') in seconds.
- Total duration of metalworking operation (operation code 'm60') in seconds.
- Total duration of metalworking operation (operation code 'm70') in seconds.
- Total duration of metalworking operation (operation code 'm80') in seconds.
- Total duration of woodworking operation (operation code 'm10') in seconds.
- Total duration of woodworking operation (operation code 'm20') in seconds.
- Total duration of woodworking operation (operation code 'm30') in seconds.
- Total duration of woodworking operation (operation code 'm40') in seconds.
- Total duration of woodworking operation (operation code 'm45') in seconds.
- Total duration of woodworking operation (operation code 'm55') in seconds.
- Total duration of woodworking operation (operation code 'm80') in seconds.
- Total duration of wood processing in seconds.
- Total time for completing wood in seconds.

- Total time for assembling wood in seconds.
- Total time for painting wood in seconds.
- Total time for packing wood in seconds.
- Total storage time of wood in seconds.

Index

Accounting-based cost pricing method, 114
Activity-based costing method (ABC method), 85
Additive manufacturing, 13
Advanced data analytics techniques, 161
Advanced manufacturing technologies (AMTs), 39, 135
Advanced planning and scheduling (APS), 24
American Customer Satisfaction Index (ACSI), 61
American Trucking Association, 51
Analogical methods, 89
Analogous recognition, 157
Analogue cost estimation (ACE), 154
Analytical approaches, 89
Application layer, warehouse architecture, 36
Applications of artificial intelligence, 26
Artificial intelligence (AI), 27, 90, 154
  applications of, 26
  techniques, 219, 220
Artificial Neural Networks (ANN), 31, 33, 90–91, 205, 219
Assembly to order (ATO), 58
Association, 21, 31
  rule learning, 31
Attitudes
  towards digitalisation and data usability of furniture's companies, 264
  towards pricing strategies of furniture's companies, 264–265
Augmented reality, 12–13
Automated order processing, 70
Automated production process, 119
Automated reporting tools, 70
Automation, 19, 69
Autonomous robots, 13
Avoiding complexity, 153

Back end of prototype, 244–246
Backwards-looking analytics, 23
Baltic Countries, similarities and differences between, 117
Baltic furniture manufacturers, 121
Baltic manufacturers' attitudes towards IT, assessment of, 137–138
Base technologies, 18
Behavioural aggregation, 222
Big data, 20, 161, 185
  analysis, 10
  research, 218
Big Data Analytics (BDA), 12, 22–26, 38
  in manufacturing, 21–22
Blow moulding, 33
Bottom-up estimation methods, 154
Buffer layer, warehouse architecture, 36
Business analytics (BA), 183–185
Business Management Systems, 141
Business representatives, 173
Business-to-business (B2B) basis, 68, 200
Business-to-consumer (B2C), 267

Calibration
  estimate, 249
  questionnaire, 108
Carbon footprint, 52

Case similarity assessments, 91
Case-based reasoning (CBR), 90–91, 219
Central and Eastern European countries (CEE countries), 182
Centralised database, 68
CEO, 111
Championing behaviour, 88
Classical knowledge management process, 241
Classical model, 223
Classification process, 21, 30
Classifier, 21
Cloud computing (CM), 13–14
Cloud integration, 68
Cluster analysis, 31
Clustering methods, 21, 31, 215
Co-design principles, 62
Cochran's *Q* test, 104, 122
Coding, 167
Command line interfaces (CLI), 248
Communication activity, 258
Communication flow diagram, 239, 243
  knowledge management cycle, 240–243
  types of knowledge, 240
Competition, 61
Competition-based approach, 135
Competition-based pricing, 87–88
Complex pricing procedure, 176
Complexity, 38–39
  in manufacturing, 152–154
  reduction strategies, 155
Complexity management, 153
  cost estimation as, 154–155
  cost estimation as complexity management tool, 154–155
  in customised furniture manufacturing, 152
  formulating strategies for managing complexity, 156–158
  methodological highlights
    coding and clustering of qualitative empirical data, 155–156
  specificity of complexity management in customised furniture manufacturing, 155
Comprehensibility, 102
Computer Numerical Control (CNC), 88
Configuration, 72
Consensus knowledge, 241
Consumers, 48–49
  consumer-driven production, 1
Contingency control measures, 175–176
Continuity-related barriers, 41
Controlling production time, 265
Conventional knowledge-based IT system, 240
Cooke's classical model, 109
Correlation analysis, 122, 126
Cost estimate, 86
Cost estimation, 218–220, 231
  approaches, 81
  classification of traditional cost methods, 84–85
  as complexity management tool, 154–155
  from costing to pricing strategies, 86–88
  expert judgement in price estimation, 91
  machine learning approach to estimate early costs of new product, 90–91
  machinery utilisation, 83–84
  managerial approach to costing, 85
  new paradigm of cost evaluation, 88–90
  practice, 130
  price evaluation strategy, 130–131
  process, 174

traditional costing methods as life cycle costing, 81–83
uncertainties regarding new product development, 85–86
Cost estimation, Benchmarking and Risk Analysis (COBRA), 220
Cost evaluation, new paradigm of, 88–90
Cost-based pricing, 87
Costing process, 131, 154
Costing systems, 82
COVID-19, 118
Creativity function, 242
Culture-based knowledge preservation strategies, 221
Custom development, 72
Custom furniture
companies, 3
manufacturing companies, 130
production, 182
Custom manufacturing, 180
Custom production, 154
Custom-made furniture, 191
Customer attitudes towards customisation, 61–62
Customer collaboration, 63
Customer engagement management, 62–64
Customer integration, 61
Customer involvement, classification of, 62–63
Customer orientation, 145, 178
Customer relationship, 157
Customer relationship management (CRM), 66, 70, 154
Customer-related barriers, 41
Customers of furniture companies, 48
Customisation, 4, 57, 62, 66–81, 72
manufacturing, 101
process, 58–59
Customise manufacturing management and engineering, 101
Customised furniture, visual analysis of production data and early pricing for, 206–208
Customised furniture manufacturing companies, 186
complexity management in, 152–156
in era of industry 4.0, 263
FI impact on, 55–56
Customised manufacturing, 56, 110, 206, 219 (*see also* Furniture manufacturing)
basic principles of, 56–57
classification of customer involvement, 62–63
customer attitudes towards customisation, 61–62
customer engagement management, 62–64
customer integration, 61
customisation process, 58–59
data specificity, 190–191
definition of, 57
knowledge-based engineering for customised product lifecycle management, 64–66
manufacturing challenges, 59
manufacturing strategies for customisation, 58
problem for, 1–3
Customised product lifecycle management, knowledge-based engineering for, 64–66
Customised production, 2, 56, 58–59, 134, 195
analysis of, 103–106
companies selected as case studies for empirical data, 110–114
historical manufacturing data collection and modelling of manufacturing processes, 106–107

inclusive governance and modelling of expert decisions, 108–109
modelling manufacturing processes based on machine learning algorithms, 107–108
outlining of methodology, 103
prototype and testing, 109
Cutting-edge techniques, 219
Cyber-physical systems, 10–11

Danish Academy of Manufacturing (MADE), 9
Data analysis, 106, 109
insights for, 188–190
instrument, 162–166
interview questions made in compliance with research instrument scheme, 164–165
Data collection, 185
Data demand, approach to, 188
Data dimensionality reduction-based visualization, 208
Data discovery, 107
Data driven culture, 38
Data exploitation, go/no-go strategy of, 196–197
Data exploration, 206
Data for testing, 259–260
Data management, 188
course of action related to avoidance of, 193–196
issues, 180
programmes, 201
strategies, 194, 196
tools for data managing and exchange, 126–128
Data mining, 244
engine, 36, 246
in manufacturing, 21–22
techniques, 22, 24, 35
Data preparation, 107, 201–203
Data presentation format, 202
Data privacy, 28
Data role in early price estimation, 160–162
Data sample, 205
Data science, 102
Data storage and retrieving, 62
Data structure for machine learning algorithm, 285–286
Data type error, 202
Data usability strategy, methodological considerations in search for, 186
Data utilisation
function, 195
goals, 194
strategies, 195–196
Data visualization, 208
early cost estimation by means of machine learning with, 200, 203–217
methods, 206, 209
Databases, 35, 246
Decision maker, 209
Decision making, 218
Decision support instruments, attitudes of manufacturers towards and other, 131–145
Decision support systems, 78
Decision tree, 30
regression, 205
Decision-making process, 203
Decision-making strategy, 227
Decisional bootstrapping, 94
Decomposition methods, 94
Deep learning, 31
Defect analysis in manufacturing, 26
Deforestation, 50
Degree of consumer involvement, 62
Delphi method, 94
Department of Statistics, 121
Derived data, 190
Descriptive analytics, 23
Descriptive functions of data mining, 21

Design, 82
solutions, 43
of user interface, 248
Diagnostic study on furniture manufacturing sector, 269–277
Digital assistance systems, 10
Digital environments, 63
Digital manufacturing, 18
Digital platforms for manufacturing, 34
components of industrial platforms, 34–37
industrial platforms, 34
Digital transformation, 40
Digitalisation, 10, 26, 41, 114
ANN, 31
applications of artificial intelligence, 26
association, 31
big data, 20
big data analytics, 22–26
classification, 30
clustering, 31
data mining and big data analytics in manufacturing, 21–22
digital platforms for manufacturing, 34–37
forming, 33–34
Machine learning in manufacturing, 31–32
machine learning models, 29
machinery, 33
ML, 27
ML pipeline and quality attributes, 27–29
moulding, 32–33
ontology, 22
regression, 29–30
in smart manufacturing, 20
welding, 32
Digitalisation process, 27
Digitisation, 26
Direct labour, 83
Direct materials (DM), 83
Direct production costs, 197
Direct visualization, 208
Diversity, 152
Domestic furniture market, 120
Double coding approach, 167
Dynamism, 152

Early cost estimation, 158
analysis methods, 167
contingency control measures, 175–176
data analyses instrument, 162–166
data preparation and transfer process, 201–203
estimation preliminary and final price, 167–169
experimental study, 205–206
from experiments to SEJ benefits of employees' engagement for, 224–232
findings from selected enterprises, 167
historical production data in furniture companies, 200–201
instrumental pricing capabilities, 173–175
managerial challenges related to uncertainty, 158–160
by means of machine learning with data visualization, 200, 203–217
methodology of qualitative analysis, 162
pricing procedure, 170
pricing processes in made-to-order manufacturing businesses, 176–180
proposal for machine learning-based forward price evaluation method, 204–205
role of data in early price estimation, 160–162

selected methods for multidimensional data visualization, 208–211
use case for visual analysis, 211–217
visual analysis of production data and early pricing for customised furniture, 206–208
Early costs of new product, 90–91
Early price assessment system, 266–267
Early prices estimation strategy, 197–198
Effective cost management, 265
Effective data management, 189
Emissions control areas, 51
Empirical data, companies selected as case studies for, 110–114
Employee involvement, 76, 79, 158, 218
Employee participation
concept, 76
in era of Industry 4. 0, 79–81
Employees' engagement for early cost estimation, from experiments to SEJ benefits of, 224–232
Empowering people in manufacturing industries, 76–81
Energy management, 19
Engagement, 79
higher scale of, 78–79
Engineer to order (ETO), 58
Engineering, 1
engineering-accounting approach, 91
value chain, 43
Ensemble forecasting, 94
Enterprise Resource Planning (ERP), 18, 66–81, 106, 153–154
automation, 70
customisation, 72
implementation, 156–157
integration, 67
security, 75
systems, 69, 73–74
Enterprises, 128
Estimates, 154
Estonia, furniture manufacturing sector, 119–120
Estonian furniture industry, 119
Estonian furniture manufacturers, 119
European Union (EU), 118, 181
EXCALIBUR software package, 109, 224, 227
Expert decisions, inclusive governance and modelling of, 108–109
Expert evaluation methodology, 253
Expert interface, 247
Expert judgement, 93, 222
availability of experts, 94–95
methodological implications of, 93–94
in price estimation, 91
validity of, 92–93
Expert opinion, 92
Expert passed prediction, 245
Expert recommendation system, 95
Expert-based knowledge, 220–221
Explicit knowledge, 240
Expressive knowledge, 240
External integration, 68

Factor analysis, 122
Fault diagnosis, 26
FedEx, 51
Financial stability, technology and, 145
Financing reporting, 85
Fixed cost, 84
Flexibility, 19
Forecast combination, 93
Forecasting by analogy (FBA), 94
Forestry, 120
Form-based interface, 248
Forming, 33–34
French manufacturing companies, The, 42
Front end of prototype, 246–247
Front-end technology, 14, 17
Frugal innovation (FI), 4, 52
concept, 53–56

impact on customised furniture manufacturing, 55–56
in manufacturing, 52
Furniture companies, 49, 263
attitudes towards digitalisation and data usability of, 264
attitudes towards pricing strategies of, 264–265
historical production data in, 200–201
Furniture industry, 2–3, 118, 200, 268
barriers in, 42–43
concept of FI, 53–55
cost estimation approaches, 81–95
customised manufacturing, 56–66
employee participation in era of Industry 4. 0, 79–81
empowering people in manufacturing industries, 76–81
ERP and customisation, 66
impact of FI on customised furniture manufacturing, 55–56
higher scale of participation and engagement, 78–79
instrumental approach towards participation, 78
regionalisation and frugal innovation in manufacturing, 52
stages of participation, 77
sustainability trends in furniture industry, 48–52
Furniture manufacturing, 120
companies, 72
course of action related to avoidance of data management, 193–196
customised manufacturing data specificity, 190–191
data usability strategies for furniture manufacturing, 199
diagnostic study on, 269–277
Estonia, 119–120
factual data representing furniture manufacturing companies, 188
historical production data analysis, 187
industry, 182
insights for data analysis, 188–190
Latvia, 120
Lithuania, 120–121
methodological considerations in search for data usability strategy, 186
processes, 82
qualitative research, 186–187
sector, 117
state of the art of data analytics in furniture manufacturing, 182–186
strategies for using data based on level of complexity of tasks customised manufacturing, 193
strategies for using furniture manufacturing data to promote industry 4.0, 180
Furniture markets, 258
Furniture production sector in the Baltic countries, 134
Furniture sector, 4

Generalisation, 21
Generating complexity, 153
Generation Xers, 48
Go/No-Go strategy of data exploitation, 196–197
Graphical user interfaces (GUI), 248
Gross domestic product (GDP), 120

Heterogeneity, 152
Hierarchical management, 87
High degree of uncertainty, 158

Historical manufacturing data collection and modelling of manufacturing processes based on machine learning algorithms, 106–107
Historical production data analysis, 187
Home window, 247
Horizontal axis, 242
Horizontal scalability, 73
Human resource diversity assessment and consideration, 60
Human resource-related barriers, 41
Human-machine interaction (HMI), 11
Hybrid pricing approach, 265

Identifiability, 242
Images, 187
  recognition research, 268
Implement role-based access control (RBAC), 74
Inclusion, 156
Inclusive governance and modelling of expert decisions, 108–109
Inclusive management, 102, 108
Independent survey participants, 172
Indication layer, warehouse architecture, 36
Indirect labour (IL), 83
Indirect materials (IM), 83
Indirect production costs, 197
Individual pricing, 185
Inductive analysis, 167
Industrial environment, 7
Industrial Internet of Things (IIoT), 12
Industrial Internet platform, 34
Industrial platforms, 34
  components of, 34–37
Industrial workers using structured expert judgment, empowering and engaging, 217–232
Industry 4.0, 3–4, 17, 52, 88, 146, 186
  barriers to implementation of, 40–43
  customised furniture manufacturing in era of, 263
  employee participation in era of, 79–81
  factors linked with success of, 38–39
  historical perspective for, 7–9
  paradigm, 7
  role of smart manufacturing in, 16–19
  strategies for using furniture manufacturing data to promote, 180–196
Industry Peer Networks (IPNs), 78
Information and communication technology (ICT), 10
Information estimate, 249
Information technology (IT), 131, 161
  adoption factors and organisational features, 133–134
  assessment of Baltic manufacturers' attitudes towards, 137–138
  organisational features and, 144
  purpose and need for, 139–141
  sector specificity and, 144
  solutions, 128, 153–154
  system, 240
  tools, 140
Injection moulding, 33
Innovation approach, 4
Input window, 247
Inquisitive analytics, 23
Instant cost estimation, 198
Instrumental approach towards participation, 78
Instrumental pricing capabilities, 173–175
Instrumentarium, 104
Integral Early Price Evaluation System, 102
Integrated early price assessment system, 3, 109, 240, 266–267

algorithm for verification of performance under laboratory conditions, 248
case study on application of price evaluation methodology, 253–257
communication flow diagram for operation of prototype, 239–243
prototype functionality and structure, 243–248
prototype operating principle for estimating cost of piece of furniture, 249–253
prototype testing conditions and challenges, 257–260
Integration, 67, 72, 102
Integrity, 28
Intelligent manufacturing
big data analytics and regionalisation, 38
complexity, 38–39
data driven culture, 38
digitalisation in smart manufacturing, 20–37
economical implication, 38
factors linked with success of industry 4.0, 38
historical perspective for Industry 4.0, 7–9
small and medium enterprises' readiness to embrace industry 4.0 possibilities, 37
Smart manufacturing, 9–19
SME-specific maturity and readiness models, 39–43
Interaction functionality, 242
Interaction-based knowledge preservation strategies, 221
Interdisciplinary approach, 102, 266
Interface of communication, 36
Internal integration, 67
International competitiveness
assessment of Baltic manufacturers' attitudes towards IT, 137–138
attitudes of manufacturers towards and other decision support instruments, 131
competences of manufacturing team, 125
cost estimation practice, 130–131
equipment and technologies used, 138–139
furniture manufacturing sector, 117–121
IT adoption factors and organisational features, 133–134
organisational features and IT adoption, 144
performance, 141–143
prevailing organisational structure in furniture manufacturing companies, 121
production and experts, 128–130
purpose and need for IT usage, 139–141
rotated component matrix, 123
sector specificity and IT adoption, 144
technology and company size, 144–145
technology and financial stability, 145
technology usage and sector specificity, 134
tools for data managing and exchange, 126–128
International Maritime Organisation, 51
Internet of Services (IoS), 11, 12, 14
Internet of Things (IoT), 11–12, 14
Interoperability, 65
Interpretation errors, 202
Interview questionnaire, 279–280
Intrusion detection systems (IDS), 74
Intuitive methods, 89
Investigate, Discuss, Estimate and Aggregate (IDEA), 129, 229

JAVA programming language framework, 109
JavaScript, 202

K-neighbour regression, 205
Key performance indicators (KPIs), 70
Knowledge, types of, 240
Knowledge accumulation strategy, 221
Knowledge base, 35–36
Knowledge cleansing cycle, 241
Knowledge discovery-oriented DMT-based applications, 25
Knowledge extraction, 22
Knowledge function, 242
Knowledge Management (KM), 183
  cycle, 240–243
  systems, 64
Knowledge preservation strategies, 221
Knowledge-based engineering approach (KBE approach), 64
  for customised product lifecycle management, 64–66

Latvia, furniture manufacturing sector, 120
LEAN system, 111
Life cycle costing, traditional costing methods as, 81–83
Linear regression method, 107, 189, 205
Liquid composite moulding, 33
Literature analysis, 158
Lithuania, furniture manufacturing sector, 120–121
Logistic regression, 29
Louvain algorithm, The, 215
Lursoft data, 120

Machine learning (ML), 27, 33, 161
  algorithms, 5
  ANN, 90
  approach to estimate early costs of new product, 90
  CBR, 90–91
  data structure for, 285–286
  early cost estimation by means of machine learning with data visualization, 200, 203–217
  historical manufacturing data collection and modelling of manufacturing processes based on, 106–107
  machine learning-based prediction, 245
  in manufacturing, 31–32
  methods, 33
  modelling manufacturing processes based on, 107–108
  models, 29
  pipeline, 27–29
  platforms, 28
  proposal for, 204–205
Machinery, 33
  utilisation, 83–84
Management decisions, 182
Managerial perspective, 78
Manual-intensive process, 204
Manufacturers, 168–169
  attitudes towards and other decision support instruments, 131–145
Manufacturing, 4
  companies, 2, 65, 86–87
  competences of, 125–130
  complexity in, 152–154
  costs, 81
  data mining and big data analytics in, 21–22
  digital platforms for, 34–37
  empowering people in manufacturing industries, 76–81
  historical manufacturing data collection and modelling of, 106–107
  machine learning in, 31–32

revolution, 9
sector, 1
strategies for customisation, 58
Manufacturing Execution System (MES), 18
Manufacturing from stock (MTS), 58
Manufacturing overhead (MOH), 83
Manufacturing to order (MTO), 58
Marketing and customer relationship, 157
Mathematical aggregation, 222
Maturity model, 40
Mean shift clustering, 31
Methodology, outlining of, 103
Micro-factories, 119
Millennials, 48
Missing data, 202
Mixed methods strategy, 223
Model evaluation module, 36
Modelling, 10
inclusive governance and modelling of expert decisions, 108–109
manufacturing processes based on machine learning algorithms, 107–108
task, 107
Modularisation, 157
Moulding, 32–33
Multi-factor authentication (MFA), 74
Multidimensional scaling (MDS), 209
MySQL database, 109

Network communications, 10
New knowledge, 241
New product development, 47
Non-convex optimization, 209
Non-linear MDS approach, 212
Non-manufacturing costs, 84
Numbers, 187

Online platforms, 78
Ontological Data Mining Agent (OntoDMA), 22
Ontology, 22
Operational data, warehouse architecture, 36
Operational modifications, 19
Organisational barriers, 40
Organisational change, 88
Organisational confidence, 88
Organisational development, 144
Organisational structures, 87, 157–158
in furniture manufacturing companies, 121
Output knowledge, 241

Pandemic, 121
Parallel tuning, 198
Parameter optimisation tasks, 25
Parametric methods, 203
Parametric models, 89
Participation, 76
higher scale of, 78–79
instrumental approach towards, 78
participation-based management, 76–81
stages of, 77
Participation Based Intelligent System, 4
Participation based management, 76–81
Performance management, 195–196
Polyurethane foam waste, 50
Positive business indicators, 175
Pre-knowledge, 241
Pre-processing (*see* Data preparation)
Predictability, 192
Prediction, 22
Prediction function, 21
Predictive analytics, 23
Predictive modelling, 161
Prescriptive analytics, 23
Price assessment tools, 232
Price calculation process, 168
Price estimation by expert evaluation, 249
Price evaluation
answers to questions for expert calibration, 253

averages of experts' predictions falling within specified ranges, 254
breaking down experts' predictions into ranges, 254
calculation of average information values, 256
calculation of calibration score, 255
calculation of information estimate, 256
calculation of lower and upper limits, 255
calculation of weights of experts, 256
case study on application of, 253–257
estimation of blended price, 257
estimation of final expert price, 256
methods, 203
strategy, 130–131
Price prediction by machine learning algorithms, 249
Price range, 192
Pricing, 66, 184
approach to, 188
complexity, 153
custom furniture, 184
custom-made furniture, 184
in made-to-order manufacturing businesses, 176–180
preconditions for making pricing paradigm shift, 265–266
process, 162, 170, 176, 179
strategies, 88
Principal component analysis (PCA), 208–209, 212
Process monitoring tasks, 25
Product and service systems (PSS), 34
Product customisation, 57
Product Data Management (PDM), 153
Product design, search for similar patterns in, 157
Product Lifecycle Management (PLM), 65, 153
Production, 82
Production business data warehouse, 36
Production data and early pricing for customised furniture, visual analysis of, 206–208
Production process, 24, 38, 106, 197
Production time, 155, 158, 198
Products/processes description, 25
Prototype, 82
back end of prototype, 244–246
communication flow diagram for operation of, 239–243
data for testing, 259–260
design of user interface, 248
front end, 246–247
functionality and structure, 243–244
operating diagram, 244
preconditions, 257–258
prototype users, 247
refining prototype, 260
selecting company, 258–259
specification, 243
structural elements, 247
and testing, 109
testing conditions and challenges, 257
user interface, 248

Qualitative content analysis of interviews, 167
Qualitative empirical data, methodological highlights coding and clustering of, 155–156
Qualitative research, 186–187
Quality attributes, 27–29
Quality classification tasks, 25
Quality management system, 110
Quantitative data, 104
Quantitative pricing methods, 203

Quantitative research survey, 110
Quantitative sociological survey research, 104
Questionnaires, 108, 269–277

Random Forest (RF) classifier, 30
Range, 192
Rating functionality, 242
Readiness models, 39–43
Reducing complexity, 153
Regionalisation, 4, 38
  issues, 263–264
  in manufacturing, 52–53
Regression, 29–30
  analysis models, 91, 154–155
Reinforcement learning, 29
Research and Innovation Staff Exchange (RISE), 5
Research methods, 102
Respondents, 104–105, 108, 170
Return on investment (ROI), 68
Root mean square error (RMSE), 205
Rule-based forecasting (RBF), 94

Scalability, 73
Scaling out (*see* Horizontal scalability)
Scaling up (*see* Vertical scalability)
Scientific knowledge, 240
Secondary functionalities, 243
Sector specificity, 144
Security, 74
Security information and event management (SIEM), 75
Selection criteria representativeness, 110
Semantic Annotation and Processing Agent model (SAPA model), 66
Semi-structured interview method, 105, 166
Semi-supervised learning, 29
Sequential pattern mining, 21
Simulations, 10, 14–15
Small and medium-sized enterprises (SMEs), 5, 40, 52, 110, 125, 159, 181
  barriers to implementation of Industry 4. 0, 40–43
  segment, 133
  smart manufacturing maturity models, 39
  SME specific maturity models, 39
  SME-specific maturity and readiness models, 39
Smart management, 57
Smart manufacturing, 9, 17–19, 160
  AR, 12–13
  autonomous robots, 13
  big data and analytics, 12
  cloud computing, 13–14
  cyber-physical system, 11
  industry 4.0, 10–11, 16–19
  IoS, 12
  IoT, 12
  maturity models, 39
  simulations, 14–15
  3D printing, 13
Smart Manufacturing Leadership Coalition (SMLC), 8
Smart product, 18
Smart supply chain, 18
Smart working, 18
Smartness dimension, 17
Spearmen's Rank correlation coefficients, 104, 122
Specialised IT tools, 128
Spring Boot framework, 109
Statistical analyses-oriented data mining techniques (SA-oriented data mining techniques), 25
Statistical calibration, 93
Structured Expert Judgment (SEJ), 221–224, 249, 281–283
  cost estimation, 218–220
  empowering and engaging industrial workers using, 217

from experiments to SEJ benefits of employees' engagement for early cost estimation, 224–232
expert based knowledge, 220–221
methodology for, 223–224
Supervised learning, 29
Sustainability Certification Label, 48
Sustainability Certification Seal of Approval, 50
Sustainability trends in furniture industry, 48–52
Symbols, 187
System for prediction, 245
System training, 244–245

t-SNE method, 209, 212
Technological barriers, 40
Technology and financial stability, 145
Technology-based knowledge preservation strategies, 221
Testing
data for, 259–260
process, 109
Texts, 187
The activity-based costing, 85
The pricing complexity, 153
3D printing, 13
Time, 170
Time management, 102
Time series analysis, 22
Total Quality Management (TQM), 77–78
Traceability, 19
Traceable costs, 84
Traditional Baltic furniture manufacturing sector, 144
Traditional cost assessment methods, 85
Traditional cost methods, classification of, 84–85
Traditional cost-based pricing strategy, 87
Traditional costing methods as life cycle costing, 81–83
Training data, 28
Training process, 205
Transfer process, 201–203
Transferability of findings, 267–268
Transformational capacity, 263

UN Framework Convention on Climate Change, 51
Uncertainty, 87
managerial challenges related to, 158–160
Unprocessed data, 201
Unsupervised learning, 29
Upfront price estimation, 204
User Access Control, 75
User interface (UI), 36, 246–248
customisation, 72
design, 248

Value-added, 84
Value-based model, 177
Value-based pricing (VBP), 86, 88, 159
Variable cost, 84
Vertical axis, 242
Vertical integration, 18
Vertical scalability, 73
Virtualisation, 10, 18–19
Visual analysis
of production data and early pricing for customised furniture, 206–208
use case for, 211–217
Visualisation process, 209, 211
Volume, variety, velocity, veracity, variability, volatility, and value concept (7V concept), 20

Warehouse architecture, 36
Weighted linear combination of opinions, 222
Welding, 32
Wood processing, 120
World Wildlife Fund, 48